D1467380

Energy and Ecology

David M. Gates
University of Michigan

QH
545
E53G38
1985
Porter

Sinauer Associates, Inc. • *Publishers*
SUNDERLAND, MASSACHUSETTS

COVER PHOTO

**The eruption of Mount St. Helens in Oregon,
May 1981. Courtesy of the U.S. Geological Survey.**

ENERGY AND ECOLOGY

Library of Congress Cataloging in Publication Data

Gates, David Murray, 1921–
    Energy and ecology.

    Bibliography: p.
    Includes index.
    1. Energy industries—Environmental aspects.
2. Power resources.   I.   Title.
QH545.E53G38   1985   333.79   84-27471
ISBN 0-87893-230-5
ISBN 0-87893-231-3 (pbk.)

Printed in U.S.A.

5 4 3 2 1

CONTENTS

PREFACE

This book is concerned with the ecological and environmental consequences of energy extraction, conversion, use and discharge. It contains accounts of world energy resources and reserves and of their rates of use and expected lifetimes. The utilization of any energy form, whether it be renewable or nonrenewable, has ecological and environmental consequences. To comprehend these consequences we must first understand the technology of each energy form: solar, biomass, coal, petroleum, nuclear and such alternative energy sources as ocean thermal energy conversion, geothermal, wind, hydroelectric, ocean tidal and ocean wave.

Since primordial times energy has flowed through the planet Earth and in the process this universal energy flow has produced a global ecosystem teeming with life. Within this living system *Homo sapiens* evolved and acquired an intelligence capable of vastly modifying, if not destroying, Earth's ecosystems. This acquired intelligence permits humans to manage the global ecosystem in a self-sustaining manner that could assure a continuing vitality in perpetuity. Or, we could "manage" Earth's destruction.

A primary difficulty we face concerning any management task is that there are many more ways to do a thing wrong than there are to do it right. This is particularly true when it comes to the management of ecosystems that are themselves extremely complex entities. An ecosystem is a unit of the landscape involving all the living organisms and physical processes within it. Destroy a single organism, eradicate a species, poison the soil or atmosphere or change the climate and the ecosystem is damaged. It becomes a different ecosystem than it was before. Is it now less stable? Is the basic fabric of the ecosystem going to unravel? Is it going to degrade until life can no longer be supported? Or, is it simply going to become impoverished, thereby sapping the vitality of the human species?

Many years ago I concluded some Congressional testimony with the statement: "We may go down in history as an elegant technological society that underwent biological degradation for lack of ecological understanding." Today the ecosystems of the world are degenerating. Soil erosion rates are greater than ever before, tropical deforestation rates are increasing, species of plants and animals are being forced into extinction, fresh water supplies are more and more contaminated with synthetic compounds, atmospheric pollution is increasing, human populations continue to grow, and vast numbers of human beings live in an ever-degrading condition. Highly concentrated forms of energy such as coal, oil, gas and uranium are being withdrawn and expended at increasing rates. Their combustion products are contaminating the air, water and soil. We can estimate lifetimes for each resource reserve within reasonable probabilities and no one can plausibly suggest that these resources will last indefinitely. Resources *will* become more scarce and more costly and the time frame is short for their depletion. One hundred years ago our energy resources were wood and coal and urban areas were badly polluted. Today our energy resources are coal, petroleum and uranium. Our urban areas are polluted and our global atmosphere is contaminated with carbon dioxide and acid rain. One hundred years from now the global climate may have changed significantly as the result of the effluent from human activities and all the world's ecosystems may be depleted irreversibly of species diversity, of nutrient enrichment, and of productive vitality.

It is perfectly clear that population growth must stop, resource conservation must become an overriding principle for society, and continued pollution of air, water and soil must become unacceptable. These goals can only be achieved by people who understand the ways ecosystems work and who know the full implications of resource depletion. This book addresses some of these issues.

Several years ago I developed a course on Energy and Ecology for undergraduate students at the University of Michigan. They have come from history, economics, law, natural resources, public health, business, all sciences, and, of course, from biology. The course is self-contained and there are no prerequisites. I have tried to make this book equally "self-contained" and hope that readers without background in mathematics and the natural sciences will not be derailed by the occasional, necessary quantitative statement. The study of energy and its ecological and environmental impacts is an ideal meeting-ground for students from humanities, social sciences, sciences, and technology. Indeed, only a truly multi-disciplinary awareness of the dangers of energy misuse and overexploitation will lead to a sensible, sensitive stewardship of Earth.

ACKNOWLEDGMENTS

Space will not allow more than a listing of the individuals to whom I express thanks for providing information and photographs. William Anderson, U.S. Geological Survey; William Avery, Johns Hopkins Applied Physics Laboratory; Sandra S. Batie, Virginia Polytechnic Institute and State University; C. B. Bottum, Jr., President, Townsend and Bottum Co.; Melvin Calvin, University of California; Judith Henderson, Louisiana State University; Paul Kilburn, Woodward-Clyde Consultants, Rocky Mountain Region; John M. Parker, Consulting Geologist, Denver; Patricia J. Rand, Atlantic Richfield Co.; Elizabeth Rodgers, Tennessee Valley Authority; James H. Stone, Louisiana State University; and James Tate, formerly of Arco Coal Co.

The Detroit Edison Company provided me with nearly 300 slides and other information concerning many aspects of energy and environmental issues. Walter J. McCarthy, Chairman of the Board, and Charles M. Heidel, President, were always helpful. Roberta Urbani of the Detroit Edison staff spent much time putting together requested information.

I wish to express my gratitude to James Ryan for drafting many of the drawings and to secretaries Meredy Vande Kopple and Janette Tobin for their endless patience with me and my manuscript.

Energy and Ecology

1

Ecology, Ecosystems and Ecological Principles

Industrial melanism. (Top) The normal, light-colored form of the peppered moth (*Biston betularia*) is difficult for predators to see against the lichen-covered bark of unpolluted trees. (Bottom) Soot from industrial smelters and power plants in England darkened the tree bark, making the light-colored moths conspicuous to predators while the dark-colored (melanic) moths are camouflaged. Photographs by H.B.D. Kettlewell.

INTRODUCTION

Ecology

ECOLOGY is the study of the relationships of organisms to their environment and of organisms to one another. It incorporates elements from all levels of biological science—from molecular, cellular, genetic, and physiological levels to species, communities, and the whole world. Ecology is clearly an extremely large subject and one about which whole books are written and courses taught. Here we can only touch upon some of the most significant principles of ecology, that is, those principles that relate directly to our discussion of the use of energy and its ecological implications.

Ecosystems

An ECOSYSTEM is an ecological unit, a subdivision of the landscape, a geographic area that is relatively homogeneous and reasonably distinct from adjacent areas. The term *ecosystem* may be applied to a meadow, a forest, a lake, a sand dune, a bog, an alpine tundra, or any other readily recognized unit of the landscape. An ecosystem is made up of three groups of components: organisms, environmental factors, and ecological processes. That is, it comprises organisms, species, populations, and communities; soil, climate, and other physical factors; and processes such as energy flow, nutrient cycling, water flow, freezing, and thawing.

The organism is the fundamental unit of ecology. Over time organisms have evolved in synergism with their environments, and as a result many internal controls keep them in dynamic equilibrium with the world around them. Some organisms, mainly plants, are rooted to their environments and must be able to survive whatever climatic variations occur, whereas others, particularly small animals, can move about to escape some of the vicissitudes of the environment.

Environmental concerns versus ecological concerns

A distinction should be made between environmental concerns and ecological concerns. Usually these terms are confused and not clearly distinguished, but there is a difference. An oil spill from a tanker will put oil on the water and a mess on the beaches. If we are concerned only with the physical presence of the oil or the aesthetic consequences, then we have an ENVIRONMENTAL CONCERN. If, however, our concern is with the effects of the oil on primary productivity by plankton, on shellfish reproduction, or on thermal regulation by birds, then we have an ECOLOGICAL CONCERN—the effects are ecological. The increase of carbon dioxide concentration in the atmosphere is expected to produce a significant climatic change in the future—a warmer climate, which is wetter worldwide but drier in certain areas, including the central United States. If we are only concerned with the resulting climatic change, then our concern represents an environmental concern. However, if the impact of climatic change on crop production or ecosystem dynamics is considered, then the issue becomes an ecological concern. A trash heap may be an aesthetic problem primarily and that is an environmental concern; but water that flows through the trash pile and becomes poisoned may pollute drinking water or streams, thereby affecting human health or animal and plant life in the stream. Iron or zinc leaching from metal cans may poison microorganisms in the soil. Concerns about these matters are clearly ecological concerns.

It is axiomatic that every environmental event has an ultimate ecological consequence, no matter how small it may be. However, our concern with a specific event may terminate with the effect of the event on the environment only or it may involve a multitude of ecological consequences.

FOOD CHAINS AND FOOD WEBS

Producers, consumers, and decomposers

Life is maintained by organisms, mainly plants, that convert carbon dioxide and water into energy-rich compounds by the use of incident solar radiation—the ultimate source of energy for all life on earth. This chemical process of assimilation of carbon dioxide by green plants is known as PHOTOSYNTHESIS. In addition to green plants, purple bacteria assimilate carbon dioxide by using hydrogen sulfide (H_2S) instead of water (H_2O) as the reducing agent. Other bacteria use organic compounds for reducing carbon dioxide. Green plants and various chemosynthetic bacteria are known as the PRODUCERS in an ecosystem. Higher plants and green algae perform most of the carbon fixation in the world, and chemosynthetic bacteria are of more significance in moving certain nutrients, such as sulfur, through the sediments of ecosystems. Because the producers fix their own food supply, they are known as AUTOTROPHS (self-feeding). Organisms that depend upon other organisms for food are known as HETEROTROPHS (other-feeding). Organisms that feed on other organisms are also known as CONSUMERS. Those that feed on plants are called PRIMARY CONSUMERS or HERBIVORES. A heterotroph that feeds on herbivores is known as a SECONDARY CONSUMER. A carnivore that feeds only on those animals that are secondary consumers is a TERTIARY CONSUMER. OMNIVORES are organisms that feed on both plants and animals.

If the world only had producers and consumers, it would not work very well because the flow of materials would be in only one direction, that is, from lower to higher order compounds. Something needs to return these compounds to more elemental forms so that they can cycle and be used over again in the food chain. This process requires a group of organisms known as DECOMPOSERS. Bacteria and fungi play this role. Bacteria generally act on animal tissue and fungi on plant tissue. Plant and animal material is degraded enzymatically and released as basic elements into the environment, where the elements are again available to the producers for reuse.

However, energy is not recycled. It moves unidirectionally through an ecosystem, being consumed at each step of the food chain.

Food chains

A field mouse may derive its energy by eating grass; then the field mouse is consumed by weasels, which in turn are taken by hawks and owls. Or the grass is eaten by grasshoppers, which are consumed by meadowlarks; and the larks may be killed by hawks. An aquatic FOOD CHAIN consists of phytoplankton (the primary producer), which are fed on by zooplankton, which in turn are eaten by fish, which are consumed by porpoises, which are eaten by killer whales. However, very often an animal may escape all predation and die of old age. Dead animals are occupied by blowflies (which lay their eggs in the carcasses), are eaten by carrion beetles, and are consumed by bacteria. Food chains are illustrated in Figures 1 and 2.

Food chains seldom stand alone but interact to form food webs.

Food webs

A FOOD WEB is the network of ecosystems through which energy flows in the entire community of plants and animals; and most food webs have a great deal of inherent stability. They will readily repair themselves when disturbed, unless the disturbance is too extreme. When the disturbance is so great that species are wiped out or entire breeding populations are decimated, ecosystem recovery to its previous form never occurs. There are

FIGURE 1. Examples of simple food chains. The sparrow eats grass seeds and is itself food for the owl. The impala feed on grass and are preyed upon by lions.

many examples of this decimation throughout the world. In areas of human activity, natural ecosystems have been replaced by urban ecosystems. Most of Europe has had its original natural ecosystems replaced by cultivated or managed ecosystems. The Great Lakes fisheries have been damaged by human activities beyond recovery to their original form.

The food web of an arctic tundra is very simple and comprises relatively few food chains and individual species, whereas a tropical forest has an enormously complex food web and includes tens of thousands of species. The number of insect species alone in the arctic tundra may not exceed two dozen, whereas in the tropical forest they may be numbered in the thousands. The arctic tundra ecosystem is much more vulnerable to disturbance than is the tropical forest ecosystem. However, human forces in the form of machinery, chainsaws, and vehicles are so overwhelming that no matter how great the diversity or inherent resilience, any ecosystem is liable to irreparable damage or at least liable to damage that may require centuries to repair.

The Great Lakes ecosystem

The Great Lakes constitute the world's largest freshwater ecosystem and, at one time, the finest freshwater fishery anywhere. However, Europeans settling the Great Lakes watershed cleared the forest to conquer the wilderness; built canals and locks to allow freighters to transport iron ore from the Mesabi and Menominee Ranges near Lake Superior to Chicago, Detroit, Cleveland, Buffalo, and beyond; fished the lakes heavily; and polluted the waters. The result was a collapse of the fishery and the eradication of one or more fish species from the Great Lakes.

Humans, as the top predator in the Great Lakes ecosystem, have been ruthless in its exploitation. Clearing the forest caused erosion of soil and siltation of spawning grounds;

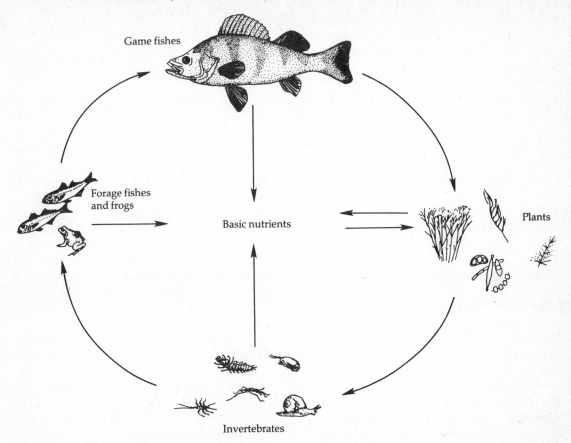

Game fishes

Forage fishes
and frogs

Basic nutrients

Plants

Invertebrates

FIGURE 2. The flow of nutrients through a food chain—from plants to invertebrates to small fishes and amphibians to large fishes. Input to the stock of basic nutrients comes from all parts of the food chain.

sunlight warmed previously shaded, cool streams. The arctic grayling, a cold-water fish once found in Michigan streams, disappeared from the region. The sea lamprey and alewife at one time were restricted to the Saint Lawrence, but these species gained access to Lake Ontario and then access around Niagara Falls to Lake Erie and finally to the upper Great Lakes through man-made canals and locks. The sea lamprey became parasites of lake trout and whitefish. The alewife, a plankton feeder, outcompeted several species of chubs for food. Sturgeon, massive bottom-feeders, were considered a nuisance by fishermen and were burned on the beaches of Lake Erie by the tens of thousands. However, sturgeon do not breed until about 15 years of age and are very slow to recover from decimation. Decades of overfishing added to the accumulated stresses on the populations of Great Lakes fishes. Finally, other human activities, such as toxic wastes, added more stress to the Great Lakes and the natural fish populations collapsed.

Today the Great Lakes fishery is very different from its original form, and what does exist must be sustained at enormous expense. The sea lamprey must be controlled by chemically treating its spawning area in the streams. The alewife cannot be controlled and will remain forever an exotic, annoying fish in the Great Lakes. The once magnificent populations of lake trout and whitefish are now a fraction of their former sizes. Coho and Chinook salmon have been introduced into the lakes and are being sustained at great expense. Sturgeon are slowly increasing in numbers once again but will never return to their former populations because of a diminished number of spawning areas.

The mismanagement and misuse of the Great Lakes ecosystem is not over. New stresses include chemical pollution by heavy metals, PCBs, and other compounds, acid rain, and thermal pollution from electric power plants. The effluents of large industrial urban areas such as Chicago, Detroit, Cleveland, and Buffalo still get into the Great Lakes, despite improved pollution abatement. Because large amounts of fresh water are required for steel making, chemical manufacturing, and electric power production, it was necessary to locate these great industries along the shorelines of the lakes.

The Great Lakes ecosystems are an example of a food web badly damaged by foolish human intervention. If wise management could have been the rule from the beginning, we might have had a highly productive fishery today, with the original food web still intact. Wide bands of the original forests should have been left along all streams entering the Great Lakes in order to prevent soil erosion and siltation of vital spawning grounds. Swamps and wetlands should not have been drained; rather, these vital areas for fish and wildfowl should have been protected. Industrial effluents should have been controlled from earliest times and commercial fish catches should never have been allowed to exceed sustainable yields.

Trophic levels

As energy flows through the food web of an ecosystem, it is temporarily stored at various stages that the ecologist defines as TROPHIC LEVELS. The producers, or green plants, always represent the first trophic level; the plant eaters or herbivores the second trophic level; the carnivores that eat the herbivores the third trophic level; and the carnivores that eat the third level animals form the fourth trophic level. Only about 10% of the energy available at one trophic level gets passed on to the next. By the third or fourth level, only about 1/100 or 1/1000 of the first level energy is available. This means, for example, that in agriculture it takes 10,000 kilocalories of corn to produce 1000 kilocalories of beef, which may be converted to 100 kilocalories of energy by humans.

A classic study of an old-field community in southern Michigan by Golley (1960) established values for the energy flow through the food chain. About 1% of the solar energy received was converted into plant material. Meadow mice were the main herbivores, and they consumed only about 2% of the energy available in the plants. The weasels, which fed mainly on the mice, utilized 30% of the annual biomass available. Although each higher level of food chain converted a higher percentage of the energy available in the previous level, it turned out that the plants used 15% of their total energy for respiration, the meadow mice 68%, and the weasels 93%. Respiration simply converts the energy to heat, in which form the energy leaves the ecosystem. The loss of energy at each higher stage of this old-field ecosystem was so great that the system could not support a top predator feeding on the weasel.

Energy flow and primary production

All ecosystems are dependent on a flux of solar energy, and all organisms above the producer level are dependent on the ability of the autotrophs in the system to capture and convert solar energy to form high-energy chemical compounds. However, not all systems of primary producers (the autotrophs) convert solar energy with the same degree of effectiveness. A tropical forest may convert about 1% of the incident solar energy to plant material; a tall grass prairie about 0.1%; and a desert 0.05% or less. An acre of corn may convert about 1.6% of the incident sunlight during its growing season, but averaged over the year the annual efficiency would be about 0.4%. For example, the net primary production of a tropical rain forest may be from 1.0 to 3.5 kg m^{-2} yr^{-1}; a temperate evergreen forest, 0.6 to 2.5 kg m^{-2} yr^{-1}; a savanna, 0.2 to 2.0 kg m^{-2} yr^{-1}; a tundra, 0.01 to 0.40 kg m^{-2} yr^{-1}; and a desert, 0.01 to 0.25 kg m^{-2} yr^{-1}. Cultivated land can produce 3.5 kg m^{-2} yr^{-1} or more, but the world average is only 0.65 kg m^{-2} yr^{-1}. The open ocean produces an average of 0.125 kg m^{-2} yr^{-1}. (More details may be found in Whittaker, 1975.)

NATURAL SELECTION

Charles Darwin introduced the concept of natural selection, a concept that can be tested by observation. NATURAL SELECTION is defined as differential reproduction and survival of an individual organism by means of inherited genetic traits. Natural selection is the process by which those individuals of a species that possess characteristics that help them adapt to their environment survive and transmit those characteristics to successive generations; during this process, those less able to become adapted tend to die out. Natural selection is also known as "survival of the fittest." As used by Charles Darwin, natural selection was a concept that helped to explain many observations he had made during the voyage of the *Beagle*. Darwin had first suggested the process of natural selection in 1844, but it was

only in the early 1900s that geneticists demonstrated it could exist, and in the 1920s that it must exist. Finally, in the 1950s an amateur lepidopterist—an M.D. by the name of H.B.D. Kettlewell, living in England and studying moths—demonstrated that natural selection does exist. He was collecting a species of moth: the peppered moth (*Biston betularia*), which inhabits dense woods. It is normally a light-colored moth with a blotchy wing pattern, but a few dark-colored forms, known as *B. carbonaria*, always existed in the moth population. Kettlewell astutely noted that a changing environment had an effect on the mixture of black and white forms in the population and that wing color was a factor affecting the survival of the moths.

Industrial melanism

MELANISM refers to dark pigmentation developed in the tissues of any organism. In moths, it refers to the dark pigmentation in the wings or parts of the body surface. The dark form of peppered moth (*B. carbonaria*) possesses the melanism trait. The region of England where Manchester and Birmingham are located (the Midlands) is heavily industrialized. Beginning early in the nineteenth century and until the 1950s, smoky emissions streamed forth from the uncontrolled industrial furnaces. These coal-blackened emissions deposited soot on rocks, soils, vegetation, and buildings, blackening everything for miles around.

The peppered moth spent the daylight hours resting on lichen-covered tree branches. Before the nineteenth century, the moth had evolved a light, mottled coloration that camouflaged it to match the natural pattern and color of the lichens and bark. These moths were preyed upon by birds, yet the adaptive coloration afforded them a high probability of not being seen by birds. Those moths that were best color-adapted survived and those that were least adapted were detected and eaten. Always among the population of light-colored peppered moths were some carrying the trait of black pigment, or melanism—a character that is genetically controlled, at least 90% of it by a single gene. As Manchester grew in size, the black smoke of industry killed off the lichens (which themselves were of light color) and blackened tree branches. Near Manchester, the black form of peppered moth was rare in 1848, but by 1900 it constituted 95% of the population. The light-colored form had been exterminated by birds. In the period 1952 to 1964, the light-colored form, once so abundant, was totally absent in the Manchester area. Air pollution controls were introduced in 1952, and a significant reduction in emissions was achieved after that time. Tall stacks installed on power plants also lifted the emissions above the local countryside. A cleaner environment around Manchester resulted, and by the 1970s the percentage of light-colored peppered moths was on the increase.

This selective effect of industrial pollution on the degree of melanism in a population of moths is known as INDUSTRIAL MELANISM. It is an ecological consequence of environmental pollution.

Polymorphism and energy-related selection factors

POLYMORPHISM exists among most plants and animals. (Polymorphism is a genetic situation in which more than one allele for a given character or trait exists within a population. The frequencies of the alleles can be affected by selection.) H.B.D. Kettlewell wrote, "The same laws which govern the adaptation of moths to a changing environment can also be applied to all other living things, including man himself." Thus, we can expect to find other examples of character shifts within a population of organisms, shifts caused by natural selection as the environment becomes impacted by the

effluents of human society. Well known are the insects that develop a resistance to DDT or the microorganisms that evolve to tolerate a drug being used to combat a disease, such as malaria. In 1948, 12 species of insects were found to contain strains resistant to DDT; then, by 1954, 25 species contained resistant strains; 137 were found in 1960, and more than 165 were found in 1967. In 1970, 224 species of arthropods were found to be resistant to pesticides; by 1980, there were 428. (If attempts are made to harvest large quantities of biomass for energy production, broadcast use of insecticides could make resistance an energy-related ecological issue).

RESPONSE TO PHYSICAL FACTORS

Organisms display a diversity of responses to the physical factors of their environments. Some organisms are highly sensitive to certain physical factors, and the range of conditions they can tolerate is severely limited. Other organisms have broad ranges of tolerance and are much less sensitive to environmental factors. An organism must have a suitable amount of heat, light, water, nutrients, and other factors in order to carry on growth and reproduction. Too much of some factors may be just as bad as too little.

Although each stage in the life cycle of an organism has a limiting-factor response, some stages are much more sensitive than others. Usually the tolerance ranges of seeds, eggs, embryos, larvae, and seedlings are much narrower than those of adults. Juveniles are almost always more sensitive to environmental factors than are adults. For example, some seedlings must grow in the protective shade of an adult. Such is the case with the Saguaro cactus of the Sonoran Desert in Arizona. The term PHYSIOLOGICAL YOUTH has been used to describe the fact that the young are most sensitive.

The tolerance ranges of all organisms change seasonally as acclimation takes place. Acclimation involves a variety of complex biochemical, biophysical, and physiological changes.

Law of limiting factors

The LAW OF LIMITING FACTORS states that the population size of any particular species may be limited by one or more factors that are present in inadequate or overabundant amounts. The law of limiting factors was first demonstrated for very simple ecosystems, such as crops. Natural ecosystems have so many interacting organisms and so many interconnections among them that a simple cause and effect relationship is often difficult to demonstrate. Nevertheless, numerous examples of specific factors limiting the abundance of a particular plant or animal species can be cited.

An interesting experiment was done at the Rothamsted experimental farm in England with a series of grass plots, which were first established in 1856. Many of the plots were unmanaged and were as nearly "natural" as possible. They were unfertilized and contained about 60 species of plants. Species diversity was high and no one plant was dominant. However, when some of these plots were fertilized with phosphorus, potassium, sodium, and magnesium, but no nitrogen, legumes (members of the pea family) became strongly dominant and other species were reduced in number. Legumes have the ability to fix nitrogen readily through microorganisms in their roots. Thus, they grew abundantly and outcompeted other plants if the other elements that they required were supplied in sufficient amounts. However, when nitrogen was supplied through fertilization, the other plant species responded vigorously and suppressed the legumes.

Sixteen different chemical elements have been identified as ESSENTIAL for the survival of most species. They are carbon, hydrogen, oxygen, nitrogen, phosphorus, potassium, calcium, magnesium, sulfur, iron, manganese, boron, molybdenum, copper, zinc, and chlorine. Several other elements are required in TRACE amounts by some species; these elements include sodium, vanadium, cobalt, iodine, selenium, silicon, fluorine, and barium. Most elements seem to be consistently low or consistently high in many ecosystems. Many soils are low in phosphorus, and the growth of plants will be limited as a result. However, an essential element present in too high a concentration may be toxic to a particular organism. Too much boron in a soil will be toxic to many plants and animals.

The conditions under which an organism lives may determine its requirement for a particular element. For example, a plant growing in the sun may require a substantial amount of zinc, whereas the same plant growing in the shade may need very little zinc. Calcium may be a limiting element for building the bones of vertebrates or the shells of mollusks. But mollusks can sometimes substitute strontium for calcium in the manufacture of shell material. (Strontium atoms follow the metabolic pathway of calcium atoms in vertebrates and are deposited in bones and teeth. This is particularly unfortunate because radioactive strontium is a byproduct of nuclear reactors and weapons.) Some plants take up and utilize more sodium when their potassium supply is inadequate.

The whole subject of limiting factors, or the law of tolerance, is extremely complex because nothing acts alone in nature. Most plant and animal responses to any given element or environmental factor are synergistic in that the response to one element or factor depends on the amounts of all the others. For example, when nitrogen is limiting to grasses, their resistance to drought diminishes. This fact became very evident when the effects of the "dust bowl" drought of the 1930s was compared with those of the 1950s and of the 1970s. Crops with good supplies of fertilizers were much hardier and could withstand extended periods of dryness much better than those without.

Several aspects of the law of limiting factors must be kept in mind when thinking about ecosystem response to human intrusion:

1. An organism may have a wide range of tolerance to some factors and not to others.
2. Organisms that are most tolerant are likely to be most adaptable to a range of habitats.
3. An organism's tolerance to a given factor may change with the availability of other factors.
4. Organisms in nature may not always be growing under optimum conditions.
5. Usually during juvenile or reproductive stages of the life cycle, an organism is least tolerant and therefore most susceptible to limiting factors.

Temperature response

All organisms are sensitive to temperature and have distinct temperature preferences. Some organisms have broad temperature tolerances and some very narrow. The antarctic fish *Trematomus benacchi* has a temperature tolerance range of 4°C (from −2° to 2°C), whereas the desert pupfish *Cyprinodon macularius* tolerates temperatures from 10° to 40°C, with a temperature preference around 20°C. The general temperature tolerances of various animals are shown in Figure 3. Most COLD-BLOODED ANIMALS, known as POIKILOTHERMS, have low metabolic rates and body temperatures quite close to the temperature of the air or water in which they are immersed. In strong sunlight, reptiles often may have body temperatures 10°C or more above the air temperature. WARM-BLOODED ANIMALS, known as *homeotherms*, have body temperatures that are regulated to remain within narrow limits—except for some animals during hibernation, when they go into an essentially cold-blooded mode. The normal body temperature for humans is about 37°C, but body temperatures as low as 22°C and as high as 43°C have been survived. Birds have the highest body temperatures of any animals, being normally 41° to 44°C; birds have recovered from 48°C. Plants generally have temperatures close to air temperature, except in strong sunlight, when leaves may be 10° to 20°C above the air temperature, or under a clear, cold sky at night, when they may be 5° to 8°C below the air nearby.

All plants and cold blooded organisms have a behavioral and growth response that increases with increasing temperature above some low temperature tolerance limit; the behavioral and growth response reaches a maximum at some optimum temperature and decreases sharply at higher temperatures, until the thermal maximum is reached. See Figure 4. The rate of egg development for most insects follows this temperature response.

The photosynthetic rate of all plants follows a similar temperature function. Figure 5 shows two photosynthetic rate curves; one for cold-adapted and one for warm-adapted plants. For example, the bristlecone pine, *Pinus aristata*, which grows at high altitudes in the mountains of California, has a temperature optimum of 15°C. Corn (*Zea mays*), which is an important food crop derived from a tropical grass, has a temperature optimum of 35° to 38°C. A desert perennial growing in Death Valley (*Tidestroma oblongifolia*) has its optimum at 45°C (not shown in Figure 5) and is still capable of photosynthesis above 50°C, a temperature at which most plants are shut down. In fact, many plants begin to sustain thermal damage when their tissue temperatures exceed 43° or 44°C. However, plants adapted to

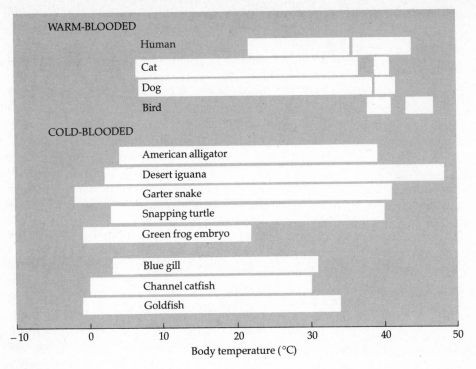

FIGURE 3. Survival temperatures and body temperatures for various animals. The shaded bands represent the normal range of body temperatures for warm-blooded animals.

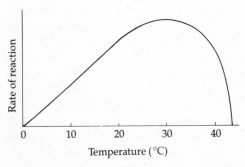

FIGURE 4. A typical temperature response curve for many organisms. For example, the photosynthetic rate of all plant leaves has this response, and the hatching success of insects also varies with temperature in this manner.

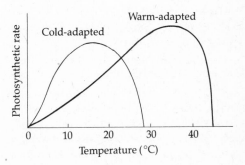

FIGURE 5. Characteristic temperature responses of photosynthetic rate in warm-adapted and cold-adapted plants.

high-temperature habitats have good resistance to thermal damage at the temperatures likely to be encountered, although they may have relatively poor cold resistance to low temperatures.

The respiration rates of all plants and animals increase monotonically with temperature until some high temperature is reached—probably around 45°C. At this high temperature, the respiration process will break down.

Algae usually optimize their photosynthetic and growth response to the temperature of the stream or lake in which they are living. However, because of temperature changes throughout a yearly cycle many bodies of water have organisms with different temperature optima. For example, many diatoms grow best when the water temperature is around 20°C, but when the temperature is near 30°C the green algae grow best and the diatoms poorly. When the water temperature is as high as 40°C, most of the diatoms will die, quite a few of the green algae will die, and the blue-green species will grow and reproduce best.

Blue-green algae are distinctive in many respects. They are believed to have originated in Precambrian times. Blue-green algae are the only prokaryotic organisms carrying out oxygen-evolving photosynthesis. Their photosynthetic apparatus is located in membranes (in the eukaryotic green algae, the photosynthetic apparatus is in the chloroplasts). Blue-green algae tolerate temperatures as high as 70° to 73°C, except when acid conditions exist. Studies of hot springs (Brock, 1973) showed that at pH values over 5 only blue-green algae were present, at pH values below 4 only green algae were present; and in the pH range between 4 and 5 both groups coexisted. Blue-green algae often produce massive blooms in polluted waters and a number of species produce toxins poisonous to fish and animals.

The temperature response of algae is of particular importance when considering the thermal effluent from power-generating plants. In streams or lakes receiving thermal effluent, increased concentration of blue-green algae would constitute a poor food source for fish, whereas the diatoms, which would be found at lower temperatures, would put energy directly into the food web. In addition, dense bacterial mats are associated with the blue-green algae in many hot springs.

The scientific literature concerning the temperature responses of organisms is very large. Important references include Gates (1980), Precht et al. (1973), and Rose (1967). These books may be consulted for further details.

Light response

PHOTOSYNTHESIS is a photochemical process requiring light to convert carbon dioxide and water to carbohydrates, starch, and sugars. Most plants respond to increasing light intensity with increasing photosynthetic rates in the manner shown in Figure 6. However, some plants require shade to grow well and must avoid direct exposure to sunlight. These shade-loving plants respond photosynthetically with more sensitivity to low levels of solar irradiance than do sun-loving plants. However, the sun plant has a greater photosynthetic rate in full sunlight than does the shade plant.

Aquatic plants respond to light in a manner similar to that of land plants. However, there is evidence of light inhibition of photosynthesis in marine phytoplankton at light intensities approaching full sunlight. There is also some evidence that ultraviolet light limits the development of aquatic plants.

Animals respond to light in a number of ways. Many organisms have a BIOLOGICAL CLOCK, which is a physiological mechanism for determining time. For some organisms,

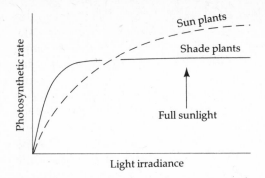

FIGURE 6. Photosynthetic rate as a function of light intensity for shade-grown and sun-grown plants. For most plants the photosynthetic rate is light-saturated when the plant is in full sunlight.

the biological clock seems to be regulated by external factors such as light, and for other organisms by internal factors. Some plants and animals possess responses having about a 24-hour periodicity. These are known as CIRCADIAN RHYTHMS. Timing mechanisms regulate endocrine changes, gonadal development, color changes in birds in spring and autumn, mating behavior in moths, insect feeding, flowering, and many other responses. Light intensity and daylength clearly are important in some of these. If an environmental change occurs in, for example, illumination of a construction site, strip-mining operation, or other industrial activity, it may have direct impact on the biological clocks of many plants and animals nearby.

Carbon dioxide response

Plants assimilate carbon dioxide in the process of photosynthesis, producing carbohydrates and other compounds. The photosynthetic rate of all plants increases with the concentration of carbon dioxide in a linear manner at normal atmospheric concentration. Only at about four times normal atmospheric concentration do photosynthetic rates become saturated with carbon dioxide. See Chapter 9 for a more complete discussion of atmospheric carbon dioxide.

PROPERTIES OF WATER

Water is essential to all life and constitutes about 70% of the weight of most plants and animals. Because of its unique physical properties, water plays a dynamic role in shaping the landscape and creating special habitats for all organisms. Water is the only substance commonly found to exist in all three physical states at the earth's surface, that is, as liquid, solid, and gas. It is colorless, odorless, transparent to visible light, and opaque to infrared radiation. Water vapor in the atmosphere allows sunlight to pass through to the earth's surface, where the sunlight heats the ground, evaporates water, or is used for photosynthesis. Water has low viscosity and flows easily. Water has its greatest density at 4°C. Therefore lakes freeze at the surface and have bottom waters at 4°C. Water expands upon freezing. Ice is less dense than water and floats. These properties are of particular value to life. Many organisms can survive the winter in the liquid cold water of deep lakes, essentially undisturbed by ice. Water evaporates readily into the vapor state and consumes heat in the process. The energy used for evaporation is carried away by the water vapor. Therefore, it is effective as an evaporative cooler or air conditioner for plants and animals. Water vapor is less dense than air. As a result moist air is less dense and more buoyant than dry air. Moist air rises in the atmosphere until cooled to the condensation temperature, at which level clouds are formed.

Plants absorb water through their roots whereas solar heating of the plant leaves removes water by causing release of water vapor into the atmosphere, a process known as TRANSPIRATION. The water stream flowing from roots to leaves transports from the soil all the vital nutrients needed by the plant for growth. Water in the leaf is necessary for photosynthesis and growth. A lack of water inhibits plant growth, whereas an adequate water supply promotes growth. Too much water can be detrimental to those plants that are adapted to using lesser amounts. Many plants are, of course, adapted to growing in streams, lakes, oceans, or wetlands. Just as animals depend on plants for food in their role as consumers, so do they satisfy some of their water needs by eating plants.

BIOGEOCHEMICAL CYCLES

Energy flows through ecosystems, whereas nutrients tend to cycle within them. Nutrient elements such as nitrogen, oxygen, carbon, phosphorus, and sulfur occur in different forms within organisms and environments. Because elements cycle within and through the biota and abiotic components of all environments, the cycles are referred to as BIO-GEOCHEMICAL. The basic elements cycle through the environment by means of chemical reactions and physical transport. Also, when formed into a variety of chemical compounds, the elements cycle through organisms in many special ways. When humans interrupt these cycles or modify them through additions of certain compounds to the environment—by interruption of flows or by harvesting certain organisms—the impact on the entire ecosystem is considerable. Not only are localized biogeochemical cycles being perturbed today, but these perturbations are becoming global in scale. In particular, the use of fossil fuel energy is changing the carbon, nitrogen, and sulfur cycles, the use of fertilizers is changing the nitrogen and phosphorus cycles, and the use of biomass for food, fiber, and fuel is modifying the hydrologic, nitrogen, and phosphorus cycles. Therefore, it is appropriate to consider these cycles in some detail.

Hydrological cycle

Water is evaporated by the sun from lakes, ponds, soils, and vegetation; it rises into the sky, where it condenses and falls back to earth as rain, snow, or dew and becomes available to the biosphere (Figure 7). The hydrological cycle is the simplest of the biogeochemical cycles. Flowing water may erode the landscape, weather the rocks, or freeze and break down both organic and inorganic structures. Over the world as a whole the evaporation and precipitation of water is in balance, although evaporation exceeds precipitation over the oceans and vice versa over the land. Only about 5% of the earth's total water is in circulation through the hydrosphere (the region of the earth occupied by water and by water vapor) and 95% is bound up in the lithosphere (the outer part of the earth composed of rock). On the average, a water molecule is transported through the atmosphere for about 10 days before precipitating out.

Plants and animals respond to a combination of water, temperature, and light. High temperatures and much sunlight cause high evaporation rates of water from soil, lakes, and oceans and high transpiration rates from vegetation. The rate of water loss from any surface depends in particular on the energy exchange with that surface by radiation and convection. The community of plants and animals present in an ecosystem depends on the availability of water. The great deserts of the world, such as the Sahara, Namib, Kalahari, Atacama, and Sonoran, are in areas where the annual precipitation is under 25 cm. Here one finds cacti, euphorbias, and other succulents. Arctic tundras, with water in frozen form more than half the year, have an annual precipitation less than 25 cm. They have sedges and grasses, but no trees. In warmer regions of the world, grasslands, savannahs, or open woodlands exist where annual precipitation is from 25 to 75 cm. Dry forests are found throughout much of North America, Europe,

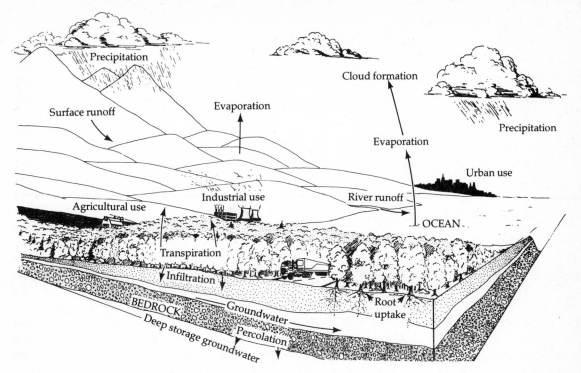

FIGURE 7. The major features of the global hydrological cycle. An average water molecule remains in the atmosphere about nine days before returning to the land or ocean surfaces.

and Asia, in areas where annual precipitation is between 75 and 125 cm, whereas rainforests are found where the annual rainfall exceeds 125 cm. The cutting of a rainforest on a Caribbean island has been shown to modify the annual precipitation from a wet to a dry condition such that the rainforest cannot return to the island. In fact, the presence of the rainforest cools the air and promotes condensation, cloud formation, and precipitation, which in turn promote growth of the rainforest.

Fresh water is not only a critical factor for the functioning of natural ecosystems, it is essential to the survival of humanity. Water is required in enormous quantities for operation of power-generating plants, mining of resources, production of steel, and conversion of shale or coal to liquid hydrocarbons. Water is essential for the irrigation of most agriculture. Massive competition for water is creating acute shortages in some regions of the world. Continuing modification by humans of the global hydrological cycle, in part or in whole, can have serious consequences on climate and in turn on the future well-being of humanity.

Carbon cycle

Approximately 49% of the dry weight of organisms is carbon, an essential element of all organic compounds. The carbon cycle is basically quite simple (Figure 8). The only source of carbon to producer organisms is atmospheric carbon dioxide, which is transformed from its gaseous state into carbohydrates and other organic materials by

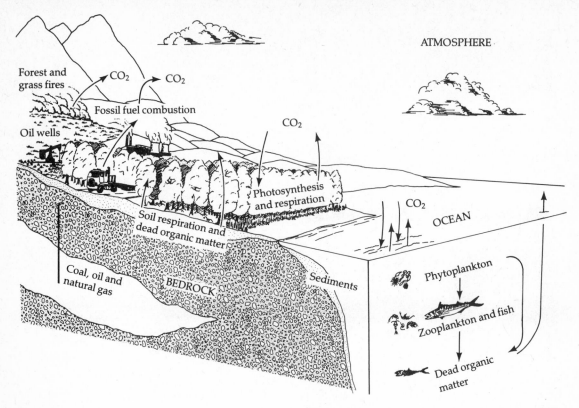

FIGURE 8. The carbon cycle, showing the sources and sinks in CO_2 exchange.

photosynthesis. Respiration of carbon dioxide by all organisms returns carbon to the atmosphere. Forest and grass fires also release carbon dioxide to the atmosphere. Carbon is deposited in soils and sequestered in sediments of streams, lakes, and oceans. Although the cycle is simple, the carbon exchanges through the food chain and the carbon chemistry of ocean surface waters are complex and not thoroughly understood.

The estimated reservoirs of carbon in the world and the carbon fluxes are shown in Figure 9. The deep ocean sediments contain approximately 32,000 Gt, the fossil fuels (oil, gas, and coal) 12,000 Gt, and the terrestrial plants and animals 1760 Gt. A gigaton (Gt) is 10^9 metric tons = 10^{12} kg. The size of the atmo-

spheric reservoir is currently just over 718 Gt, a quantity that represents a concentration of 339 parts per million (ppm) or 0.0339% by volume. The atmospheric concentration of carbon in the form of carbon dioxide gas is increasing at a rate in excess of 1 ppm per year as the result of the burning of fossil fuels, the manufacture of cement, and the cutting of forests. This flux of carbon is estimated at 5.0 to 7.0 Gt/yr. This human activity, which is significantly affecting the atmospheric carbon reservoir, and through it the earth's climate, is one of the great environmental issues of our time. It is described in Chapter 9.

Prior to human intervention, the carbon fluxes in and out of the atmosphere may have been in equilibrium, at least for short periods

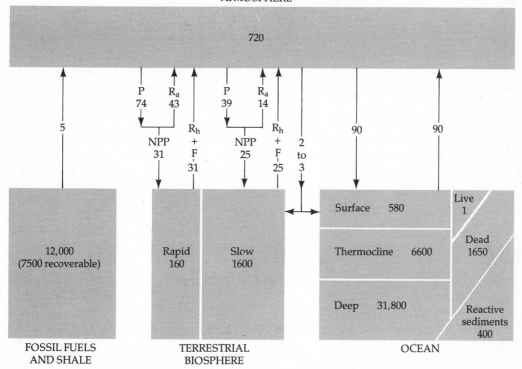

FIGURE 9. The reservoirs of carbon in the world in gigatons (Gt) and the annual fluxes of carbon in Gt/yr. 1 Gt = 10^9 metric tons = 10^{12} kg of carbon. P, Gross primary production or production by photosynthesis; R_a, carbon released by respiration of autotrophs (mainly the green plants); NPP, net primary production ($P - R_a$); R_h, carbon released by heterotrophic respiration (by fungi, bacteria, and animals); F, carbon released by fires.

of time (less than 1000 years). It is now estimated that the atmospheric concentration of carbon dioxide prior to the industrial revolution may have been about 268 ppm. There is evidence that the concentration was as low as 150 ppm during the peak of the last ice age, about 50,000 years ago.

It is clear from Figure 9 that the largest fluxes of carbon flow between the oceans and the atmosphere. The fluxes between the land surfaces and the atmosphere are somewhat smaller. It is thought that there may be a small net flux from the terrestrial biosphere to the atmosphere of about 2 Gt/yr from the cutting and burning of forests, and a net flux to the oceans of about 2.5 or 3.0 Gt/yr. Our understanding of these fluxes is not sufficient to make a more precise statement about them. The oceans may be the only major net sink for carbon. Therefore, a more careful consideration of the oceanic carbon cycle is in order.

The principal forms of carbon in the ocean are inorganic, occurring as dissolved carbon dioxide plus carbonic acid ($CO_2 + H_2CO_3$), bicarbonate ions (HCO_3^-), and carbonate ions (CO_3^{2-}) in percentages of 1, 89, and 10, respectively. The precise amounts of each of these compounds depends on the pH of the water. Calcium carbonate rock is formed in the oceans by a combination of carbon ions with calcium present in the water. The surface waters of the ocean (upper 70 m) are supersaturated in CO_2 with respect to calcium carbonate ($CaCO_3$). The deep ocean waters are undersaturated because of higher pressure, lower temperature, and increasing acidity due to the release of CO_2 by respiration from organisms. The various reactions involving carbon compounds in seawater are completely reversible and may be written as

$$CO_2 + H_2O \rightleftharpoons H_2CO_3 \rightleftharpoons H^+ + HCO_3^-$$
$$\rightleftharpoons H^+ + H^+ + CO_3^{2-}$$

If the concentration of CO_2 increases in the atmosphere, and by diffusion in seawater, the reaction will drive to the right and more bicarbonate and carbonate will result. However, the real limiting factor to the rate of such a process is the time it takes for CO_2 to diffuse across the atmosphere–sea surface interface. It is estimated that it would require hundreds of years or longer for CO_2 emitted into the atmosphere by the burning of fossil fuels and the cutting of forests to diffuse into the oceans.

Nitrogen cycle

Nitrogen is an essential component of protein, particularly of amino acids, and is a highly important element in biogeochemical cycles. It is the most abundant element in the atmosphere, 78% by volume for a total mass of 3.9×10^{18} kg. Nitrogen (N_2), unlike CO_2, cannot be used directly from the atmosphere by plants but is absorbed by plant roots from the soil in the form of nitrate (NO_3^-), nitrite (NO_2^-), ammonia (NH_3), or ammonium (NH_4^+). Animals obtain their nitrogenous compounds by eating plants and passing them through the food chain.

The nitrogen cycle is fairly complex; nitrogen moves through the biosphere along several pathways (Figure 10). Most of the nitrogen incorporated into the tissues of organisms arrives there only after FIXATION (also commonly referred to as REDUCTION) by numerous microorganisms, some of which exist in the roots of plants. Nitrogen fixation is the process by which gaseous nitrogen is combined with other elements, in this case with hydrogen to form ammonia (NH_3) or ammonium (NH_4^+). Industrial fixation of nitrogen produces fertilizers, which are used in agriculture. Small amounts of nitrogen move from the atmosphere to the soil as NH_4^+ and NO_3^- in rain and are then absorbed by roots. This NH_4^+ comes from industrial burning, volcanic activity, and forest fires; and NO_3^- arises from oxidation of N_2 by O_2 or by O_3, which is produced by lightning and solar ultraviolet radiation. NO_3^- is also produced by ocean salt spray that ends up in rainwater.

Decaying plant and animal remains and manure from animals return nitrogenous compounds to the soil. Most of these compounds are insoluble and not directly available for plant use. However, soil bacteria make great use of these compounds in metabolism, and in the process they make reduced nitrogen again available to plants. The most important processes in the nitrogen cycle are AMMONIFICATION, NITRIFICATION, and DENITRIFICATION, each of which involves organisms.

AMMONIFICATION is the process by which ammonium (NH_4^+) is produced during the decomposition of nitrogenous organic matter. Heterotrophic (other feeding) bacteria, actinomycetes, and fungi, occurring in soil or water, utilize organic (nitrogen-rich) compounds for metabolism and in the process convert some of this to NH_4^+. The ammonification process is defined as exothermic because it releases energy that the organisms use for metabolism. The ammonia released by the organisms is simply their way of getting rid of excess nitrogen.

NITRIFICATION by bacteria is a two-step

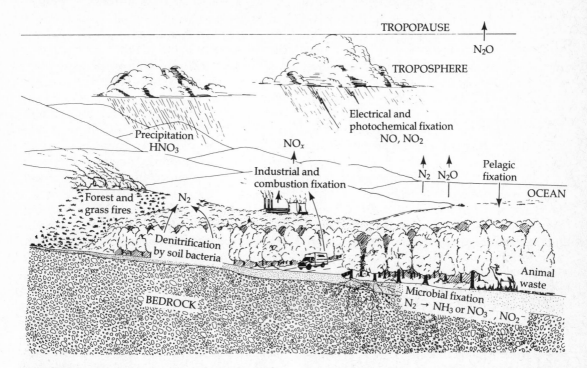

FIGURE 10. The nitrogen cycle. The troposphere is the lower part of the atmosphere and its temperature decreases with increasing height; the stratosphere has an increasing temperature with increasing height; the tropopause is the boundary between the two regions. Pelagic refers to the open ocean and specifically to its surface water. NO_x represents various oxides of nitrogen such as NO and NO_2.

process. NH_4^+ is converted to NO_2^- by *Nitrosomonas* bacteria, and NO_2^- is converted to NO_3^- by *Nitrobacter* bacteria. The pH of the soil is critical for these reactions, which work best under warm, moist conditions with about neutral pH (pH 7.0). Nitrification may also occur in water. In many acid soils, nitrifying bacteria are less abundant; under these circumstances NH_4^+ becomes a more important nitrogen source than NO_3^-. Many forest trees absorb most of their nitrogen as NH_4^+ because forest soils tend to be slightly acidic. Acid rainfall will exacerbate the situation and make it even more difficult for vegetation to absorb nitrogen. Because of its positive charge, NH_4^+ tends to adhere to soil particles and is therefore not as readily absorbed by plant roots as NO_3^-, which does not adhere.

DENITRIFICATION is the process by which molecular nitrogen is released as an end product and returned to the atmosphere. This process also releases energy to be used by soil organisms for metabolism. Nitrate (NO_3) is reduced by NO_2^-, N_2, or N_2O by bacteria such as *Pseudomonas* and by fungi. The oxygen released in the presence of glucose or phosphate provides the high energy compounds for life. Denitrification occurs generally under anaerobic conditions and therefore is most likely to occur in oxygen-poor soils (poorly aerated) or where there is much organic decay, a situation of high oxygen demand. The surface waters of lakes are well stirred by wind and are aerobic. Hence, one would not expect denitrification to occur under such circumstances. On the other hand, one might expect denitrification to be taking place in the bottom waters of many lakes, where much organic matter accumulates and essentially ANAEROBIC conditions develop.

It is clear from the description of these three processes of ammonification, nitrification, and denitrification that fungi and various microorganisms play an extremely critical role in the nitrogen cycle and therefore in primary productivity. Many of these organisms are free-living in the soil or water, but others, particularly some of the bacteria, live symbiotically with some vascular plants in nodules formed on their roots. Vascular plants on which root nodules form are the legumes (peas, beans, alfalfa, acacia, clover, and soybeans) and some species of pine and alder. When these plants are present in any natural system of mixed vegetation, the nitrogen availability goes way up and the productivity of the other plants present increases very strongly. The response of the whole ecosystem may be quite striking to see. The alders growing along lakes or streams may greatly affect the productivity of the waters themselves. Nitrogen compounds are highly soluble in water, do not readily bind to soil particles, and therefore are easily carried by ground waters into lakes and streams. Nitrogen, used in fertilizers spread on lawns or agricultural fields, dissolves in rain or irrigation water, and the runoff enriches streams and lakes. The nitrogen enrichment produces a rapid growth of algae and higher aquatic plants. These plants then grow abundantly; when they die, much of the available oxygen is consumed during the decay process and the

lake becomes ANAEROBIC (lacking oxygen). This often occurs in lakes surrounded by great expanses of fertilized lawns.

It is evident that the nitrogen cycle is absolutely essential to life and that any human activities that interfere with or perturb it will affect the living systems. Automobiles and power plants release massive amounts of oxides of nitrogen (NO_x) to the atmosphere, from which they are returned to the earth's surface by diffusion and mixing or by precipitation, a process contributing to acid rain. To understand the impact of the oxides of nitrogen on ecosystems, one must thoroughly understand the nitrogen cycle. Unfortunately our understanding of the global nitrogen cycle is far from complete, particularly with regard to the fluxes of nitrogen compounds to or from each part of the global ecosystem.

The largest reservoirs of nitrogen are rocks, sediments, and the atmosphere, with 1.9×10^{20}, 4×10^{17}, and 3.9×10^{18} kg N, respectively. Dissolved nitrogen in the oceans is estimated to be 2.2×10^{16} kg. All other reservoirs are small by comparison. Plant biomass, for example, has about 1.4×10^{13} kg N.

Combustion used in human activities, one of the better-known fluxes, puts about 19×10^9 kg N yr^{-1} into the atmosphere. The global biological fixation rate in terrestrial ecosystems is about 139×10^9 kg N yr^{-1} and in oceanic ecosystems about 20 to 120×10^9 kg N yr^{-1}. Industrial activities throughout the world generate about 36×10^9 kg N yr^{-1}. [Further information about the nitrogen cycle may be found in Svensson and Soderlund (1975), Clapham (1973), and Kormondy (1976).]

Sulfur cycle

All organisms require some sulfur. Inorganic sulfate (SO_4^{2-}) is a major source of sulfur for organisms. Sulfur is used in the folding of amino acids to form protein molecules in protoplasm. Plants and microbes can reduce sulfate for protein synthesis, but animals obtain sulfur-containing ions from the amino acids in their food.

The sulfur cycle is shown in Figure 11. The transfer rates given are in 10^9 kg S yr^{-1}. Figures in lighter type are the rates of transfer of sulfur believed to have occurred prior to human influence on the cycle and the boldface figures show the estimated inputs from human activities.

Sulfate is abundant in nature; it is leached from the soil and is replenished by rain. Consequently, vegetation can usually meet its needs. Volcanoes eject sulfur into the atmosphere. Overcropping of vegetation can deplete sulfate availability so much that sulfate must be replaced by use of fertilizers. Humans are providing another source of sulfur to the world by burning fossil fuels, which release sulfur as oxides of sulfur or as sulfate. When these compounds are mixed with rain, the rain becomes dilute sulfuric acid (H_2SO_4). (Acid rain is discussed in Chapter 10.) Sulfates are much more soluble than phosphates and move readily between land and ocean, where they accumulate. Sulfur is 240 times more abundant in seawater than in fresh water. Sulfur deposits accumulate where seawater evaporates, and gypsum (calcium sulfate crystals) remains on the surface. The sedimentary or mineral phase of the sulfur cycle is also important. Sulfur is precipitated in sediments in the presence of cations of elements

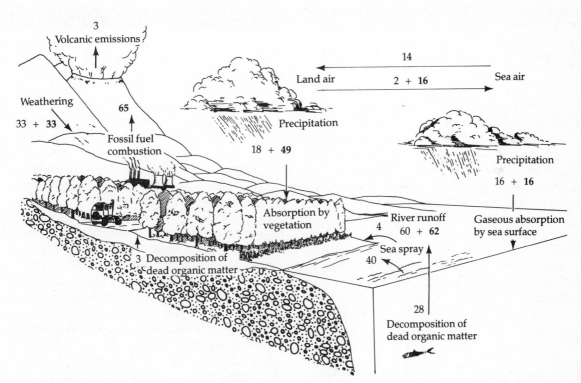

FIGURE 11. The sulfur cycle. The numbers in lighter type are the estimated transfer rates for the natural global ecosystem prior to human interference; those in boldface represent the human inputs to the ecosystem. All transfer rates are expressed as 10^9 kg S per year.

such as iron (Fe) and calcium (Ca) to form ferrous sulfide (FeS), ferric sulfide (Fe_2S_3) (which is iron pyrite), and calcium sulfate ($CaSO_4$).

Plants absorb sulfate through their roots and to some extent through their leaves. It then is passed to animals through the food chain and returned to the soil by excretion or by death and decay. Sulfur is released to the atmosphere as sulfur dioxide (SO_2) by burning vegetation. Sulfur occurs in living tissue as the sulfhydryl molecule within amino acids. Organic matter decomposes through the activity of bacteria or microbes, which convert the sulfhydryl molecule into hydrogen sulfide (H_2S). Under aerobic conditions, some chemosynthetic bacteria oxidize the hydrogen sulfide to sulfate (SO_4^{2-}) and use the energy liberated in the oxidation to obtain carbon from carbon dioxide by reduction. The sulfate produced is then reused by green plants, which absorb it through their roots. Under anaerobic conditions, such as in the bottom waters of some lakes, it is impossible to oxidize hydrogen sulfide, but certain photosynthetic bacteria use hydrogen sulfide to manufacture carbohydrates and in the process oxidize hydrogen sulfide to elemental sulfur or to sulfate. The presence of large quantities of hydrogen sulfide in some deep lakes, such as below 200 meters in the Black Sea, probably is the reason for a lack of fish or other animals there.

As with other biogeochemical cycles, the sulfur cycle is far more complex than we have indicated. If we are to understand the impact of human activities on the sulfur cycle, all aspects of the cycle must be understood.

Phosphorus cycle

Phosphorus is a compound of major importance to organisms, playing key roles as high energy phosphorylated compounds such as deoxyribonucleic acid (DNA), ribonucleic acid (RNA), the energy-carrying molecule adenosine triphosphate (ATP) and as phosphate salts in bones and teeth. Phosphorus is actually in a higher ratio to other elements in most of these compounds than are other key elements.

Plants must assimilate phosphorus from inorganic phosphate, generally as orthophosphate ions. Organisms other than plants, that is, the consumers and decomposers, get their phosphorus in organic form from plants or secondarily from animals. The phosphorus cycle is illustrated in Figure 12.

Phosphorus does not occur naturally in the atmosphere in any form except to a small extent in dust. In waters where there is abundant dissolved oxygen, phosphorus is oxidized and forms insoluble compounds, which precipitate and are not available to aquatic plants. Eventually those precipitates form phosphate rocks. Those phosphate rocks, when uplifted by geologic processes, erode and eventually return phosphorus to the global ecosystems. Large quantities of phosphate rocks are strip-mined in Florida, and the phosphorus is used for fertilizers. Guano, the excrement of birds, is rich in phosphorus. Enormous deposits of guano are found along the west coast of Chile and Peru, where they are being mined for fertilizers.

In soils, phosphorus comes from decaying organic matter, from which phosphate is released as inorganic, ionic phosphate. In this form it can be taken up directly by plants. However, it may also be bound to soil particles. The sedimentation part of the phosphorus cycle is quite complicated. Phosphate, which is not very soluble, reacts chemically with either aluminum, calcium, iron, or manganese to form insoluble inorganic compounds. Phosphate can also be incorporated into the crystal structure of clay minerals. Acidification of soils through acid precipitation helps to promote the unavailability of phosphorus to plants.

Phosphorus in lakes occurs in three forms: inorganic, particulate organic, and dissolved

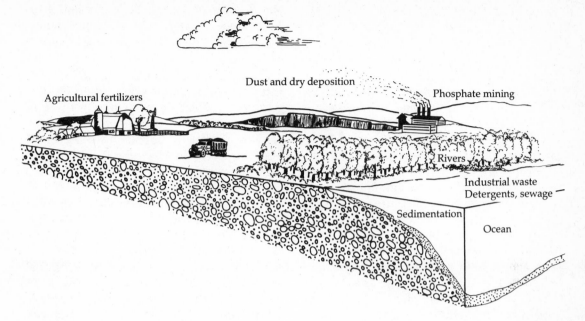

FIGURE 12. The phosphorus cycle. Phosphorus does not occur naturally in the atmosphere except on dust particles. Phosphorus is found mainly in phosphate rocks, which formed from sediments. Humans mine these for fertilizers.

organic. It cycles through all these forms. The inorganic phosphorus is orthophosphate; the particulate organic phosphorus is in suspension in living and dead protoplasm and is insoluble; and the dissolved organic phosphorus comes from particulate matter that has been excreted and decomposed.

Phosphorus entering streams or lakes as phosphate from lawn fertilizers or from household or industrial detergents may produce undesirable nutrient enrichment. The result is often a rapid growth of algae and higher plants, followed by increased respiration, reduced oxygen concentration, and fish die-off. The stream or lake becomes a stinking mess.

Lake Balaton is the largest lake in Central Europe. It is located about 100 km from Budapest. Although a shallow lake, it is an important resort region for Hungary.

Phosphorus entering from sewage and agriculture sources has polluted Lake Balaton. Heavy algal blooms occur, and the lake turns a dark green color. When this happens, the lake has an offensive odor and in general people will not swim in it. When the upper layers of water are opaque to sunlight—because of the algae—the plant life in the lower layers dies off. The decaying plant material consumes oxygen, and the reduced oxygen levels kill the fish. (Massive fish kills have resulted.) When the lake becomes anaerobic, there is a further release of phosphate from the bottom mud, and this further exacerbates the problem.

Every summer 2 million tourists each spend an average of a week on Lake Balaton. Each tourist puts about three grams of phosphorus into the lake each day, half of it from detergents and half from excrement passing through sewers. There is no organic pollution of the lake from sewage, but there is phosphorus pollution because phosphorus is not removed by primary or secondary treatment (removing phosphorus requires special and expensive tertiary treatment).

Hungary has not banned the use of phosphate-containing detergents, as have many countries. (Swiss tourists buy large quantities of phosphate-containing detergents in Hungary to take back to Switzerland.) In addition, fertilizers are being used in increasing amounts in Hungary. The amount of phosphorus, nitrogen, and potassium in the fertilizers used in the agricultural land around Lake Balaton is now about 400 kg per hectare—about seven times higher than it was 20 years ago. A good deal of the problem could be reduced by controlling soil erosion into Lake Balaton and by eliminating phosphate-containing detergents.

Lake Washington near Seattle had a serious phosphate problem. It was solved by building a circumferential sewage line at a cost of $85 million. But Lake Balaton is three times larger, and the cost today for installing such a circumferential sewer would be staggering. Many farm ponds and small lakes in America are becoming eutrophic from agricultural runoff and the use of lawn fertilizers. Many resort developments surrounding lakes encourage expansive green lawns, which they fertilize regularly. The fertilizer runs into the lake and the lake becomes eutrophic and repulsive. Minimal use of fertilizers and proper planting of vegetation along the lake shore can often limit the input of phosphorus to the lake to tolerable amounts. In New York State, a sharp reduction in the concentration of phosphorus in Lake Onondaga near Syracuse followed the implementation of legislation limiting the percentage of phosphorus allowed in detergents.

POPULATIONS AND COMMUNITIES

Any ecosystem has a large number of individual organisms, which belong to many different species. A POPULATION is defined as that assemblage representing the total number of individuals of a single species in a specified area. The number of sugar maple trees in a forest would constitute a population. Another population would be made up of the number of white birch trees, or the number of American beech. All of the populations representing all of the species in the forest would make up a unit referred to as a COMMUNITY. A plant or animal community has a population distribution in which the most prevalent species are those with intermediate abundances and the very rare and extremely common species are fewer in number, as shown in Figure 13. Note that the scale for the number of individuals per species doubles for each unit, which makes it logarithmic. The distribution shown, which is bell-shaped, is called a LOGNORMAL DISTRIBUTION. The data of the upper curve represents the relative abundances of diatom species in a natural stream community. The lower curve represents the diatom species abundances in a polluted river. Note that for the polluted water a few species are extremely common (that is, have large populations), that more rare species are present, and that the total number of species is much lower than for the unpolluted, natural stream. Most stressed ecosystems will exhibit this response. A few species will be favored even though most will be disadvantaged, whether the stress is thermal, nutrient, or acid stress. The lognormal distribution does not hold for some communities of organisms, such as those on islands. (For further reading on this topic, see Ricklefs, 1979.)

Ecological succession

Every community of organisms is subjected to a changing and fluctuating environment, and, in turn, is itself changing in composition. There are diurnal, monthly, and annual cycles, long-term trends, and short-term dis-

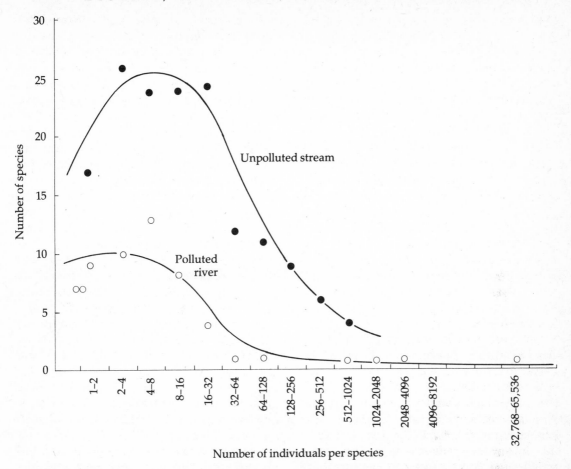

FIGURE 13. Distribution of the relative abundances of diatom species in a natural, unpolluted stream (upper curve) and in a polluted river (lower curve). The scale for the number of individuals per species doubles with each unit, making it a logarithmic scale. These curves are known as lognormal distributions. (Redrawn from data by Patrick, 1963.)

turbances. The rates of flow of energy, nutrients, and water change. Plants and animals die and are replaced by new generations. However, overall a community of organisms will appear relatively stable while at the same time containing a mosaic of patchiness caused by fires, blow-downs, floods, or other disturbing events. The apparent long-term constancy of a community structure is an indication of its relative stability. The rapidity with which a community recovers from a disturbance is also an indication of its relative

stability. The gypsy moth may defoliate a forest community but many properties of the community can help it recover. Trees have a great deal of resilience to damage and can survive one or two years of defoliation by insects. On the other hand, some insect infestations may kill off certain age classes of trees. The pine bark beetle may kill most of the old trees of a stand while younger, more vigorous trees survive. Woodpeckers, finding an abundant food supply of beetles and good nesting sites in the dead trees, will increase rapidly in numbers. Then over the next few years the population of bark beetles will diminish, and finally the number of woodpeckers will diminish as their food supply is depleted. The community of trees and animals will continue to repair and age until some time in the future when another beetle infestation may occur.

When a community of plants and animals is severely disturbed, such as by fire, lumbering, or both, there will occur great changes in the composition of the community and ecological succession will result. ECOLOGICAL SUCCESSION is an orderly process of community development that involves changes in species structure and community processes with time and culminates in a stabilized (steady-state) ecosystem. What was once a mature virgin forest may, after fire or lumbering, become an open field of grasses and herbs, followed by shrubs and young trees of certain species, followed by other tree species. Eventually, after hundreds of years, the mature forest may once again occupy the site. To be more specific, let us consider this example. A mature climax forest of red and white pine is lumbered and burned, thereby leaving a vast open field. The first year many grasses and fireweeds invade as seeds blow into the area. During the next few years these grasses will increase in numbers, but soon bracken fern, honeysuckle, and shrubs will come in and begin to shade out the grasses. Then young aspen shoots will appear (aspen being clonal, propagates by root extension); the bracken and aspen will suppress many of the grasses, and an aspen community will dominate for many decades. This aspen community will contain birch, cherry, maple, and many other trees and shrubs. However, if seed sources from mature pines are available, these will blow into the area or be carried in by animals (squirrels), and young red pine and white pine trees will begin to grow beneath the aspen. The process of pine invasion will be slow, but eventually there will be many

pines persistently rising through the aspen. Aspen is a tree that is shade intolerant and self-pruning. As they grow older and taller, they have leaves only on branches that are near the top and project into the sunlight. But the pine is shade tolerant and will grow well in the shade of the aspen until it is taller than the aspen, at which time the aspen will die off. The red and white pines will continue to grow and flourish until the forest becomes a mature climax pine forest.

At each stage of this preceding succession, the microclimate of the site was modified by the plants themselves. At first it was an open, exposed, sunlit site where the soil was readily warmed and dried, but as grasses and bracken fern covered the surface, the soil became cooler and more protected against moisture loss. After the establishment of aspen, the soil was even more shaded, cooler, and more moisture retentive. The wind moving over the site was completely different during each stage of succession. Thus, we see that the entire character of the ecosystem changed progressively during succession. The community structure changed and the physical environment changed.

On dry sandy soils, the mature red and white pine forest, once established, would be relatively stable over many hundred, if not several thousand, years and would be what the ecologist refers to as a CLIMAX FOREST. However, on more moist, richer soils, the pine community might be succeeded by a community of hardwoods made up of birch, sugar maple, beech, and maybe red oak. Eventually this hardwood forest would be the climax type.

In many parts of the world, the climate is not suitable for forests, and climax in the successional sequence might be a grass community. This situation is found in prairie and tundra ecosystems. Prairie grasses can establish extremely deep root systems and thereby pump energy and nutrients to considerable depths below ground. It is this deep root system, whereby the underground biomass may exceed the aboveground biomass by a factor of 10, that makes this community stable against the vicissitudes of drought and wind. It is also why the prairie soils are rich and productive for agriculture throughout the wheat belt and part of the corn belt of the United States. We will discuss prairie communities in more detail when we consider the western coal fields.

Succession does not always lead to climax communities, nor is succession only limited to plants. It occurs among all organisms, plants and animals. The insect or bird populations of the open field are grasshoppers, tiger beetles, sparrows, and nighthawks; of the aspen forest, wood boring beetles, woodpeckers, chickadees, and nuthatches; and of the pine forest, beetles, pileated woodpeckers, vireos, warblers, and chickadees. The number of insect and bird species is generally greater at some intermediate stage of succession than during the climax stage. Carcasses of dead animals provide food for a large number of scavenging and detritus-feeding organisms. The African savannah, which comprises grasslands and acacia-type trees, has a huge diversity of small and large animals. When a lion makes a wildebeest kill and feeds on the carcass, it is immediately followed by vultures, flies, and microorganisms of decay. The nutrients contained in the carcass pass through the food web and eventually return to the soil.

It is important to realize the dynamics that occur within ecosystems and the patchiness of communities. There is always disturbance in nature—fire, wind, thaw, drought, flood, earthquake—and the result is a certain pervasive patchiness to any community. There are always openings in a mature climax forest, whether pine or hardwoods, and in these openings grow aspen, birch, bracken fern, and grasses. The abundance of such openings, and of previous openings in various

stages of succession, is great, so that when any new disturbances occur there are adventitious plants and animals ready to move into the area.

Wherever there is disturbance by humans, there will be opportunity for succession. Whether it is the cutting of a forest for the use of its biomass for fuel, the stripping of the land surface for coal mining, the damming of a valley for hydropower, or the mining for oil shale, there is the potential for succession to follow. We will follow this story later with various energy sources.

Ecosystem stability

What about the stability of ecosystems? The concept of stability is a difficult one in ecology and one that is hard to prove because it requires long-term studies. I think of the STABILITY of an ecosystem as its ability to resist change or to minimize change in the character of the ecosystem when subject to a perturbation. The perturbation might be an extended period of drought, a heavy insect infestation, or a fire. For example, a grassland has high stability against fire because most of its biomass is beneath the ground surface.

Ecologists generally believe that increased diversity and stability increase with succession, although there are exceptions to this rule. Many ecosystems are more diverse during intermediate stages of succession than they are during early or late stages of succession.

A tropical forest is known to have a much greater diversity of species than does a monoculture of lodgepole pine in Yellowstone Park, for example. One also infers that the tropical forest is much more stable and can recover from disturbance much more quickly than might the lodgepole pine forest. We think of the arctic tundra as having very low diversity and reduced stability plus an extremely slow recovery rate from disturbance. Generally ecosystems characterized by extreme environmental factors, such as deserts or tundras, have reduced diversity. A tropical climate is much more steady and has variations of less amplitude than do either temperate, alpine, or arctic climates.

Humans, in order to supply the food demands of a burgeoning population, have planted vast acreages of monoculture, for example, corn, wheat, and sorghum. However, they can only maintain these monocultures at the expense of huge energy and nutrient inputs. We know that these monocultures are highly unstable and vulnerable to disturbance. The corn blight invasion in the midwest in 1972 is an excellent example of this vulnerability. In 1969 and 1970, corn inbreds and hybrids having the "Texas"-type cytoplasm for male sterility were introduced throughout the corn belt of the central and southern United States. This was thought at the time to be a marvelous advance in corn agriculture because by virtue of the male sterility the corn did not need to be detassled to prevent cross-pollination. However, it turned out that this "Texas"-type corn was unusually susceptible to the southern corn blight fungus. The blight began to appear along the Gulf Coast early in the growing season. Its spread into the midwest was so rapid during the summer months that its damage to the corn crop took on epidemic proportions. The damage might have been worse; as it was, 15% of the United States corn crop in 1970 was damaged. This disaster was the result of extensive monoculture. The next year would have been worse had not the use of this type of corn been discontinued.

Leave any one of these monocultures to its own existence and it soon will be overrun by natural systems. An outbreak of spruce budworm infestation is far more devastating to large contiguous stands of balsam fir than when the fir is in mixed stands with many other species. We have seen the loss in America of most of our mature American elms as a result of a fungal infection known as Dutch elm disease, which is transmitted by a bark beetle. The elms were planted throughout most of our towns and cities, as well as in the countryside. From these experiences and observations of natural ecosystems, we infer that all monocultures are highly unstable and that the less the diversity of species within an ecosystem, the less the stability of the ecosystem.

Growth of populations

If an organism has an adequate supply of nutrients and ample space and if other requirements are well met, then, starting with a few individuals, its population will grow exponentially (Figure 14A). If a pair of houseflies produced 100 offspring (50 females) and all lived to create the next generation and if the next generation did likewise, the increase in the number of houseflies would be exponential (Table 1). The first pair gives rise to 31.25 billion flies in the sixth generation. This number constitutes the population, assuming each adult dies after laying eggs. Some adults will live more than one generation and will lay more eggs. We also assume all eggs are fertile and all juveniles mature to reproduce more eggs. These assumptions are unrealistic in nature. This scenario does show, however, that exponential growth is a powerful function. Many populations tend to follow exponential growth over short periods of time, but clearly such a growth pattern cannot continue indefinitely. Before proceeding to some of the more realistic situations, let us first express exponential growth in mathematical form.

The number of individuals in a population is N. The number of births B is taken to be proportional to the number of individuals. Hence,

$$B = bN \tag{1}$$

where b is the birth rate.

The number of deaths D is also considered to be proportional to the number of individuals. Hence,

$$D = \delta N \tag{2}$$

where δ is the death rate.

The increase, or decrease, of a population per unit time is the change in number of individuals ΔN in a time interval Δt, which clearly is the number of births less the number of deaths occurring in the same interval of time. Hence,

$$\Delta N/\Delta t = B - D = bN - \delta N$$
$$= (b - \delta)N = rN \tag{3}$$

where $r = b - \delta$, the intrinsic rate of increase of the population.

If the increments ΔN and Δt are infinitesimal, then by the notation of calculus, $\Delta N/\Delta t = dN/dt$, a differential of N with time. Equation 3 can be integrated with respect to time to give

$$N = N_0 e^{rt} \tag{4}$$

where N_0 is the initial number in the population at time $t = 0$. This is the analytical expression for EXPONENTIAL GROWTH. A plot of N versus t is shown in Figure 14A. An assumption made here is that b and δ are constant with time. This is seldom true in nature. A special case occurs when the birth rate just equals the death rate, for example, $b = \delta$ and $r = 0$. Then $N = N_0$, which says that the population just replaces itself and does not grow or diminish in size.

TABLE 1. Production of houseflies[a]

Generation	Number of flies
1	100
2	5,000
3	250,000
4	12,500,000
5	625,000,000
6	31,250,000,000

[a]Assume that each female lays 100 eggs and that half of these eggs become females, each of which is capable of laying 100 eggs.

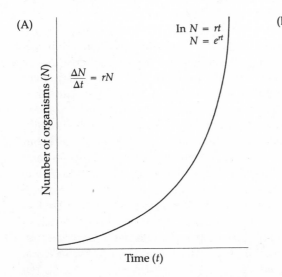

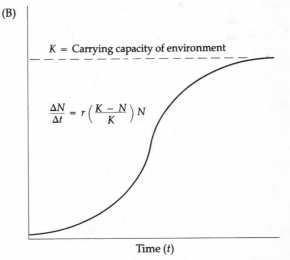

FIGURE 14. Two simplified forms of population growth; (A) exponential growth and (B) logistic growth. (Redrawn from Wilson and Bossert, 1971.)

The exponential law given by equation 4 is also known as the COMPOUND INTEREST LAW. We shall meet it again in Chapter 5 in which we shall describe the demand for energy and the growth of demand. It is exactly the same compound interest law that applies to our bank accounts.

As a population grows in size, the death rate will probably increase as a result of a reduction in available food and space. So instead of δ remaining constant, it may increase at a linear rate with N and be expressed in the following form:

$$\delta = \delta_0 + k_\delta N \qquad (5)$$

where k_δ is the rate of increase of the death rate or the slope of the straight line increase. Clearly $k_\delta = (\delta - \delta_0)/N$.

The birth rate may decrease with increasing population size and may be expressed by

$$b = b_0 - k_b N \qquad (6)$$

where k_b is the rate of decrease of the birth rate or the slope of the straight line decrease.

If we substitute equations 5 and 6 into equation 3, we get

$$dN/dt = (b_0 - k_b N) - (\delta_0 + k_\delta N) \qquad (7)$$

This equation looks a little more complex than the simple exponential equation. It is known as the LOGISTIC EQUATION FOR GROWTH. This growth form is shown in Figure 14B. Note that the slope of the curve, which is dN/dt, is zero when $N = 0$ and when $b_0 - k_b N = \delta_0 + k_\delta N$. This latter condition occurs when the birth rate just equals the death rate, when the population maintains a stable size, or when it is just able to sustain itself. When we solve this equality for N, we get

$$N = \frac{b_0 - \delta_0}{k_b + k_\delta} \qquad (8)$$

This value of N is called the CARRYING CAPACITY of the environment. This particular value of N is traditionally given the symbol K. If we let $r = b_0 - \delta_0$ and substitute K for N in equation 8, then, rewriting the logistic equation, we get

$$\frac{dN}{dt} = rN \left(\frac{K - N}{K} \right) \qquad (9)$$

Just as the exponential equation can be integrated over time, so can the logistic equation. The number of individuals N in the population as a function of time is

$$N = \frac{K}{1 + [(K - N_0)/N_0]\, e^{-rt}} \qquad (10)$$

where N_0 is the initial population density at $t = 0$ and K is the value of N when $t = \infty$. The graphic form of this equation is shown in Figure 14B.

The logistic growth, or sigmoid, curve is the form we intuitively know must, more or less, fit any increasing population over time. Exponential growth can never continue indefinitely. Eventually limiting factors of some kind must slow the growth rate and eventually establish an equilibrium population. These limiting factors produce a decreasing birth rate and an increasing death rate, responses that represent a negative feedback system. In ecology, the effectiveness of the feedback mechanisms in stabilizing a population is extremely important. Not all populations become stabilized at a high density, but having achieved it temporarily, they may then die back to a very low density. Plants that pioneer an open field may reproduce quickly, have a high r value, and in short order cover much of the field. Such species generally have the ability to rise rapidly to a high density when they have found a suitable habitat. However, their density will decrease as other more competitive species take over the habitat. Then they may migrate to other suitable habitats where they can operate at the pioneer stage and repeat their rapid growth performance, or they can survive in low densities among their stronger competitors in this first habitat. A population of a species with these characteristics is known as an r-ADAPTED SPECIES.

In contrast to the *r*-adapted species is another type—referred to as a *K*-ADAPTED SPE-CIES—that lives in relatively constant environments. Plant species in an old climax forest are good examples of *K*-adapted species. Here the plant species present are able to maintain stable equilibrium population densities at or near the carrying capacity of the habitat. The intrinsic rate *r* of natural increase tends to be low. A species in this situation only needs to be able to replace losses from its population, losses that are occurring slowly, rather than expand its population density quickly. Trees and shrubs that can deny sunlight to other plants and whose root systems can out-compete other plants for moisture and nutrients are *K*-adapted species.

The field of population biology is very complex. The ideas presented in this chapter are just the very simplest of the strategies used by plants or animals to survive in natural ecosystems. Many variations of these adaptations exist. Further information may be found in Clapham (1973), Kormondy (1976), Ricklefs (1979), Wilson and Bossert (1971), and Boughey (1973).

2

The Atmosphere

A low-pressure cell with counter-clockwise circulation located off the east coast of the United States, as seen from a weather satellite. Courtesy of the National Center for Atmospheric Research.

INTRODUCTION

The physical properties of the atmosphere determine our weather and climate; they also affect the destination of pollutants within the global ecosystem. Therefore, we must understand the characteristic properties of the atmosphere if we are to appreciate the consequences of human activity on the biosphere. Or, if we are to harness sunlight, wind, or waves for energy, we must understand the dynamics of the atmosphere and the oceans.

Earth is a unique planet in our solar system because it has a climate propitious for life. However, the weather and the climate of Earth depend in a critical manner on the relation between Earth and the sun; that is, the uneven heating of Earth by the sun is a fundamental factor in our weather and climate. Note from Figure 1 that the solar rays striking the surface of Earth at the equator illuminate a much smaller surface area than they do at high latitudes. Because of this difference in illumination, the atmosphere and the ground surface at the equator receive much more solar heat per unit area than they receive at high latitudes. In addition, at the equator the solar rays pass vertically through the atmosphere and therefore pass through the shortest distance and least amount of air between outer space and the ground. Those solar rays traversing the atmosphere at higher latitudes travel greater distances and pass through greater quantities of air between outer space and the ground. Because the air absorbs some sunlight, the greater the path length through the air the greater the amount of sunlight absorbed and the less sunlight at the ground. Hence, more sunlight reaches the equatorial surface than reaches the polar surface.

The atmosphere and the ground surface

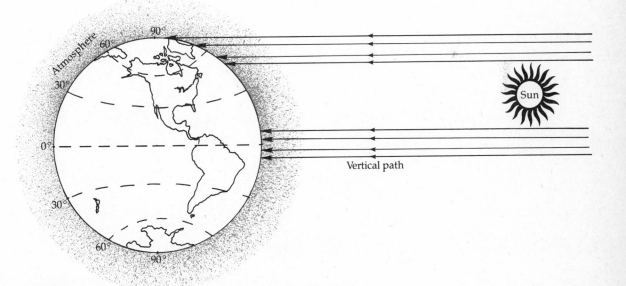

FIGURE 1. Incidence of solar radiation on Earth at the times of the spring and fall equinoxes. Note the longer path length through the atmosphere of the solar rays arriving at higher latitudes compared with the solar rays at the equator.

radiate heat to outer space day and night. However, it turns out that on an annual basis polar regions radiate more heat to space than they receive from the sun and tropical regions receive more heat than they emit. This fact accounts for the temperature difference between the warm equatorial and the cold polar regions on Earth.

GLOBAL CIRCULATION

The temperature difference between the equator and the poles creates a thermodynamic engine. Warm air is less dense than cold air, and, as a result, warm air rises and cold air

sinks. If matters were only this simple on Earth, then warm air would ascend in the tropics, flow toward the poles, and sink, returning toward the equator as cold air near the ground surface. However, instead of one large convective cell existing in each hemisphere, three convective cells exist in each hemisphere, as shown in Figure 2. The warm air in the tropics rises and flows poleward. However, by the time the air has reached 30° north or south of the equator, it has cooled so much by radiating heat to space that it becomes more dense and sinks to the ground, returning to the equator. Meanwhile, the cold air at the poles is moving toward the lower lat-

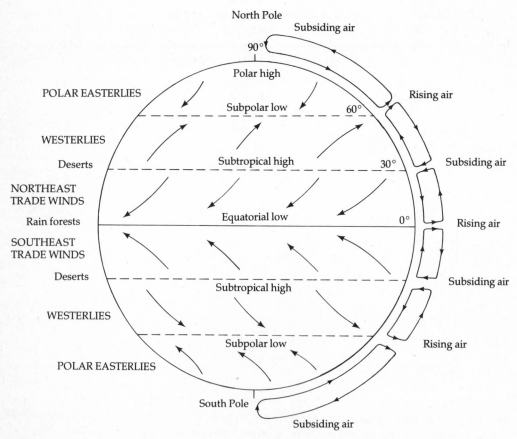

FIGURE 2. General atmospheric circulation patterns for the rotating Earth.

itudes. However, in the process, it is gaining heat as it flows over Earth's surface. This polar air becomes warmer and less dense and finally rises at about 60° north or south of the equator. Between latitudes 30° and 60°, another convective cell is formed, with rising air at 60° and sinking air at 30°.

The air currents at the adjoining boundaries of two cells must be similar. They must flow either upward together or downward together, but they cannot flow contrary to one another.

Atmospheric pressure exerted by warm air and by cold air

Warm air is less dense than cold air, and warm air holds more moisture than cold air. Therefore, warm air and cold air exert different pressures. ATMOSPHERIC PRESSURE is essentially the weight of an air column that extends from the ground to the outer edge of the atmosphere. If a column of air is cold and dense, then it is heavy and exerts high pressure at the Earth's surface. If the column of air is warm and light, it exerts low pressure at Earth's surface.

Atmospheric pressure exerted by moist air and by dry air

Moist air is less dense than dry air. This is because the molecular weight of dry air, made up of only nitrogen and oxygen gases, is 28.966. Water molecules (H_2O) have a molecular weight of 18.016. When water is added to dry air, the water molecules displace some of the dry, heavier air molecules and the average density of the air goes down. Because moist air is less dense than dry air, it is more buoyant. Moist air rises over dry air—a phenomenon responsible for clouds rising high in the sky. In addition, because a column of moist air is less dense than a column of dry air, the column of moist air exerts a lower pressure.

Dry, cold air masses typically constitute HIGH-PRESSURE CELLS, and moist, warm air masses constitute LOW-PRESSURE CELLS. Therefore, as shown in Figure 3, low-pressure regions occur where warm air is ascending and high-pressure regions occur where cold air is sinking.

Angular momentum

An object traveling in a straight line has LINEAR MOMENTUM, which is equal to the product of the object's mass and velocity. When an object is rotating, it has ANGULAR MOMENTUM, which is equal to the product of the object's mass, velocity, and distance from the axis of rotation. An ice skater who is spinning has angular momentum.

An object in motion continues in motion unless acted on by an outside force. The skater spinning at a particular rate has a certain amount of angular momentum. When the skater extends his or her arms, the rate of rotation slows. By pulling in the arms, the skater increases the rate of rotation (see Figure 4). This phenomenon has been described as the conservation of momentum. When the skater's arms are extended, the average distance of the skater's mass from the spin axis is increased. However, the angular momentum of the skater remains constant. Therefore, as distance increases, velocity, or rate of spin, decreases.

Easterlies, westerlies, and trade winds

The same thing that happens to the skater happens to the air on a spinning planet. The Earth is rotating from west to east. When cold air is moving from the pole toward lower latitudes (a north–south movement), the spin radius of the air is increasing. Because the angular momentum of the atmosphere remains constant, the air spins more slowly as it moves toward the equator. That is, at lower latitudes the ground underneath the air

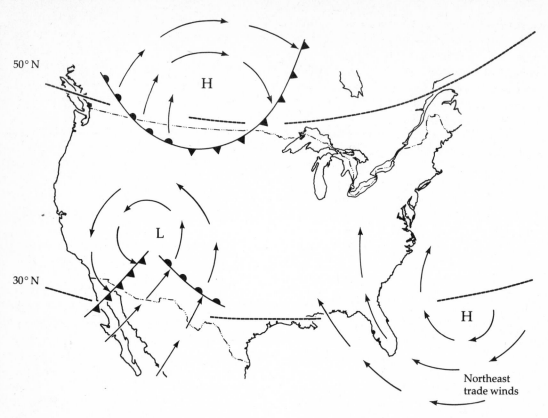

FIGURE 3. Typical high- (H) and low-pressure (L) systems and their associated fronts. A cold front is represented by triangles pointing in the direction of movement and a warm front by semicircles. In the northern hemisphere, air circulates clockwise around a high-pressure cell and counterclockwise around a low-pressure cell. Warm air rises where low atmospheric pressure exists, and cold air sinks where subsiding air occurs.

moves west-to-east at a speed greater than that of the air. This slower spinning of the air relative to the spinning ground produces, for an observer at the air–ground interface, a westward flow of these polar air masses that are moving equatorward. These flows are called the POLAR EASTERLIES, because the air is moving from the east.

Near the equator, the air also is moving equatorward from 30° north or south latitude. The NORTHEAST and SOUTHEAST TRADE WINDS are the result of the same westward flow effect described in the preceding paragraph.

However, air starting at 30° latitude and moving toward 60° latitude is reducing its spin radius. As a result, the rotational speed of the air increases. Therefore, this poleward moving air in the middle latitudes moves from the west more rapidly than the Earth's surface, and the observer on the surface detects a west-to-east movement of the air over the surface. The west-to-east-moving air masses are called the PREVAILING WESTERLIES and occur at the middle latitudes.

The polar easterlies, prevailing westerlies, and northeast and southeast trade winds are shown in Figure 2.

Coriolis effect

Surface temperature differences cause a movement of air poleward or equatorward; and changes in angular velocity move the air eastward or westward relative to the Earth's surface. However, much of the action in the atmosphere is the result of air moving from regions of high pressure to regions of low pressure. This movement would be strictly in straight lines at right angles to isobars (lines of constant pressure) if it were not for a special effect that occurs over a rotating planet. This effect is called the CORIOLIS EFFECT, and it causes the wind in the northern hemisphere to be deflected to the right of a straight path, and, in the southern hemisphere, to the left. In other words, the Coriolis effect gives a twist to the air movement.

This phenomenon can best be understood by considering what would happen to a rocket that is aimed straight at its target. For example, on the map, Atlanta, Georgia is nearly due south of Detroit, Michigan. A rocket fired from Detroit toward Atlanta would hit the Earth far west of Atlanta. Hence, looking along its path from Detroit toward Atlanta, we would observe that the rocket seems to be deflected to the right. This deflection occurs because the linear rotational velocity of the Earth's surface is greater at low latitudes than at high latitudes. The linear rotational velocity of the surface is 1040 mph at the equator and 0 mph at each pole. The linear rotational velocity difference between Atlanta and Detroit is about 90 mph. So while the rocket is enroute, Atlanta moves out from under it. The rocket should be aimed east of Atlanta in order to impact at Atlanta. If the rocket is fired from Atlanta toward Detroit, it would need to be aimed west of Detroit in order to hit its target because the rocket would have the rotational velocity of Atlanta instead of the slower-moving Detroit.

Air flowing over the Earth's surface is subject to the same effect as is the rocket. The

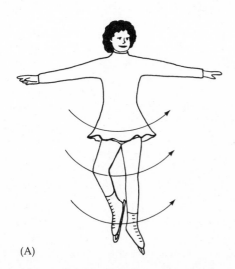

(A)

(B)

FIGURE 4. (A) When the skater's arms are extended, the average distance of the skater's mass from the spin axis is increased and the rate of rotation slows. (B) By pulling in the arms the skater increases the rate of rotation. As distance increases, velocity, or rate of spin, decreases; however, the angular momentum (see text) of the skater remains constant.

Coriolis effect causes the air to flow in a circular motion. In the northern hemisphere, air flows clockwise and outward around a high-pressure cell and counterclockwise and inward around a low-pressure cell, as shown in Figure 3. In the southern hemisphere, the flows reverse direction. Low-pressure cells are known as CYCLONES and high-pressure cells as ANTICYCLONES. All high-pressure cells and low-pressure cells are carried around the world with the prevailing winds.

FRONTS AND WEATHER

A high-pressure cell of cold, dry air has a COLD FRONT advancing along its leading edge (Figure 3). In North America this cold air is flowing from the north. On the back side of a high-pressure cell is a WARM FRONT with the air flowing from the south toward the north. A low-pressure cell of warm, moist air has a warm front advancing along its leading edge. In North America this warm air is flowing from the south. On its back side will be a cold front descending from the north. On a weather map a cold front is shown as a line with sharp triangles along its leading edge and a warm front is a line with semicircles along the leading edge. The triangles or semicircles point in the direction of air movement (see Figure 3).

The cloud cover and precipitation associated with a cold front and with a warm front are shown in Figures 5 and 6, respectively. Along a warm front, warm, moist air moves over colder, drier air ahead of it. As the warmer air overrides the colder air, a warm-front rain may result. This warm air tends to be unstable and gusty. We are forewarned of an advancing warm front by the appearance

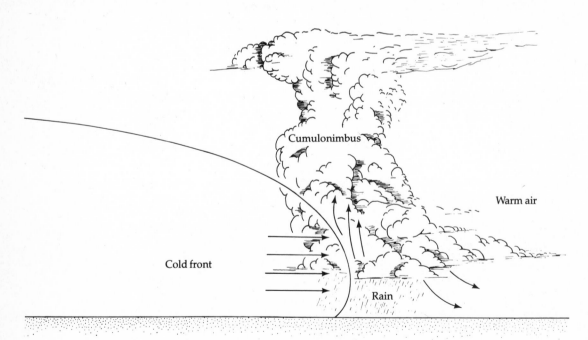

FIGURE 5. Typical cloud types associated with an advancing cold front that is moving in under warmer air.

of high cirrus clouds followed by cirrostratus, altostratus, stratus, and nimbostratus, from which rain will fall. Warm-front precipitation is more widespread and of longer duration than is the more sudden rainfall associated with a cold front. Along a cold front, cold, drier air moves underneath the warm, more moist air ahead of it. The cold front tends to be sharp and fast moving, and cold-front precipitation is likely to be more violent, that is, a cold front produces more sudden thunderstorms than does a warm front.

Continents are usually dominated by high-pressure systems in the winter and by low-pressure systems in the summer. However, things are never quite this simple, and there often is a mixture of high-pressure and low-pressure systems across the continent at any time of year. High-pressure systems are usually strongest in the winter and have cold air advancing on strong winds with calm air at the center of the high-pressure cell. Summer high-pressure cells are weakest, are slow moving, and often produce hazy skies. High pressure generally indicates fair weather and low pressure foul weather. As with the high-pressure cells, the low-pressure cells are stronger in autumn, winter, and spring than they are in the summer.

Extratropical low

At latitudes between 30° and 60°, a LOW-PRESSURE TROUGH, known as an EXTRATROPICAL LOW, may form in the atmosphere (Figure 7). The trough being formed in Figure 7 is out over the Pacific Ocean. One that formed earlier has become a full-blown low-pressure cell and is located over the east central United States. Extending to the southwest out of this low-pressure cell is a cold front and to the northeast is a warm front. The cold front may

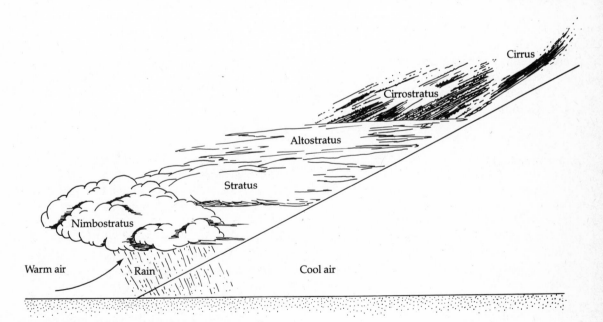

FIGURE 6. Typical cloud types associated with an advancing warm front that is moving in over colder air.

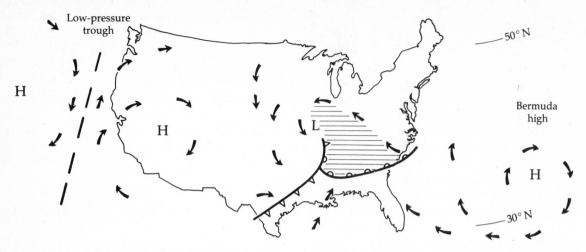

FIGURE 7. High- and low-pressure systems characteristic of summertime conditions over and near the continental United States. An extratropical low-pressure trough will often form between two high-pressure cells, as shown here off the West Coast. The Bermuda high-pressure cell affects the East Coast weather significantly during the summer.

move faster than the warm front and catch up with it to form an occluded front (which would be shown by alternate triangles and semicircles on a weather map). The direction and strength of the wind changes abruptly and often more than 90° across these fronts. Rain is likely to spread ahead of the occluded and warm-front parts of the low-pressure system.

Bermuda high

For those of us residing in North America, the Bermuda high shown in Figure 7 is of particular significance. It is this high-pressure cell that contributes to the northeast trade winds over the Caribbean much of the year and to the southerly flow of air into the central midwest of the United States during the summer. This southerly flow is putting warm moist air into the Gulf states and as far north as Iowa, Illinois, and Ohio to form the summer climate conditions characteristic of the corn belt. It was this warm, moist air flow that carried a highly infective fungus into the corn crop of the midwest in 1970. (See the section on ecosystem stability in Chapter 1 for a full description of the near disaster involving the hybrid corn crop and this fungus, which suddenly mutated and spread northward on the Gulf breezes.)

TERRAIN FEATURES

Air flow with respect to a variety of terrain features is of particular importance when considering air pollution and various environmental effects. For example, the differential heating that occurs between sea and shore produces an onshore breeze in the daytime and an offshore breeze at night. Sunlight warms the land more rapidly than it warms the sea, with the result that the air over the land rises and pulls cooler air from the sea toward the land. At night the process reverses as the land cools more rapidly than the sea and cold air drainage produces an offshore breeze. Often masses of air pollutants will drift offshore during the nighttime from a coastal city and then return the next day to add their burden to the new pollutants being generated within the urban area.

Mountain valleys have some very special climates and meteorology. All valleys trap cold air in the valley bottom because cold air is heavier than warm air. When pollutants are released into such a stable air situation, they become trapped and cannot escape. This is what happened at Donora, Pennsylvania in late October 1948 when a smothering blanket of smog developed from the steel and zinc mills located in a mountain valley. Twenty people died in three days from the air pollution. The normal death rate for Donora was two people in a three-day period. In addition, about 42% of the total population of Donora became ill. A similar situation occurred in December 1930 when the Meuse Valley in Belgium was blanketed with industrial smog. Sixty-three people died, and more than a thousand became ill.

Los Angeles and Denver are both located in a valley or a depression. Los Angeles is between the sea and the San Gabriel and San Bernardino mountains. Cold air settles in the Los Angeles basin and cannot escape to the east. Pollutants accumulate in the air, various photochemical reactions change the chemical composition of the smog, and the result is an eye-burning, lung-choking smog. Denver, Colorado is also located in a basin. It is bounded on the west by the Rocky Mountains and on the east by the high plains that rise from the Platte River valley in which Denver is located. The result is that cold drainage air from the mountains becomes trapped in the Denver basin. Denver prior to 1955 was virtually an unpolluted city. During the next three years the freeways of Denver were built and many high-rise office buildings were located downtown. Denver quickly became a seriously polluted city. Today it is the second most polluted city in America,

second only to Los Angeles. Recent studies in Denver show that motor vehicles contribute only 27% of the particulates that are the primary constituent of the Denver brown cloud. Coal and oil combustion contribute 34% of the particulates, wood burning 18%, natural gas combustion 12%, and other sources about 9%.

The nature of air flow up and down valleys and around or over hills, buttes, or escarpments is of particular importance in terms of air pollution. Many coal-burning electric power plants are being constructed in mountainous and canyon regions of the western United States. Even when abatement equipment is installed to reduce emissions, there are nevertheless serious residual emissions that will cause deteriorating air quality in this region.

STRUCTURE AND COMPOSITION OF THE ATMOSPHERE

Temperature structure

The Earth's atmosphere has both a temperature structure and a composition structure that are more complex than we might generally assume. The vertical temperature structure of the atmosphere is shown in Figure 8. The temperature of the atmosphere becomes colder with height at a rate of about 6.5°C per kilometer. Above 12 km, the air temperature remains relatively constant at −58°C until the height of 20 km. Above this height the air temperature increases until approximately 50 km, where it is about 0°C (nearly as warm as at the surface). For another 5 km or so the air temperature remains fairly constant and then again cools until at 80 km a minimum temperature of nearly −90°C is reached. Above 80 km the air temperature is constant for another 10 km, above which warming occurs again. This thermal structure has given us the opportunity to distinguish various regions of the at-

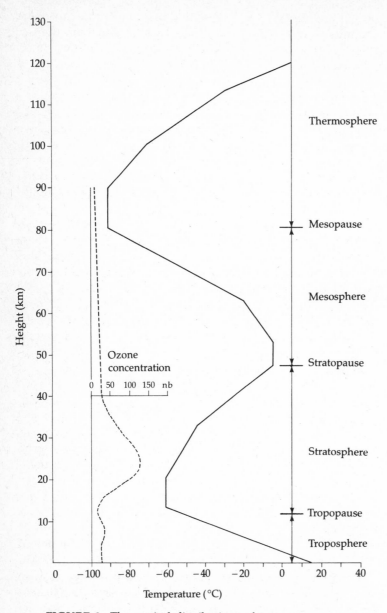

FIGURE 8. The vertical distributions of temperature and ozone in the atmosphere. The various regions of the atmosphere and the boundaries between them are defined. The terms tropopause, stratopause and mesopause apply to those boundaries where the temperature is constant with height.

mosphere. That just above the ground surface up to 12 km (actually 9 km in polar regions and 16 km in equatorial) is called the TROPOSPHERE. The next region is the STRATOSPHERE, the next the MESOSPHERE, and last the THERMOSPHERE. The zones that separate the various regions are the TROPOPAUSE (at 12 km), the STRATOPAUSE (at 50 km), and the MESOPAUSE (at 80 km).

Atmospheric composition

Earth's atmosphere had evolved from primordial times in synergism with the evolution of life, and in the process it has struck a delicate balance between the incoming solar radiation and the outgoing heat radiation. The balance of these radiations maintains a terrestial temperature that is conducive to life. The visible rays of sunlight are necessary for photosynthesis and for the production of all life on Earth, yet the short-wave ultraviolet rays from the sun are destructive to life. Fortunately most of the ultraviolet radiation is filtered out by the atmosphere. This convergence of factors, which gives just the right environmental conditions to permit life to prosper on Earth, is truly amazing; these conditions do not exist on the other planets of our solar system. What are these characteristics of atmospheric composition that make our atmosphere so unique?

Earth's atmosphere comprises 78% molecular nitrogen (N_2), 21% molecular oxygen (O_2), 0.93% argon, 0.03% carbon dioxide (CO_2), and 0.04% of a variety of rare gases. These gases are uniformly mixed in the troposphere and stratosphere, except at very great heights. Above 80 km, in the thermosphere, separation by gaseous diffusion, photodissociation, and ionization occurs.

Pollutants

Pollutants emitted at the ground–atmosphere interface are dispersed by the winds into regions other than those in which they originate. Sulfur dioxide, oxides of nitrogen, and particulates emitted by automobiles, home furnaces, industry, and power plants may be carried over great distances, but they generally remain in the troposphere. Normally the tropopause acts as an effective "lid" on the troposphere because cold air at the base of the stratosphere is underlying warmer air at greater heights (a stable situation). In the troposphere, warm, moist, or dusty air can be carried by advection to great heights above the ground and up to the tropopause "lid." Usually they will not go further upward. However, there are breaks in the tropopause through which pollutants may move into the stratosphere. In addition, pollutants may get into the stratosphere by gaseous diffusion, or thunderstorms may penetrate the tropopause and carry pollutants above it. Large volcanoes often eject dust and gases well up into the stratosphere. It is known that fine dust from the Krakatoa, Indonesia eruption in 1883 produced red sunsets for several years. The eruption of Mt. Agung on Bali in 1963 produced dust in the stratosphere that was detectable until 1973 by photometers at Mauna Loa, Hawaii. The Volcan de Fuego, Guatemala eruption of 1974 produced colorful twilight glows for many months. The eruption of Mt. Saint Helens in Oregon in May 1980 was mostly sideways and as a result very little debris was ejected into the stratosphere. Local, low-level winds spread heavy amounts of ash several hundred kilometers east-northeast from the volcano. The volcano El Chichon in southern Mexico erupted in April 1982 and ejected massive amounts of fine dust (20×10^6 metric tons) and sulfur dioxide (50×10^6 metric tons) into the stratosphere. This is believed to be the largest eruption since Krakatoa and may result in a colder global climate for several years. Solar radiation incident at the ground has already been reduced at several stations by the volcanic cloud.

Ozone

Ozone is a molecule of particular importance to the thermal structure of the atmosphere and to life at the surface because it acts as an ultraviolet filter. Its distribution is shown in Figure 8. It is apparent that it is not uniformly mixed in the atmosphere because its peak concentration is at a height of about 24 km.

The heating that takes place in the stratosphere is largely caused by ozone because ozone absorbs the solar ultraviolet radiation very strongly. The uppermost regions of the vertical ozone distribution, where the ozone concentration is low, absorb the incident ultraviolet rays of the sun very well. Once absorbed, there is not much energy remaining at these wavelengths for further absorption in lower regions of the atmosphere, even though the ozone concentration is greater at 25 km than it is at 50 km. Because of this differential absorption of solar ultraviolet energy, the atmosphere is strongly heated at 50 km and much less so at 25 km.

Ozone, a minor constituent of the atmosphere, is formed by the photodecomposition of oxygen molecules. Ultraviolet rays from the sun break down molecular oxygen (O_2) to atomic oxygen (O). These atoms then combine with other oxygen molecules to form the ozone molecule, O_3. The chemistry of this reaction is

$$O_2 + h\nu \rightarrow O + O \qquad (1)$$

$$O + O_2 + M \rightarrow O_3 + M \qquad (2)$$

where M is a catalyst, another molecular species that aids in the reaction but at the same time is not changed by the reaction. The quantity $h\nu$ represents the quanta of light entering the reaction, where h is a constant and ν is the frequency of the light.

Ozone is destroyed photochemically by sunlight or by collision with atomic oxygen. The reactions are

$$O_3 + h\nu \rightarrow O + O_2 \qquad (3)$$

$$O + O_3 \rightarrow 2O_2 \qquad (4)$$

Ozone exists at those points in the atmosphere at which the rate of formation balances or exceeds the rate of destruction. Other reactions in the atmosphere also contribute to the destruction of ozone, and, in the lower atmosphere near the ground, these are generally so effective that little ozone exists in the normal unpolluted air. In addition, rain is effective in removing ozone from the atmosphere. These other destructive reactions can be written in the following form:

$$X + O_3 \rightarrow XO + O_2 \qquad (5)$$

$$XO + O \rightarrow X + O_2 \qquad (6)$$

The net result is

$$O + O_3 \rightarrow 2O_2 \qquad (7)$$

and the molecule X has played a catalytic role; therefore it still exists as a free molecule to destroy more ozone. X may be H, HO, NO, Cl, or Br.

Atmosphere ozone is destroyed by Freon gas that is released into the atmosphere by human industrial activity. Freon is a chlorofluorocarbon, such as $CFCl_3$. It is used in spray cans in many parts of the world but has been banned for such use in the United States. Freon also is used in refrigeration and as a foaming agent in the manufacture of seat cushions. The $CFCl_3$ molecule is transported by diffusion and mixing into the stratosphere, where it is photodissociated. The reaction is

$$CFCl_3 + h\nu \rightarrow Cl + CFCl_2 \qquad (8)$$

After dissociation of Freon, chlorine atoms (Cl) are free to participate in the destruction of ozone through reactions 5 and 6, where X = Cl. A reduction of the ozone concentration in the stratosphere means that increased amounts of ultraviolet radiation will penetrate the atmosphere and reach the earth's surface.

RADIATION

The atmospheric gases have interesting properties with respect to the passage of radiation through them. They are all transparent to visible light and partly transparent to ultraviolet and infrared radiation. VISIBLE LIGHT is that part of the spectrum to which the human eye is sensitive, essentially from the edge of the ultraviolet at 400 nm to the edge of the infrared at 700 nm. In between these wavelengths we perceive all the colors of the spectrum—violet, indigo, blue, green, yellow, orange, and red.

Ultraviolet radiation

Radiation of wavelengths shorter than 400 nm is called ULTRAVIOLET RADIATION. We know that ultraviolet radiation is destructive to life; specifically to some of the molecules most critical to life's vital functions. Ultraviolet rays can produce sunburn in humans. The most essential part of a cell from the standpoint of damage is the genetic material—deoxyribonucleic acid, or DNA. Because DNA carries the genetic messages that control cellular growth and reproduction, any damage to this macromolecule is potentially serious. Increased ultraviolet radiation at Earth's surface means increased numbers of skin cancers to humans and a variety of damaging effects in plants and animals.

Infrared radiation

Radiation of wavelengths greater than 700 nm is called INFRARED RADIATION. All objects at temperatures above absolute zero emit infrared radiation in the wavelength range 4 to 20 μm (4000 to 20,000 nm). Infrared radiation is emitted by a radiator heating a room; it is radiated by the human body; and it is emitted by Earth's surface to outer space. Sunlight contains wavelengths of infrared radiation from 700 nm to 2.5 μm. Liquid water absorbs nearly all incident infrared radiation and transmits none. Some atmospheric gases absorb only certain infrared wavelengths and transmit others well. Neither nitrogen gas nor oxygen gas absorbs energy in the infrared, but water vapor, carbon dioxide, and other polyatomic molecules do absorb infrared radiation.

The greenhouse effect

The energy balance of the planet Earth is a balance of the energy received from the sun and the flow of infrared radiation to outer space. The result is a mean planetary surface temperature of about 288°K or 15°C. The earth's surface radiates heat to outer space according to the radiative laws of physics. The ground surface radiates an amount of energy proportional to the fourth power of its absolute surface temperature. All of this radiation is at infrared wavelengths, most of it in a broad band from 4 to 20 μm, centered at 10 μm. We see from Figure 9 that some of the absorption bands associated with carbon dioxide and water vapor can interfere with the free passage of this longwave radiation to outer space. Carbon dioxide gas and water vapor in the atmosphere absorb part of the outgoing stream of infrared radiation from the ground. Then these molecules reradiate this energy both outward toward space and back downward toward the ground. This returns some of the radiant flux of thermal energy to the ground, thereby raising the surface temperature and contributing to a warmer climate. This whole process has been dubbed "the greenhouse effect" because the glass roof of a greenhouse lets sunlight in while blocking the passage of infrared radiation out.

The spectral characteristics of the radiation returned to the ground from the atmosphere is shown in Figure 9. The carbon dioxide absorption band centered at 14 μm is responsible

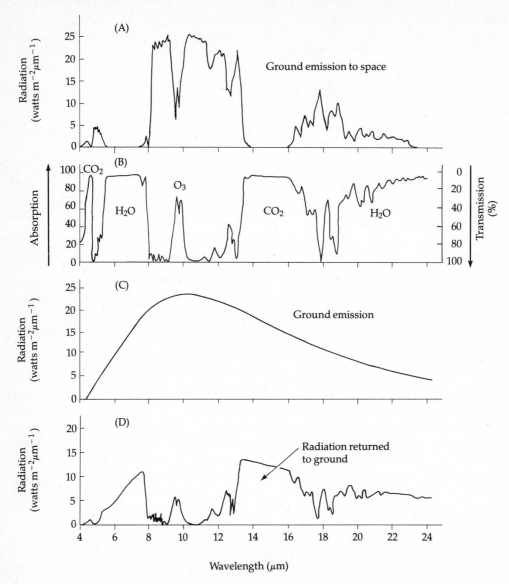

FIGURE 9. The infrared spectral distribution of thermal radiation and atmospheric transmission. (A) Radiation from the ground that reaches outer space; (B) transmission (and absorption) by atmospheric gases; (C) the "blackbody" radiation emitted by the ground surface; and (D) radiation received at the ground from the atmosphere.

for much of this return radiation as is also the absorption and emission by water vapor in the band, which begins at about 16 μm and continues to very long wavelengths. In addition, the water vapor band centered at 6.3 μm and the ozone band at 9.6 μm also contribute to the radiation returned to the ground surface. If the atmosphere were completely transparent to infrared radiation, all of the ground-emitted radiation would escape directly to space and Earth would be much colder than it is.

Because the Earth's surface and its atmosphere delicately balance radiation from the sun and to outer space, any disturbance of the composition of the atmosphere will affect this balance and potentially change the Earth's climate. If the amount of carbon dioxide, water vapor, or ozone in the atmosphere is changed, there will be an effect on climate. If new molecular species such as chlorofluorocarbons are added to the atmosphere, the climate may change. The effect of the increasing atmospheric carbon dioxide concentration on climate is described in Chapter 9. When substantial amounts of dust are added to the atmosphere from volcanoes, dust storms, or human activities, climate changes result. Light-colored dust increases the reflection of sunlight to space, thereby reducing the amount of solar energy absorbed in the atmosphere or at the ground surface. This energy loss results in lower temperatures. On the other hand, dark-colored dust increases the absorption of sunlight and produces a warming climate. Dust also causes radiation to undergo multiple scattering between the ground surface and the sky, resulting in more trapped radiation in the Earth–atmosphere system and warmer temperatures. Dust interferes not only with the passage of sunlight through the atmosphere, but also with the flow of the infrared radiation emitted by the Earth's surface to outer space. Useful references to the nature of the weather and the atmosphere are Campbell (1979); Lutgens (1979); and Roberts and Landsford (1979).

3

Energy

Modern society uses enormous amounts of energy, as illustrated by this nighttime picture of Detroit, Michigan. Courtesy of the Detroit Edison Company.

INTRODUCTION

All activities in the universe require energy. Nothing moves, changes, grows, or decays without the expenditure of energy. Energy is essential to all life. Energy flows downhill from high-potential sources like the stars (our sun) at millions of degrees temperature to the cosmic cold of outer space at a few degrees above absolute zero. The planet Earth intercepts some of this energy flux from the sun; captures, transforms, and stores some of it; and reradiates the rest back to outer space. This energy flux is necessary for all natural and human-developed systems on Earth.

During much of the 5.0 billion years since Earth was formed, the natural systems of the planet evolved slowly and progressively under the impelling forces of radiative energy flow. Life evolved, atmospheric composition and structure changed, and complex ecosystems of millions of organisms developed. Indeed, a system formed on Earth that reversed the second law of thermodynamics, a law suggesting disorder from order in the universe—a running down of the universe. Instead, life produced order out of disorder. Simple molecules formed more complex molecules, and these in turn formed self-replicating structures capable of storing and utilizing energy. Simple one-celled plants or algae evolved; these organisms captured sunlight and produced carbohydrates. These plants, and more complex ones to follow, formed the food supply for other organisms in the evolutionary chain of events. Plants and animals multiplied, diversified, and continued to evolve into higher forms. Complex food webs formed and ecosystems comprising thousands of organisms, linked together through energy and nutrient flows, carpeted the earth. Some organisms such as plants formed the base of the food chain and others were intermediate, and still others became "top" predators. At each stage of the food chain, energy was degraded into heat as high-energy compounds of the previous stage were used for food. Humans evolved and became the ultimate voracious "top" predator, with an insatiable appetite for energy.

During millions of years of birth, growth, death, and decay of organisms, a very small fraction of the solar energy flowing through the earth's ecosystems was left behind as inanimate organic compounds to form great biogenic reservoirs of coal, oil, and gas. Humans, subsisting for several million years on the energy supply of the natural web of food and fiber, began to discover the potential energy locked up in those organic reservoirs. With the release of this energy and the creative consequences of an inventive mind, humans constructed a complex society of machines, buildings, chemicals, and synthetics. Cheap, available energy made possible the development of great centers of civilization on Earth and gave humans enormous mobility, including the exploration of outer space.

Industrial civilization upset the energy balance of the natural systems and through chemical releases poisoned the organisms forming the natural food webs. Human populations grew, consumption of materials proliferated, ancient hydrocarbon reserves were drawn down at alarming rates, organisms died, species disappeared, ecosystems degraded, and global climate was probably changed by the impact of human effluent. Gradually it has dawned on this intelligent "top" predator that something is not right on Earth—that continuing growth of population and growth of the consumption of resources will result in eventual degradation of the global ecosystem under the impact of the industrialized society; that the industrialized system is not self-sustaining; that there is a limit to resources; that more people on Earth means less food, energy, and material goods per capita in the long run; that human welfare is dependent on species diversity, ecosystem stability, and a more or less "normal" climate.

NET ENERGY

If one had an infinite supply of energy compared to the rate of consumption, it would last indefinitely. That is essentially the situation for the natural ecosystems of the Earth, which draw upon solar energy. Even though the sun's nuclear furnace will eventually diminish and the earth will grow cold, such a time is so distant that it has no meaning to the lives of organisms and ecosystems, which are measured in years or even thousands of years. However, human consumption of energy is so rapacious, not of solar energy but of limited reserves of coal, oil, gas, and uranium, that the lifetimes to extinction are startlingly short.

It always requires energy to mine, refine, transport, and use coal, oil, gas, or uranium. The more depleted the source, the deeper the mine or well; the more offshore it is, or the colder the climate where the resource is located, the greater the amount of energy spent to procure more energy and the less the net gain. What really counts is net energy, not gross energy. Many sources of energy are low grade or dispersed. As we dig deeper in the earth for coal, we expend more energy in the extraction process and the net energy decreases. Solar energy at the sun is concentrated, but when it reaches the Earth it is highly dispersed by virtue of the inverse square law of the distance. To capture this dispersed energy spread over Earth's sunlit surface requires an enormous network of machinery, whether mirrors, solar cells, or harvesters. It is extremely costly in terms of energy to gather in this solar energy and use it productively. The net energy gain is not as great as one might wish. Fortunately, there is a net energy gain in the use of solar energy,

and there is a promising, though somewhat limited, future for this energy form.

When new oil fields or gas reserves are discovered, such as the enormous reserves on the north slope of Alaska or the North Sea reserves of gas, only the number of barrels of oil available or cubic feet of gas are mentioned. No one mentions what the energy expenditure will be to recover these reserves nor the available net energy. The energy expenditure includes the energy inputs for exploration, such as road building, housing, drilling, machinery, vehicles (including airplanes); for extraction including pipelines, pumping, road vehicles, machinery, and housing; for transportation involving pipelines, trucks, and rail and ship tankers or both; for storage including tanks and maintenance; and finally for refining and conveyance to the user. One can readily understand why offshore oil or Alaskan reserves become decidedly less attractive from a net energy standpoint than the great Permian basin fields of Texas and other areas of the central United States or the Middle East fields where procurement is comparatively easy. Energy reserves that seem to offer many years of supply usually are figured as gross energy rather than net energy and as a result developers may estimate that they will last longer than they actually will.

One reason why we do not hear about net energy is that the energy subsidy to explore, extract, transport, refine, and use a new source is coming from a very different source. The energy supplied to explore and develop the Prudhoe Bay oil and gas field came from coal, oil, gas, or nuclear energy from elsewhere, energy that was simply a part of America's total energy consumption.

In other words, developing Prudhoe Bay was not looked upon as an energy drain elsewhere. Cheap, readily available energy in Texas or the Middle East allowed the development of less available, more "energy"-expensive fuel from Prudhoe Bay. The same reasoning applies to the development of oil shale and other synthetic fuels.

ENERGY AND ECOSYSTEMS

There is much about energy use to be learned from the science of ecology. Ecologists have long studied the way plants and animals utilize energy, how they compete among themselves for the available energy, and how they live within the confines of limited energy flows. Through the study of ecosystems, we can learn much about the most effective and efficient ways to use energy. Natural ecosystems have evolved a certain inherent stability as they approach the climax stage or steady-state conditions. During the early stages of succession, when few species occupy an open field, for example, energy may be used wastefully but in such a manner as to provide for rapid growth. However, these plants and animals of the early successional stage may not be able to compete with the more energy-efficient species that invade the ecosystem later.

Ecologist H. T. Odum has written extensively about the use of energy by human society and natural ecosystems. Odum (1973) has enunciated various principles that he believes society might use to improve the management of their own affairs. I paraphrase some of them here.

1. The true value of energy to society is *net energy*.
2. Societies win and dominate that maximize their useful total power from all sources and flexibly distribute this power toward needs affecting survival.
3. During times when there are opportunities to expand one's power inflows, survival (and competition) is based on rapid growth, even though there may be waste.
4. During times when energy flows have been tapped and there are no new sources, those systems win that use all available energies in long-staying, high-diversity, steady-state works.
5. The successfully competing economy must use its net output of richer quality energy flows to subsidize the poorer quality flow so as to maximize total power.
6. Increasing energy efficiency with new technology is not always an energy solution since most technological innovations are really diversions of cheap energy into hidden subsidies of energy-expensive structures.
7. Without energy subsidy there is no yield from the sun possible beyond the yields of forestry and "natural" agriculture.
8. During periods when the expansion of energy sources is not possible, the many growth-promoting policies of government become an energy liability.

Society has been near steady state during most of human history; only during the last two centuries has there been an enormous growth of human society as a result of readily availably energy supplies. Reflect on the incredible rate at which wilderness America has been settled and industrialized. Our economic system has been predicated on continued growth. As energy becomes more costly, as net energy gains become smaller, as mineral resources grow more scarce, as population densities increase, as climate changes adversely, the people of the world are going to find it necessary to adopt principles for survival that optimize the use of energy.

ENERGY—WHAT IS IT?

It is necessary to go to the science of physics to get the definition of energy, for in physics a whole framework of concepts has been constructed based on certain fundamental ideas. These ideas have arisen from many centuries of observation by humans of the natural physical world.

ENERGY is the ability to do work. But what is work? WORK is defined as what happens when a force acts through a distance. When you push a cart in a supermarket, you are doing work by exerting a force on the cart and moving it through a distance. When you walk, you exert a force against the floor and move your body a certain distance, thereby doing work and expending energy. When you pick up a suitcase from the floor and lay it on a table, you have done work and expended energy. When you drive an automobile, you are moving it and yourself over a considerable distance as the result of a force exerted through the transmission and through the wheels to the road surface. In the process work was done and energy was expended. The energy required to do this work came from burning gasoline in the engine.

We sometimes classify energy into different forms. We define MECHANICAL, ELECTRICAL, CHEMICAL, MAGNETIC, NUCLEAR, and HEAT energy, among others. We also define KINETIC ENERGY and POTENTIAL ENERGY. The energy contained in the gasoline in your automobile is in the form of chemical energy, but when exploded in the engine cylinder it is rapidly converted to mechanical energy as the expanding vapor drives the piston. The piston, linked to the crankshaft, delivers this mechanical energy through the transmission to the wheels. Perhaps some of the energy from the automobile engine turns an electric generator to deliver electrical energy to the lights. The electrical energy heats the filament of the light bulb to incandescence and the energy is emitted as heat and light. At every stage in the automobile where energy was transferred from one form to another some energy was lost in the form of heat. We do not get something for nothing. As we shall learn later, there is no such thing as a perfect machine. With every event, no matter how slight, there is a loss, an expenditure, of energy. Observers of all these things have also concluded that energy can neither be created nor destroyed, but can only be transformed.

Kinetic energy

An object in motion possesses KINETIC ENERGY by virtue of its speed. A bullet fired toward a target has kinetic energy. A train in motion has kinetic energy. A waterfall has kinetic energy in the descending water, and this energy may be used to drive an electric turbine or turn a waterwheel. The kinetic energy content of a moving object is proportional to the mass m of the object and to the square of the speed v. The relationship is written as

$$\text{kinetic energy} = 1/2 \times \text{mass} \times (\text{speed})^2$$
$$= (1/2)mv^2$$

Hence, when the speed of an object is doubled, its kinetic energy content is increased fourfold. It would require four times more energy to accelerate an automobile from rest to 20 miles per hour than to 10 miles per hour, not accounting for frictional losses. Water thundering over Niagara Falls contains an enormous amount of kinetic energy, which is available for generating electric power.

Potential energy

An object at rest possesses POTENTIAL ENERGY by virtue of its height above some reference level. The quantity of potential energy is proportional to the mass of the object and to the height h. A proportionality constant, the acceleration of gravity (g), converts the mass to a weight or force and allows us to write for potential energy

$$\text{potential energy} = \text{mass} \times \text{gravitational acceleration} \times \text{height}$$
$$= mgh$$

The value of g is 9.80 m sec^{-2} or 32.3 ft sec^{-2}. A book resting on a table possesses potential energy relative to the floor by virtue of its height above the floor. Displace the book so that it falls off the table and that potential energy will convert to kinetic energy of motion. Just before it strikes the floor all of the potential energy will have converted to kinetic energy, except for a very small amount lost to frictional heating of the air. When the book strikes the floor, all of its kinetic energy is converted to heat energy, which warms the book and the floor and then is conducted and is radiated away. All of the original energy can be accounted for, despite any number of transformations. The kinetic energy of the water at the base of Niagara Falls is derived from its potential energy at the top of the falls prior to its drop. The great height of the falls gives the water an enormous amount of potential energy, which in turn makes for a large amount of kinetic energy.

Laws of mechanics

Fundamental to our understanding of energy is a description of the laws of motion. More than three centuries ago, Galileo discovered experimentally the laws of motion, but it was Sir Isaac Newton who put them into useful form. These laws are known as the law of inertia, the law of force, and the law of action and reaction, otherwise known as Newton's first, second, and third laws. They are

NEWTON'S FIRST LAW.
A body at rest will remain at rest or a body in motion will continue with that motion unless acted on by an external force.

NEWTON'S SECOND LAW.
If the motion of an object is changed in speed or direction, the object will have undergone acceleration as the result of a force acting on it. (Newton discovered that for a given force, the acceleration of an object varies inversely with its mass.)

NEWTON'S THIRD LAW.
Whenever one object exerts a force on another object, the second always exerts an equal force back on the first.

From these three laws, all of classical mechanics was derived, including our concepts of ENERGY and WORK. Because energy is the ability to do work and because work is done when a force acts through a distance, we can write

work = force × distance = energy

Newton's second law of motion gives us the definition of a FORCE. Hence,

force = mass × acceleration

Power is a concept of great importance and describes how rapidly energy is used. POWER is defined as the time rate at which energy is expended. Hence,

power = energy/time

Systems of units

It is not sufficient only to have a concept of the term *energy*; we must also be able to measure it and quantify it, using a set of units. Two systems of units are in general use today. We must be able to work with both systems. The METRIC SYSTEM of units is being adopted throughout Europe, Asia, Canada, and to some degree in the United States. The BRITISH SYSTEM is used by engineers in the United States and by some English-speaking countries elsewhere. By all logic, we should drop the British system of units immediately, but it is too ingrained in the habits of our engineering profession to make the change easy. Therefore, throughout this book, both systems will be used.

In metric units

force = mass × acceleration
newtons = kilograms × meters per second per second

work = force × distance
joules = newtons × meters

power = energy/time
watts = joules/seconds

In British units

$$\text{force} = \text{mass} \times \text{acceleration}$$
$$\text{pounds} = \text{slugs} \times \text{feet per second}$$
$$\text{per second}$$

$$\text{work} = \text{force} \times \text{distance}$$
$$\text{foot-pounds} = \text{pounds} \times \text{feet}$$

$$\text{power} = \text{energy/time}$$
$$\text{foot-pounds}$$
$$\text{per second} = \text{foot-pounds/seconds}$$

Because power is energy expended per unit of time, we can turn this around and express energy expended as the product of power × time. This rearrangement suggests that suitable energy units could be written as

$$\text{energy} = \text{power} \times \text{time}$$
$$\text{watt-hours} = \text{watts} \times \text{hours}$$

Often for the use of electrical energy, a unit of KILOWATT-HOUR (kWh) is used, which is 1000 watt-hours. In British units, the HORSEPOWER (hp) is used.

$$1 \text{ horsepower} = 550 \text{ foot-pounds/second}$$

It would be nice if the confusion of units simply ended here. But such has not been the tradition of physics. Scientists studying heat, which is energy, had to have their own set of units. In the metric system of units, the CALORIE is used to measure heat and in the British system, it is the BRITISH THERMAL UNIT (Btu). The definitions of these are

One calorie equals the amount of heat required to raise 1 gram of pure water from 14.5° to 15.5°C at standard atmospheric pressure. (One gram is the mass of 1 cubic centimeter of water.)

1 British thermal unit equals the amount of heat required to raise 1 pound (slug) of pure water from 59.5° to 60.5°F at standard atmospheric pressure. (A pound mass is defined as 0.4536 kg.)

It is important to be able to convert from one type of unit to another. Careful laboratory experiments have established the various energy equivalents. Some of these are

$$1 \text{ calorie} = 4.186 \text{ joules}$$
$$1 \text{ Btu} = 778.2 \text{ foot-pounds}$$
$$= 252 \text{ calories} = 1055 \text{ joules}$$
$$1 \text{ kWh} = 3412 \text{ Btu}$$

ENERGY CONTENT OF FUELS

Very large quantities of energy and fuel are used by society. The units defined above are fundamental, but often they are inconvenient to use because of their small size. It requires too many powers of ten to express in joules or calories the amount of energy consumed in a year by the United States. Although these expressions increase the confusion of terminology, the energy industry nevertheless expresses energy demand or consumption in terms of *tons of coal equivalent*, *barrels of oil*, or *quads*. Therefore, we need to establish values for them. It is important to do so here in order to have this information for the chapters that follow.

1 metric ton (10^6 grams) of coal equivalent
(mtce) $= 7 \times 10^9$ calories
$= 29.3 \times 10^9$ joules
$= 27.8 \times 10^6$ Btu

This assumes a heating value for coal of 7000 calories per gram or 12,000 Btu per pound. Eastern bituminous coal has a heating value of 13,280 Btu per pound and western subbituminous coal a value of 7252 Btu per pound. The amount of energy in an actual ton of coal depends on the type of coal and the particular mixture used. However, an equivalent unit for the energy content represented by a "hypothetical" ton of coal can be defined as given here.

1 barrel (bbl) of oil (42 gallons)
$= 1700 \text{ kWh}$
$= 6.12 \times 10^9 \text{ J}$
$= 194 \text{ W-yr}$
$= 5.80 \times 10^6 \text{ Btu}$ (Exxon uses 5.55 $\times 10^6$ Btu, for example.)

1 cubic foot (ft^3) of natural gas
$$= 0.29 \text{ kWh}$$
$$= 1.05 \times 10^9 \text{ J}$$
$$= 0.033 \text{ W-yr}$$
$$= 1000 \text{ Btu}$$

1 metric ton of oil equivalent (mtoe)
$$= 1.43 \text{ mtce}$$
$$= 39.68 \times 10^6 \text{ Btu}$$

1 short ton of U_3O_8
$$= 400 \times 10^9 \text{ Btu} \quad \rbrace \text{ in burner}$$
$$= 14.4 \times 10^3 \text{ mtce} \quad \rbrace \text{ reactors}$$
$$= 30 \times 10^{12} \text{ Btu} \quad \rbrace \text{ in breeder}$$
$$= 1.08 \times 10^6 \text{ mtce} \quad \rbrace \text{ reactors}$$

1 quad $= 10^{15}$ Btu
$$= 172 \times 10^6 \text{ bbl of oil}$$
$$= 10^{12} \text{ ft}^3 \text{ of gas}$$
$$= 36.0 \times 10^6 \text{ metric tons of coal}$$

Quad is an abbreviation of quadrillion and is used as a measure of the national and global demands for energy.

In the United States coal amounts are measured in short tons, whereas on the world coal market, quantities are measured in metric tons.

1 short ton $= 2000$ lb
$$= 907.2 \text{ kg}$$
$$= 0.9072 \text{ metric tons}$$

It is possible to compare some of these measures of energy. For example, one metric ton of coal equivalent is worth the energy value of 4.78 barrels of oil and 2.77×10^4 ft^3 of natural gas.

ENERGY LAWS

There are a number of laws of physics that determine limitations on our use of energy. They are known as the zeroth, first, and second laws of thermodynamics.

ZEROTH LAW OF THERMODYNAMICS
Two systems in equilibrium with a third one are in equilibrium with each other. (In general, all spontaneous processes tend toward equilib-

rium. Once the system is at equilibrium, no useful work can be obtained from it. A bar of metal with one end hot and the other cold will come to a uniform temperature given sufficient time. A ball will roll downhill, and, in the process, one can extract energy and do work. At the bottom of the hill the ball is in equilibrium with its surroundings—likewise for the water flowing over Niagara Falls.)

FIRST LAW OF THERMODYNAMICS.
Energy can be neither created nor destroyed, only transformed. (The first law of thermodynamics is simply a statement of the conservation of energy. Chemical energy can exist in a battery, but when it is used to put an electric current through a wire it is transformed to heat because the wire gets hot. When the electric current flows through the filament of a light bulb, the energy appears as light and heat. When the electric current causes a motor to rotate, the electrical energy is transformed to mechanical energy. No matter which form it takes, the total energy in the system is a constant.)

SECOND LAW OF THERMODYNAMICS.
No process involving an energy transformation will spontaneously occur unless there is a degradation of energy from a concentrated form into a dispersed form.

It is impossible to obtain work by using up the energy in the coldest object present.

The universe, or any section thereof, is tending toward maximum disorder or maximum entropy. (Each of these statements expresses the second law in a different way. The second law is a very fundamental principle that governs the way the energy world works; in particular, it governs the efficiency of any machine or energy-converting device.)

Entropy

The concept of ENTROPY represents the amount of energy in a system that is unavailable for doing work. It is related to the random motions within the system at the molecular level, which depend on temperature. The energy lost (q) in these random motions is the product of the absolute temperature (T) and the entropy of the system (S). Hence, $S = q/T$. The lower the absolute temperature, the larger the entropy and the greater the fraction of the total energy that is unavailable to do work. All systems tend toward equilibrium, or, in other words, run down. The transfer of energy is toward an ever less available and more dispersed state. In our solar system the flow of energy is such that the trend is toward low-grade, dispersed heat energy distributed at uniform temperature. This represents a slow running down of our solar system.

It would appear that life on Earth violates the second law because living things produce more order where formerly there was disorder. However, life on Earth is not a closed system. For life to create more order out of disorder, it must use the flow of solar energy that is coming from the high-temperature sun (high degree of order) and is passing to the low-temperature cold of outer space (low degree of order), in this case after striking Earth's surface. Increased order is produced among the molecules on Earth that make up organisms at the expense of increasing the amount of disorder in the stream of solar energy. Life on Earth is not an equilibrium system.

Carnot cycle

A thermal engine, such as a steam engine or a power plant, is a machine that utilizes a substance, such as water, and carries it from a high temperature to a low temperature. In the process, energy is extracted from the hot water and work is done. The French military engineer Sadi Carnot, who thought about the efficiency of machines, made the first significant contribution to our understanding of the second law of thermodynamics in 1824.

Carnot's main purpose was to understand the conditions necessary for operating a steam engine at maximum efficiency. He knew that no real engine could perform as well as a highly idealized engine, and saw that an understanding of this would give him an idea of the upper limit for the efficiency of an engine. A steam engine takes energy from a heat source (the boiler), passes it through a cylinder where it drives a piston, and then condenses the spent steam by transferring energy to a heat sink (the condenser). All of the energy lost between the steam and the condensed water is not available for doing work because some of the energy is lost to the environment as heat. Carnot imagined an ideal engine that was perfectly insulated from its surroundings. The fluid in his engine absorbed heat at a high temperature T_h and condensed at a low temperature T_c. This process is known as a CARNOT CYCLE. Carnot found the efficiency η of such an engine to be given by

$$\eta = \frac{T_h - T_c}{T_h}$$

where the temperatures are expressed in degrees absolute (see the next section for the definition of absolute temperature).

If the hot steam is at 500°K and the condensed steam at 300°K, then the efficiency would be

$$\eta = \frac{500 - 300}{500} = \frac{2}{5} = 0.4$$

This means that only two-fifths of the energy in the steam could be used for useful work and that three-fifths of the energy cannot be recovered. This represents an efficiency of 40%. In practice an actual steam engine operating between these same two temperatures would have frictional losses and heat energy would flow to its surroundings. The actual efficiency might be about 30%.

It is clear that the higher the temperature of the steam, the higher the efficiency of the engine. If $T_h = 600\,°K$ and $T_c = 300\,°K$, then $\eta = 0.5$ or 50%. This is the reason that the steam in electric generating plants is raised under pressure to temperatures as high as 900 °K, to give an idealized efficiency of 66%. Even efficiencies of 40% for steam electric plants are seldom realized in practice.

Temperature scales

Different temperature scales are used in the metric and British systems of units. There are advantages to both systems. The metric system uses the CENTIGRADE scale, with the freezing point of water defined as 0 °C and the boiling point as 100 °C. The British system uses the FAHRENHEIT scale, with the freezing point of water at 32 °F and the boiling point at 212 °F. There are 180 divisions on the Fahrenheit scale and 100 divisions on the centigrade scale. Thus, $1\,°C = 9/5\,°F$ or $1\,°F = 5/9\,°C$.

Conversion from the centigrade to the Fahrenheit scale and vice versa is very easy:

$$T(°F) = 9/5\,T(°C) + 32$$

$$T(°C) = 5/9\,[T(°F) - 32]$$

It is often useful to know the values on both scales for a few temperatures of common experience. For example, 10 °C = 50 °F, 20 °C = 68 °F, 30 °C = 86 °F, and 40 °C = 104 °F.

A temperature of absolute zero has been determined to be 273° below 0 °C. Absolute zero is the temperature at which all random motion of particles ceases. It is the temperature at which any molecule is in its lowest energy state. The absolute temperature scale in the metric system, the KELVIN scale, is defined as having the same size units as the centigrade scale, except that it begins at absolute zero rather than at the freezing point of water. Hence,

$$T(°K) = T(°C) + 273$$

In the British system, the absolute scale of temperature is the Rankine scale. It has the same size units as the Fahrenheit scale, but zero is 492° below 0 °F. Hence,

$$T(°R) = T(°F) + 492$$

Radiation laws

Earth is warmed by solar radiation but loses heat to outer space by emitting infrared radiation. It is the balance between the incoming and outgoing streams of radiation that keep Earth's surface at a mean temperature of about 15 °C. Your body radiates infrared radiation, a loss of energy that helps to maintain your body temperature at its normal level. Your home loses infrared radiation, an energy loss that you pay for in your heating bill. But what is this law of physics that causes all objects to lose energy by radiation?

Physicists discovered more than a century ago that all objects at a temperature above absolute zero radiate energy at a rate proportional to the fourth power of their absolute temperature. This is known as the STEFAN-BOLTZMAN LAW and is written

$$R = \epsilon\sigma\,(T + 273)^4$$

where $\sigma = 5.67 \times 10^{-8}\ \mathrm{Wm^{-2}\ °K^{-4}}$ and ϵ is the emissivity of the surface of the object. The temperature T is in °C but it is converted to absolute temperature by adding 273 °C to it. A perfect or ideal black surface radiates with an emissivity $\epsilon = 1.0$. All other surfaces radiate with less efficiency and for them $\epsilon < 1.0$. Your skin may radiate with an emissivity of about 0.97.

The Stefan-Boltzmann law is also known as the blackbody law of radiation. The fourth power exponent in the radiation law shows that the amount of radiation emitted is strongly determined by the surface temperature of the object. If an object had a surface temperature of 300 °K, it would radiate 459 $\mathrm{Wm^{-2}}$; but with a surface temperature of 600 °K, the radiation emitted would be 7348 $\mathrm{Wm^{-2}}$ or 16 times the amount emitted at 300 °K.

If the mean temperature of Earth is 15°C or 288°K and if its surface is a perfect black object, then it would radiate 390 Wm^{-2}. The Earth's surface receives on the average from the sun about 340 Wm^{-2} (see Gates, 1980). To balance this amount of incoming energy at a surface temperature of 15°C, the average emissivity of Earth is calculated to be about 340/390 = 0.87.

Objects at the ambient temperature of Earth emit only infrared wavelengths of radiation. A person does not see this radiation because our eyes are insensitive to infrared wavelengths. However, you can feel the warmth emitted by a radiator in a room if you stand next to it, even though you cannot see the heat. As the temperature of an object increases, it emits energy of shorter wavelengths; and if it becomes sufficiently hot, it will emit visible light. If you were to sit around a wood fire and watch the coals cool as the fire dies away, you would notice that the visual sensation changes with time. At first there would be a flickering yellow flame, the coals would be "white" hot, become orange and then dark red as they cooled. A few hours later you might not see any light from the coals, but you would feel with your hand the warmth emitted as infrared radiant heat. Although you may not consciously feel the low levels of radiant heat emitted by the walls of the rooms in your home, you can experience the significance of this radiation. Just sit in a room where the air temperature is warm, but the walls are cold, and see how you feel. Or try sitting for an extended period near a cold window in winter and experience the reduced level of infrared radiation that you receive from the cold window; contrast this to the warmth you receive if you sit near a warm inside wall.

The laws of radiation are fascinating and very important. Radiation flows throughout the universe. Stars and galaxies exchange energy by radiation. Planets are heated or cooled by the flow of radiation coming to or going from them. All objects on the surface of the Earth are heated and cooled by radiation. Life on Earth is formed by the flow of radiation. Radiation is electromagnetic energy that is made up of many wavelengths. We have names for these wavelengths, depending upon their size, and we classify regions of the spectrum according to the ultraviolet, visible, infrared, microwave, and radio regions.

4

Energy Resources

Coal is still the major source of energy for the United States. Here at Superior, Wisconsin on Lake Superior, coal is being delivered to a freighter bound for Detroit, Michigan. Courtesy of the Detroit Edison Company.

INTRODUCTION

The long-range well-being of the human population of the planet Earth will depend upon our ability to understand the location and quantity of MINERAL FUELS within the earth and how they might be extracted and used. We also must understand the demography of the human population and how it must be limited because ultimately all problems of resource depletion, environmental pollution, and ecological degradation relate to population size.

The information in this chapter is taken mostly from Schurr et al. (1979). These authors depended upon various government agencies, international organizations, and private corporations to provide them with the basic data base for making the estimates given in their report. Estimating global resource reserves is difficult at best. Furthermore, our knowledge concerning the amounts and distributions of various resource stocks is highly variable. Coal, its origin and formation, seems to be understood, and the distribution of coal beds seems to be well known. We do not completely understand the origin of petroleum, and therefore our understanding of its distribution is incomplete. Our knowledge about the distribution of uranium is even less adequate.

A discussion of mineral fuel requires a careful definition of terminology. The total amount of a mineral fuel in the earth is considered fixed, but how much of it is recoverable using current technology is one thing or how much is recoverable eventually using an unknown future technology is another matter entirely. The total natural stock of a mineral fuel is called the RESOURCE BASE. That part of the resource base that is estimated with at least a small degree of certainty and may be eventually recoverable is termed RESOURCES. Resources are classified by two criteria: the geological plausibility of their existence and the economic feasibility of their recovery. That portion of resources that is known to be present with a high degree of certainty and that is economically recoverable with current technology is referred to as RESERVES. That portion of resources that is contained in relatively unexplored extensions of known reserves and that would be economically recoverable with current technology is referred to as INFERRED RESOURCES. The accuracy of estimates of inferred resources is considerably less than that of estimates of reserves. The term HYPOTHETICAL RESOURCES applies to undiscovered quantities that might exist in known producing regions and extensions of these, but that are more costly and difficult to recover than are the reserves. The estimates of resources given in the following tables are believed to be on the conservative side because they exclude the so-called SPECULATIVE RESOURCES, which may exist outside known producing regions and under unknown geological conditions. The term CONVENTIONAL, used in the tables, refers to sources of mineral fuels that are commercially used today. Oil shale, for example, is not currently in great use and is classified as an UNCONVENTIONAL RESOURCE.

UNITED STATES RESERVES AND RESOURCES

Table 1 gives the amounts of reserves and resources within the United States for coal, oil, gas, and uranium.

Coal

The estimates given in Table 1 for coal are based on compilations by the U.S. Geological Survey. The United States is fortunate in having massive reserves and resources of coal. The total quantity is estimated at 1.8×10^{12} short tons, with an energy content of about 37,800 quads. This quantity represents a heat equivalent of 21×10^6 Btu per short ton of

TABLE 1. U.S. recoverable reserves and resources of conventional mineral fuels[a]

Fuel	Identified		Undiscovered hypothetical resources	Total	Quads of Btu equivalent[b]
	Reserves	Inferred resources			
Coal (billion short tons)	260	648	895	1803	37,863
Oil (billion barrels)	34	23	82	139	806
Natural gas liquids (billion barrels)	6	6	16	28	115
Gas (trillion cubic feet)	209	202	484	895	917
Uranium (thousand short tons)	890	1,395	1,515	3800	1,140 (LWR) 68,400 (FBR)
Total (quadrillion Btu)	6163	14,391	20,287	—	40,841 (LWR) 108,101 (FBR)

[a]From Schurr et al. (1979).
[b]LWR, Light water reactor; FBR, fast breeder reactor.

coal. (A short ton is equal to 2000 pounds or 907.2 kg.)

There is a considerable difference in the heat content of coal, depending on whether or not it is anthracite, bituminous, subbituminous, or lignite. Because coal in the eastern United States is mainly anthracite and bituminous and west of the Mississippi River it is mainly subbituminous and lignite, the energy and sulfur content of coal available in these different regions is quite different. Western coal is of lower energy and sulfur content than eastern coal. Environmental regulations for low sulfur emissions from the burning of coal is placing increasing demand on western coal. In the years ahead, much of the coal near the surface will be mined first; but as much deeper coal is taken the cost of extraction will become higher and higher. Recovery factors for coal are 60% for reserves, 50% for inferred resources, and 40% for hypothetical resources. For coal, the inferred resources are at relatively short distances from known deposits and at depths of less than 1000 feet, whereas hypothetical resources are at greater distances and at depths from 1000

to 6000 feet. The economics of recovery will differ greatly in the two cases.

Oil

The reserves and resources of oil within the United States are decidedly much smaller than are those for coal (see Table 1). Total undiscovered resources of oil are estimated at 82×10^9 barrels (the range of estimates is 50 to 127×10^9 barrels). Adding this to reserves (34×10^9) and inferred resources (23×10^9), we get a total of 139×10^9 barrels, with an energy content of 806 quads. A heat equivalent of 5.8×10^6 Btu per barrel is used for this conversion. Recently there has been a tendency to revise these estimates upward because they were based on lower prices for oil. Because of the higher price incentive, an estimate is made of crude oil resources and reserves of 200×10^9 barrels for an energy content of 1160 quads, somewhat greater than that given in Table 1. There are also natural gas liquids associated with the production of natural gas. These add 28×10^9 barrels of oil with an energy content of 115 quads.

The major oil shale deposits of the United

States are in Colorado, Utah, and Wyoming. The total amount of shale oil is estimated to be as great as 170×10^{12} barrels. However, much of this may prove to be extremely difficult or impossible to recover. Schurr et al. (1979) give a very much lower amount (e.g., 330×10^9 barrels) for the "in-place" quantity. Of this they estimate that 60% will be recoverable and will yield 198×10^9 barrels or 1148 quads of energy. These authors go on to say that if economics were to allow the extraction of oil from shale with an oil content of 15 gallons per ton instead of 30 gallons per ton, the potential total recoverable oil becomes 1×10^{12} barrels or 5800 quads of energy. Hence, the amount of oil to be derived from shale is 1.4 times that recoverable from conventional oil using the lower figure and is 7.2 times using the larger figure. Taken at face value these are impressive amounts, but no consideration has been given here to net energy gain or to the environmental impacts of the extraction process.

Gas

Conventional natural gas reserves and resources are estimated at a total of 895×10^{12} ft³, with 917 quads of energy (see Table 1). (The range of estimates is 733 to 1066×10^{12} ft³.) The heat equivalent of gas is taken to be 1025 Btu per cubic foot.

A lot of unconventional gas resources are to be found in sands, shales, coal seams, and geopressurized zones. The use of these resources will augment very substantially the limited conventional oil and gas resources of the United States. A major question has to do with recovery efficiencies and the net energy gain. If 25% of the gas in tight sands and black shales were recovered, these sources would produce 198×10^{12} and 71×10^{12} ft³, respectively, for a total of 269×10^{12} ft³ of gas, with an energy content of 276 quads (203 + 73 quads) (see Table 2). Coal seams may yield 87 quads of natural gas, and geopressurized zones could give

a minimum of 308 quads. Some estimates indicate as much as 16 times this amount. There are enormous uncertainties connected with these estimates, largely due to the lack of technology. The technology for production from geopressurized zones is much less developed than for recovery from tight sands and black shales. Geopressurized zones exist under the Gulf of Mexico as well as under the adjacent continental coasts. In addition to natural gas production from these zones, there would be very high pressure hot water that would be available at the wellhead and could be used for the hydraulic generation of electricity and then used for process heat.

Both oil and natural gas proven reserves in the United States have been declining since 1970. New oil discoveries and gas discoveries have been declining since 1965. In the case of natural gas, the reserves-to-production ratio fell from 20 in 1960 to 11 in 1976—about the lowest level considered acceptable for normal production.

Uranium

It is much more difficult to estimate the uranium reserves and resources of the United States than to do so for the fossil fuels. Nuclear power is a new industry, and the search for nuclear fuels has been less thorough than for other fuels.

Identified reserves and inferred resources of uranium are given in Table 1 as 890×10^3 and 1395×10^3 short tons, respectively. Another 1515×10^3 short tons of undiscovered hypothetical uranium resources are recognized for a total of 3800×10^3 short tons. A wide range of other estimates are possible. These are discussed by Schurr et al. (1979). The heat value of the total uranium reserves and resources would be equivalent to 1140 quads if burned in light water reactors (LWR) and to 68,400 quads if burned in fast breeder reactors (FBR). Light water reactors utilize

TABLE 2. U.S. unconventional oil and gas resources[a]

Resource types	In-place quantities	Recovery factor %	Recoverable quantities	
			Quantity recovered	Quads of Btu equivalent
Oil (billion barrels)				
Oil shales	330	60	198	1148
Heavy crude oils	150	20	30	174
Total	480		228	1322
Gas (trillion cubic feet)				
Tight sands	793	25	198	203
Black shales	284	25	71	73
Coal seams	850	10	85	87
Geopressurized zones	3000	10	300	308
Total	4927		654	671

[a]From Schurr et al. (1979).

only about 0.7% of uranium 235, but breeder reactors may use all of the uranium isotopes. Hence, in principle, the fast breeder is 140 times (100%/0.7%) more effective than the light water reactor. However, in actuality this is not the case. For various reasons not given here the ratio of 60 to 1 is used, rather than 140 to 1.

Schurr et al. (1979) describe in detail the calculation of the heat content of a short ton of uranium when utilized in a light water reactor.

A 1,000-MW LWR reactor with a burnup of 30,000-megawatt-days per metric ton of uranium fuel, a capacity factor of 65 percent, and a busbar efficiency of 32 percent needs (1,000 × 365 × 0.65)/(30,000 × 0.32) = 24.7 tons of enriched uranium (U) each year. At 0.2 percent tails, 3 percent enrichment, and 15 percent conversion and fabrication losses, 202 short tons of natural uranium (U_3O_8) are required to produce this annual core replacement. Thus, evaluating electric energy at the equivalent thermal input of 10,665 Btu per kilo-watt hour, 1 short ton uranium (U_3O_8) equals 1,000 × 365 × 24 × 0.65 × 10,665 × 1000)/202 = 300 billion Btu. Using a comparative ratio of 60 to 1, breeder reactor technology would result in a heat content of 18 trillion Btu per short ton uranium (U_3O_8).

The breeder reactor will extend uranium and thorium resources much further into the future than if they were used only in light water or similar reactors. Also low-grade resources would become more valuable because of the energy they could provide.

WORLD RESERVES AND RESOURCES

There are enormous disparities around the world in the amount of effort that has gone into exploration for fuel resources. Therefore, the uneven distribution of these resources as presently understood could be in part the result of a lack of exploration in some regions. For example, the United States appears to contain about 20% of the world fuel resources, but has only 7% of the world's land mass. Africa, with 20% of the world's land mass, appears to have 1% of the global fuel resources. It is very likely that huge new fossil fuel resources, including oil and gas, will be discovered in many geologically promising areas in Africa, Asia, and polar regions. The distribution of global conventional mineral fuel recoverable reserves and resources is given in Table 3. The numbers from China are questionable. For uranium the numbers are not available from China, the Soviet Union, or other Communist countries.

Coal

The amount of coal available in the world is truly impressive. The easiest way to make the comparison of coal amounts with other re-sources is to compare the numbers in the bottom line of Table 3, because these are given in energy units. Over 5.0 trillion metric tons of coal or 140,600 quads of heat value are available. This is eight times the energy of oil and gas resources combined. Only the energy available through breeder reactors greatly exceeds the energy available from coal. However, the quantities of radioactivity generated by breeder reactors creates an enormous dilemma for decision-makers and one that has not yet been resolved.

If current rates of global energy consumption were constant for the future (e.g., 276 quads per year), the world coal resources would last another 500 years. Ninety percent of the world's coal resources are in China, the

TABLE 3. World recoverable reserves and resources of conventional mineral fuels[a]

Region or nation	Coal (billion metric tons coal equivalent)		Oil (billion barrels)	
	Reserves	Resources	Reserves	Resources
United States	178	1285	29	110–185
Canada	9	57	6	25–40
Mexico	1	3	16	145–215
South and Central America	10	14	26	80–120
Western Europe	91	215	24	50–70
Africa	34	87	58	100–150
Middle East	—	—	370	710–1000
Asia and Pacific	40	41	18	90–140
Australia	27	132	2	—
Soviet Union	110	2430	71	140–200
China	99	719	20	—
Other Communist areas	37	80	3	—
Total	636	5063	642	1450–2120
Quintillion (10^{18}) Btu	17.7	140.6	3.7	8.4–12.3

[a]From Schurr et al. (1979).

Soviet Union, and the United States. The United States alone has 25.6% of the estimated coal reserves and resources of the world. Also of particular significance for coal is the high resource to reserve ratio of 8, and the high reserve to production ratio of almost 200 at the present time. For example, the world ratio between oil resources and oil reserves is only 3 and the oil reserves to oil production ratio is 30. Clearly there will be increased demand for America to ship coal abroad because of our rich endowment of this resource. Coal is not an entirely nonhazardous, environmentally benign resource. The risks, disadvantages, and environmental consequences of burning coal are discussed in Chapter 8. It is just possible that the technology and economics of coal liquefaction and gasification will improve in the future to the point where the huge transportation and pipeline system for oil and gas delivery may be utilized. However, it is unlikely that these developments will be very significant until well into the twenty-first century. The possibility of such a development is highly speculative.

Oil and gas

The world's total oil and gas reserves are given in Table 3 as 642×10^9 barrels (3700 quads) and 2502×10^{12} ft^3 (2600 quads), respectively. These are known quantities and are reasonably accurate. The quantities of esti-

TABLE 3. (*Continued*)

	Gas (trillion cubic feet)		Uranium[b] (thousand metric tons U)	
	Reserves	Resources	Reserves	Resources
United States	205	730–1070	643	1696
Canada	59	230–380	182	838
Mexico	32	350–480	5	7
South and Central America	81	800–900	60	74
Western Europe	143	500	87	487
Africa	186	1000	572	772
Middle East	731	1750	—	—
Asia and Pacific	89	—	45	69
Australia	31	500	296	345
Soviet Union	910	2850	n.a.	n.a.
China	25	—	n.a.	n.a.
Other Communist areas	10	—	n.a.	n.a.
Total	2502	8710–9430	1894	4288
Quintillion (10^{18}) Btu	2.6	8.9–9.7	7.4 (LWR) 443.2 (FBR)	16.7 (LWR) 1003.4 (FBR)

[b]n.a., Information not available from Communist countries; LWR, light water reactor; FBR, fast breeder reactor.

mated oil and gas resources are much less accurate and are thought by some experts to be distinctly on the conservative side. The midrange of the values shown in Table 3 for estimated total global oil and gas resources are 1800×10^9 barrels of oil (10,440 quads) and 9070×10^{12} ft^3 of gas (9300 quads). The figures include the recent Mexican oil and gas developments. On the basis of the present world annual consumption of roughly 22×10^9 barrels (127.6 quads), the oil resources might last over 70 years. Consumption does not remain constant but has grown annually at rates as high as 7.5% and more recently at 2.5%. The lifetimes estimated for oil and gas resources are substantially less than 80 years at those annual growth rates.

The great inequity in the global distribution of oil creates considerable international tension, and chances are that these tensions will not reduce in the future. The United States, for example, is drawing down its domestic reserves and resources at a disproportionate rate and thereby is becoming increasingly dependent on the availability of foreign supplies. The Soviet Union, on the other hand, would appear to be in a favorable position with respect to their domestic oil and gas supplies.

Professor Thomas Gold, an astronomer at Cornell University, has a radical new idea that the Earth may contain within its core an almost unlimited supply of methane. Professor Gold is interested in the interior dynamics of planets. He believes that there are massive quantities of abiogenic methane gas beneath the Earth's crust and that this gas slowly leaks to the surface through fissures in the rock structure. This gas has a purely physical origin and was not formed biologically as were conventional coal, oil, and gas resources. Petroleum companies are taking great interest in the idea and will be researching a technology to explore for this gas. However, once discovered, it will be excessively expensive to recover.

Uranium

The world uranium reserves and resources are given in Table 3. The amount of energy available from the uranium reserves and resources is dependent on the method of burning this fuel. Two estimates are given: one for the uranium when burned in a light water reactor and another when burned in a fast breeder reactor. Information concerning quantities of uranium located within Communist countries is not available. For a discussion of the energy calculation for uranium, see the section in this chapter on the United States uranium supply.

5

Energy Demand

Energy production requires the use of a considerable amount of land. Shown is the Dow Chemical refinery at Sarnia, Michigan. Courtesy of the Detroit Edison Company.

INTRODUCTION

All organisms require a supply of resources to sustain themselves at a viable level. Many animals devise schemes for affording themselves some protection from the vicissitudes of the environment around them. These schemes involve burrowing, house-building, nest-making, and other materials-intensive activities that require the expenditure of energy over the minimum metabolic energy required for survival. However, the human species has developed an appetite for homes, machines, transportation, and clothing far beyond the minimum need for survival. Furthermore, the demand for these materialistic extravagances is grossly uneven in human society, with a few people demanding thousands of times more than the portions of others. For example, the Concorde aircraft serves only an infinitesimal elite. The people of the United States in 1975 used 29% of the world's commercial energy resources although they constituted only 5% of the world population.

Since World War II many of the developed nations of the world have had a growth economy and have experienced a growth of population, a growth in demand for materials, and a growth of energy consumption. This insatiable appetite of developed nations for energy and materials is spreading to lesser developed regions of the world, where populations are increasing at unprecedented rates. Everyone wants more of everything. "Bigger is better" has dominated the economic philosophies of corporate leadership—an attitude that also prevails in many national governments. Cultures throughout the world are being homogenized. Traditions are being traded for synthetic schemes that promise a better life. Diverse cultures that have been locally independent are becoming simplified and are becoming regionally or globally interdependent.

With human flight into outer space, a new perspective of the planet Earth was gained. The world is now seen as a finite ecosystem of limited space and resources. But world population continues to grow at 1.8% per year, as does also the demand for resources. There is a choice, however. The people of the world may allow populations to grow and resource depletion to occur until the boundaries of the ecosystem in all its dimensions limit and degrade the human spirit to some degenerate level, or they may perceive and plan a rational scheme of conscious population control and resource utilization for optimum human activity indefinitely into the future.

Population growth and resource demand rates rarely remain constant; rather they are everchanging. These rates of growth and demand are an inherently complex phenomenon. However, certain simple concepts allow us an insight into various futures that, within fairly broad limits, must be highly meaningful. These concepts are the substance of this chapter.

COMPOUND INTEREST

Everyone is interested in how fast their bank account can grow. We are concerned with whether or not we are drawing 5%, 7%, or even 10% interest on our money. We know that if we can leave the interest acquired to accumulate in the account at a fixed rate of interest we will make more money each year. However, few of us have ever computed how long it would take to double our money at a given rate of interest if we should let the interest accumulate. It so happens that when our demand for energy increases each year at a constant rate we have a growth in energy consumption similar to the growth of our bank account. The compound interest law that applies to our bank account applies to energy consumption. When a population of people, houseflies, bacteria, or any other life form increases at a constant rate, it can be shown that the law of compound interest accounts for the growth rate.

When a resource demand is growing at a constant percentage per year, the growth is exponential. A particular property of such exponential growth is that the time required for the quantity to increase its size by a fixed fraction is a constant. For example, the time necessary for the quantity to double in amount would be a constant. The easiest way to understand this is to look at it analytically or in mathematical form.

An amount A grows at an annual rate of interest i, where i is the fractional rate of change of A per unit time and may be written as

$$i = \frac{1}{A} \cdot \frac{dA}{dt} \qquad (1)$$

Or to state this another way, the rate of change in the amount A is proportional to the amount A, hence

$$\frac{dA}{dt} = iA \qquad (2)$$

In order to obtain the growth of A with time, we must sum or integrate this with time. Rewrite equation 2 in the following form:

$$\frac{dA}{A} = i \, dt \qquad (3)$$

Integrating, we get

$$\ln A = it + C \qquad (4)$$

or

$$A = e^{it+C} = e^{C}e^{it} \qquad (5)$$

The quantity e^{C} is a constant that can be evaluated by considering that at time $t = 0$, $A = A_0$. Hence,

$$A = A_0 e^{it} \qquad (6)$$

This is the so-called EXPONENTIAL LAW, also known as the COMPOUND INTEREST LAW. Compound growth at various rates is illustrated in Figure 1.

The same law can be written using bases other than base e. For example, we could use base 10, and equation 6 would be written as

$$A = A_0 10^{it} \qquad (7)$$

Doubling time

It is often convenient to know the time it will take for the quantity A to double. If the initial

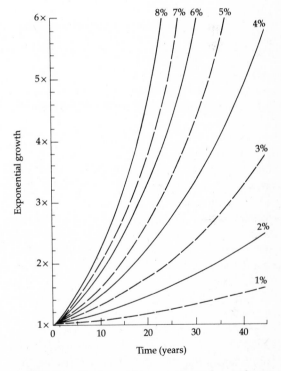

FIGURE 1. Compound or exponential growth as a function of time (in years) for annual rates of increase of 1 through 8%. Note the very short time required for a quantity to double or triple its value when the annual growth or consumption rate is 3% or greater. For example, at 4% it will require only 18 years to double a quantity and at 7% a quantity will double in only 10 years.

value of A is A_0, then at what time $t = T_2$ will $A = 2A_0$? Substituting into equation 6, we get:

$$2A_0 = A_0 e^{iT_2} \qquad (8)$$

$$\ln 2 = iT_2 \qquad (9)$$

$$T_2 = \ln 2/i = 0.693/i \qquad (10)$$

Here i is the interest rate given as a decimal fraction, not as a percentage. For example, a 5% interest rate has a value for i of 0.05. To use percentage p instead of interest i as a decimal fraction, we can substitute p for i given that $p = 100i$, or $i = p/100$. In terms of p, equation 10 becomes

$$T_2 = 69.3/p \qquad (11)$$

At an annual interest rate of 5%, the doubling time $T_2 = 69.3/5 = 13.8$ years. If $p = 10\%$, $T_2 = 6.93$ years. If $p = 15\%$, $T_2 = 4.62$ years. It does not take a very high interest rate for a quantity to begin to double very quickly when compound interest is involved. At a 5% inflation rate prices will double in 13.8 years.

Also consider what two doubling times will do to a quantity. Take the case of a 5% growth rate which has a doubling time of 13.8 years. The quantity growing at this rate will double again in another 13.8 years. Therefore, it will increase fourfold in 27.6 years. In three doubling times, or a period of 41.4 years, the initial amount will have increased eightfold.

Bartlett (1978) reminds us of some of the basic facts concerning the exponential law. As Bartlett relates it, "The game of chess was invented by a mathematician who worked for an ancient king." As a reward for the invention of chess the mathematician asked the king to place a grain of wheat on one square of the board, double it on the next square, double that amount on the next square, and so on until he completed the process on the sixty-fourth square. The last square should have had 2^{63} grains of wheat, a staggering quantity, and the total amount of wheat grains on the board should then have been $2^{64} - 1$. Note that when eight grains were placed on the fourth square eight was greater than the total of seven grains already on the board. The number of grains placed on any given square was always one more than the total number on all previous squares.

President Carter told the American public on 18 April 1977, "And in each of these decades (the 1950s and 1960s), more oil was consumed than in all of man's previous history combined." A growth rate that has a doubling time of 10 years is 6.93%, but the truly impressive thing is that for each doubling time more energy is consumed than in all of previous history.

Modest growth rates can amount to massive numbers in very short order. If the world had 4 billion people in 1975 and the population was growing at 1.9% per year, the population doubling time should be 36.5 years. Thus, there should be 8 billion people by the year 2010 and 16 billion by 2047. These are sobering statistics. It is unlikely that the world population will stabilize under about 10 billion, even though everyone would be better off if it would stabilize at a much lower number.

Bacteria grow by division, such that one bacterium becomes two, then two become four, then four divide to yield eight, and so forth. Bartlett (1978) elaborates this wonderful example of exponential growth. He asks us to imagine a hypothetical strain of bacteria that has a doubling time of one minute. Then imagine placing one bacterium in a bottle at 11:00 a.m. It is observed that the bottle is full at 12 noon—truly exponential growth in a finite environment. Bartlett says, "This is mathematically identical to the case of the exponentially growing consumption of our finite resources of fossil fuels."

Table 1 is an enumeration of the state of the bacteria during the last few minutes in the bottle. At 11:54 a.m., the bottle has become

1/64 full. Certainly no conscious bacterium would conceive of running out of space when looking about at the 63/64 empty space in the bottle. Now to quote Bartlett's description of the final minutes in the bottle.

> Suppose that at 11:58 a.m. some farsighted bacteria realize that they are running out of space and consequently with a great expenditure of effort and funds they launch a search for new bottles. They look offshore on the outer continental shelf and in the Arctic, and at 11:59 a.m. they discover three new empty bottles. Great sighs of relief come from all the worried bacteria, because this magnificent discovery is three times the number of bottles that had hitherto been known. The discovery quadruples the total space resource known to bacteria. Surely this will solve the problem so that the bacteria can be self-sufficient in space. The bacterial Project Independence must now have achieved its goal.

How long can the bacterial growth continue if the total space resources are quadrupled? The answer is clearly only two more doubling times or two minutes, since the doubling time is one minute. With four times the space available the bacteria will now run out of space at 12:02 p.m. This is the fate of any population of organisms that grows exponen-

tially. In reality, exponential growth will not continue to the last second because limiting resources will slow down the rate so that the last tortured moments of crowding will be extended; but that prospect is not a very pleasant one. It amounts to growth of population limited by strangulation.

RESOURCE DEMAND AND CONSUMPTION

The exponential law may now be applied to RESOURCE CONSUMPTION. The demand for resources is growing exponentially and therefore the annual rate of consumption $r(t)$ is also growing exponentially. According to our earlier discussion, we can write, in the same manner as equation 6, the following:

$$r(t) = r_0 e^{kt} \tag{12}$$

where k is the fractional change of the consumption rate per year and r_0 is the present consumption rate.

The total consumption C of a resource between the present ($t = 0$) and a future time ($t = T$), when the annual rate of consumption $r(t)$ is growing exponentially, is

$$C = \int_0^T r(t)dt$$

$$= r_0 \int_0^T e^{kt}dt = (r_0/k)e^{kt} \Big|_0^T \tag{13}$$

$$= (r_0/k)(e^{kT} - 1)$$

Exponential expiration time

If the total amount of a resource available is R, one can determine the EXPONENTIAL EXPIRATION TIME (EET) by finding the time T_e at which consumption equals the available resource, for example, $C = R$.

$$R = (r_0/k)(e^{kT_e} - 1) \tag{14}$$

$$EET = T_e = (1/k)\ln(kR/r_0 + 1) \tag{15}$$

TABLE 1. Ideal multiplication of bacteria in a bottle[a]

Time	Fraction full	Fraction empty
11:54	1/64	63/64
11:55	1/32	31/32
11:56	1/16	15/16
11:57	1/8	7/8
11:58	1/4	3/4
11:59	1/2	1/2
12:00	1	0

[a]Modified from Bartlett (1978). Bacteria in this scenario have a doubling time of one minute.

The determination of EET may be looked upon as unreasonable because, as a resource is consumed and nears depletion, the rate of consumption will no doubt be reduced. However, the EET gives us some idea of the minimum length of time a resource will last.

Examples

In Chapter 4, we presented tables of the estimated recoverable reserves and resources of conventional mineral fuels in the United States. Table 1 shows the United States to have 139×10^9 barrels (estimated) of recoverable oil, the equivalent of 806 quads. If the United States consumes this oil at a rate $r_0 = 4 \times 10^9$ barrels per year = 23.2 quads per year and if the consumption rate were to increase at a modest 1% per year ($k = 0.01$), we can calculate the exponential expiration time by using equation 15:

$$\text{EET} = (1/0.01) \ln [(0.01) (806)/(23.2 + 1)]$$
$$= 29.8 \text{ years}$$

If the rate of consumption were constant at 23.2 quads per year, then the 806 quads of oil would last 34.7 years. These are very sobering numbers. Thirty or thirty-five years is not much time.

Estimated recoverable oil from shales is given in Table 2 in Chapter 4 as 1148 quads. This would give a total recoverable oil reserve of 1954 quads. Using this number, we can calculate the EET for increased rates of consumption of 0, 1, and 2% per year. We get an EET of 84.2, 61.1, and 49.4 years, respectively. These numbers are still very sobering. In fact, this means that if the United States uses all the domestic oil available, at a modest consumption rate, the maximum possible lifetime of our oil reserves is well under 100 years. People alive today will witness the extinction of our oil reserves in their lifetimes.

Now if we take into consideration the estimated recoverable reserves and resources of coal and natural gas in addition to oil, not counting oil shale, we get from Table 1 in Chapter 4 the following totals: 37,863 + 806 + 115 + 917 = 39,701 quads. Now adding unconventional oil and gas resources, including oil shale, we get 39,701 + 1,993 = 41,694 quads. This is seen to be the maximum possible reserve of domestic fossil fuel energy available. The present total annual energy consumption in the United States is about 80 quads per year. When annual growth rates in consumption of 0, 1, and 2% are used, the following EETs are obtained: 521, 183, and 121.8 years. These are extremely modest growth rates. However, see the next section for estimates of these rates over the next two decades. If the growth rates in fossil fuel consumption are 3 and 4%, the EETs become 93.7 and 77.1 years. If imports and nonfossil energy sources can be used, these expiration times may be stretched out. The United States is in a favorable energy reserve position in that there is a vast supply of coal within its boundaries. However, its inexorable appetite for energy consumption does not auger well for a safe energy future unless a number of vital decisions are made early, decisions that involve conservation and the use of nuclear, solar, and other nonfossil energy sources.

UNITED STATES ENERGY DEMAND

The health of our society has always been predicated on economic growth. If continued economic growth also means continued growth of energy demand, then from what we have just seen we are going to be in increasing difficulty with regard to our energy future. It is a matter of judgment as to whether or not 25, 50, or even 100 years is a comfortable amount of time to make a transition from a petroleum-based economy to one based on solar and nuclear energies.

The Exxon Corporation makes continuing

estimates of United States energy demand projections for the next couple of decades. They do not regard their projections to be future forecasts—an important distinction. The following information is from *Exxon Company: USA's Energy Outlook 1980–2000* (Exxon, 1979).

> The increasing costs of meeting social goals, in addition to higher energy costs (which will cause investments to be made to save energy) will divert a greater amount of capital from investment in output-producing equipment, resulting in lower than historic gains in labor productivity. This factor, along with projected slower growth rates in the labor force, will result in slower long-term growth in GNP on the average of just under 2.7% annually through 2000 compared with 4.1% over the 1960–1973 period.

> While continued growth in total free world oil supplies is projected through the 1980's, a plateau in production appears likely around the turn of the century.

Consuming sector

Table 2 represents the U.S. energy demand by consuming sector. The nonenergy category represents the use of oil, gas, and metallurgical coal as feedstock or raw material rather than as fuel in the manufacture of such products as petrochemicals, asphalt, lubricants, and steel. United States energy demand is projected to increase from 82.0 quads per year in 1980 to 106.6 quads per year in the year 2000. Energy demand growth is expected to average 1.1% per year in the years 1980–1990, in contrast with 4.1% during the years 1960–1973 and 0.6% between 1973 and 1980. However, during the period 1990–2000, it is projected to increase to 1.6% per year, principally because of the energy required for synfuel production and a lack of any other gains in energy efficiency (such as auto mileage improvement after 1985).

Supply forms

Table 3 gives the energy supply forms of the United States as hydroelectric, geothermal, solar, nuclear, coal, gas, and oil. Note in particular the expected growth of coal and nuclear, the slight decline of gas and oil, and the small expected contribution of hydro, geothermal, and solar during the next two decades. Oil will remain the primary source of energy for the United States until the end of the century when coal will have an equal share. The oil share of the total U.S. energy consumption in terms of percentage will drop from 46% in 1980 to 38% in 1990 and 32% in 2000, but the absolute volume of oil consumed remains almost constant. Coal was 19% of the total in

TABLE 2. U.S. energy demand (quads/yr) by consuming sector

Consuming sector	1960	1980	1990	2000
Nonenergy	4.0	6.8	7.8	8.9
Industrial	16.9	26.7	31.3	39.4
Transportation	11.4	21.0	19.9	21.2
Residential/ commercial	14.0	27.5	32.0	37.1
Total	46.3	82.0	91.0	106.6

TABLE 3. U.S. energy supply forms (quads/yr)

Energy supply form	1960	1980	1990	2000
Hydroelectric, Geothermal, Solar	1.7	3.2	3.8	5.3
Nuclear	—	3.6	8.5	13.4
Coal	10.2	15.9	23.9	34.7
Gas	13.7	21.6	20.1	18.5
Oil	20.7	37.7	34.7	34.7
Total	46.5	82.0	91.0	106.6

1980 and is estimated to be 33% in 2000. More and more coal will be used to make synthetic oil and gas. Nuclear energy is limited to electric utility use. Nuclear energy is expected to grow from 5% of the total energy supply in 1980 to 9% in 1990, to 13% by the year 2000. The absolute volume of natural gas consumed and its share of total energy demand are expected to decline steadily.

Electric utility supply

Table 4 presents the electric utility demand of the United States and its forms of supply. Electric utilities account for about 9% of total energy demand in 1980. Their share is expected to grow to about 10.6% in 1990 and to just over 11% in 2000. On an input basis, electric utilities require about 30% of the total energy consumed in this country and will use 36% by the year 2000. Of the total electric utility energy demand, coal accounts for 49% in 1980, and this percentage remains nearly constant for the remainder of this century. However, the nuclear share of energy input for the electric utilities goes from 15% in 1980 to 27% in 1990 to 34% in 2000. Very few new nuclear plants are being installed at the

present time. It may very well be that there will occur a serious nuclear lag because of the complexities connected with new nuclear construction and with public attitudes. Hydroelectric can never pick up a large share of the electric utility generating needs.

Coal supply

The coal supply of the United States is given in Table 5. The coal reserves in the United States are abundant and account for 90% of all known domestic fossil fuel resources. Annual coal production is expected to grow from 0.734 billion short tons in 1980 to nearly 1.3 billion short tons in 1990 to 2.2 billion short tons in 2000. These figures include coal exports that will rise from 47 million short tons in 1980 to 143 million short tons annually by the year 2000. The largest user of coal is the electric utility industry, which consumed about 75% of the domestic coal used in 1980. However, its share of the total coal market will decrease to about 45% by the year 2000 (Exxon, 1979). Note the changing share of the coal market between the eastern and western coal reserves. The eastern mines are deep and expensive, but close to the markets.

TABLE 4. U.S. electric utility fuels[a]

Fuel	1980	1985	1990	2000
Nuclear capacity (1000 MW)	68.0	123.0	146.0	211.0
Coal (million short tons/yr)	542	695	802	970
Oil (million barrels/day)	1.4	0.8	0.4	0.3
Gas (trillion cubic feet/yr)	2.6	1.6	1.8	0.6
Other (million barrels/day oil equivalent)	1.5	1.6	1.8	2.5

[a]From Exxon Company (1979).

Western coals are shallower, are less expensive to extract, and have lower sulfur and Btu contents than eastern coal. Gas and liquid fuels produced from coal are expected to become commercially available in the mid to late 1980s.

Gas supply

The gas supply of the United States from 1960 to 2000 is given in Table 5. Natural gas production in the United States has been declining since 1973 when it peaked at 22 trillion cubic feet (tcf) per year. By the year 2000, natural gas production is expected to be only about 11.3 tcf or about half the 1973 production. Gas production on Alaska's North Slope is expected to start in the late 1980s. More than 35% of U.S. natural gas production in 1990 and nearly 50% in 2000 must come from reserves yet to be discovered. Rapid exploration of the petroleum reserve potential of the Outer Continental Shelf is necessary if the decline in natural gas production is to be offset.

Oil supply

The oil supply in the United States from 1960 to 2000 is given in Table 5. Domestic oil production peaked in the early 1970s and is declining. North Slope production was 1.5 million barrels per day in 1979. Domestic oil supply is estimated to decline from 10 million barrels per day in 1978 to 6 million barrels per day in 1990. By the year 2000 about 65% of conventional domestic oil supplies must come from reserves yet to be found. Most new domestic sources will be either in Alaska or on the outer continental shelf. Synthetic oil from shale and coal will enter the market

TABLE 5. Expected U.S. supplies of coal, gas, and oil[a]

Supply source	1960	1980	1990	2000
Coal production (million short tons per year)				
Western	22	232	581	1170
Eastern	412	502	704	1049
Total	434	734	1285	2219
Gas supply (trillion ft^3 per year)				
Domestic conventional	12.8	18.2	15.1	11.3
Synthetics	—	0.4	1.0	3.1
Imports	0.2	1.4	2.6	2.8
Total	13.0	20.0	18.7	17.2
Oil supply (million barrels per day)				
Domestic conventional	8.1	9.9	6.1	6.3
Synthetics	—	—	1.0	4.6
Imports	1.8	8.1	9.3	5.5
Total	9.9	18.0	16.4	16.4

[a]From Exxon Company (1979).

in the late 1980s and will gain a greater share of the future market. By the year 2000 they may contribute 10 to 20% of the U.S. oil supply. If they fail to do so, imports will need to increase.

Energy future

The people of the United States have often been accused of being the most profligate users of energy in the world. The National Research Council's Committee on Nuclear and Alternative Energy Systems (CONAES) set up a Demand and Conservation Panel to consider some energy demand scenarios for the United States. CONAES (1980) describes the details of this study with particular emphasis on the low side of possible energy demand.

> The Panel concluded that whatever driving forces might be involved, it will be technically feasible in 2010 to use roughly a total amount of energy as low as that used today and still provide a higher level of amenities, even with total population increasing 35 percent.

Several questions were asked and answered by the panel. For example, they asked and answered this question: Is a lower-energy future a bleak one for the United States? Our national well-being can be improved while energy growth is constrained by higher energy prices and perhaps by more regulation. Improvements in the efficiency of energy use can be made in the United States of about 1% per year for several decades. If population growth slows even more than it has, present amenity levels may be maintained and then slowly expanded without a major increase in energy use.

Is a low-energy future a low-technology future? The opposite is probably true. Technological innovation will be necessary to achieve a low-energy future. Computers and communications equipment today require far less energy than they did many years ago.

Technologies that exist today can be applied to improve energy efficiency. Time is needed to make changes in capital equipment, particularly by replacing it as it becomes outmoded. United States energy consumption could actually peak during the next 20 years. The various factors that drove the high-energy demand path of the past are no longer with us.

WORLD POPULATION

Human populations increased very slowly during the first 3 million years or longer of human existence. At about A.D. 1 there were between 200 and 300 million people. Table 6 and Figure 2 show the growth of the human population since A.D. 1 and show the time required to double the population. The most startling feature of Table 6 is the fact that the doubling time has been getting shorter. It took the entire span of human existence to bring the population to 1 billion people by 1825. Then, in the next 102 years, it doubled to 2 billion, and in another 48 years to 4 billion. The present population of the world is about 4.7 billion and going up with a doubling time of 37 years. Some projections indicate 8 billion people by the year 2013. The momentum of population growth is very strong; although some suggest that the global population will not exceed 8 billion, it seems

TABLE 6. World population and doubling time

Year (A.D.)	World population (billions)	Doubling time (years)
1	0.25	1650
1650	0.50	200
1850	1.1	80
1930	2.0	45
1975	4.0	38
2013	8.0	—

unlikely to stop short of 10 billion or more.

The population growth rates are very uneven throughout the world. In 1950, one-third of the population lived in the developed nations in Europe, North America, the Soviet Union, Japan, and Australia, and two-thirds lived in the developing countries. The world population increased 76% from 1950 to 1980, from 2.5 to 4.4 billion people, with fewer than 100 million added to Europe. The Soviet Union and the United States each grew a little less than 50% during this 30-year period. The developing countries almost doubled, increasing from 1.7 to 3.3 billion. The increase in Latin America was 125% and in Africa it was 114%. The Asian population increased 85%. It is expected that population growth rates will slow throughout most of the world during the next 20 years, except in Africa. A doubling time of 38 years is given in Table 6 and represents an annual growth rate of 1.84% for the world population. No one can be certain as to exactly what the rate will turn out to be because there are too many factors affecting it. Population projections by the United Nations indicate that until the year 2000 Europe will increase less than 8%, the United States and the Soviet Union about 17%, and all the other developed countries about 12.5%. Less developed countries are likely to increase 50% or more. Africa is expected to increase 76%, Latin America 65%, southern Asia about 55%, and eastern Asia 24%.

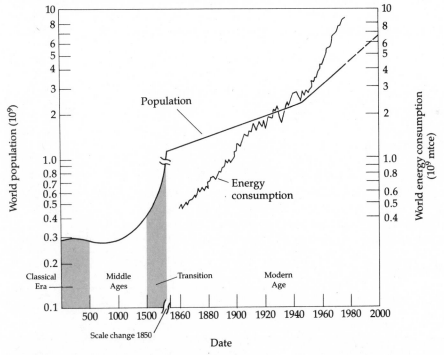

FIGURE 2. The world population growth (in billions of people) with time and the world energy consumption (in metric tons of coal equivalent). Note the change in the time scale following the transition period that ended with the year 1800. The time scale from 1800 to 2000 is much expanded over that up to 1800.

WORLD ENERGY DEMAND

Just contemplating the global energy demand created by 8 or 10 billion people is staggering. A close relation exists between energy consumption and population growth as shown in Figure 2. We see that the rate of growth of energy consumption is greater than the rate of growth of population. The total world energy consumption is approximately 276 quads. For a world population of 4.7 billion, the average annual world energy consumption today is 2.1 metric tons of coal equivalent per capita. However, the spread in per capita energy consumption between the haves and have-nots continues to increase. The highly industrialized nations have annual energy consumptions of 4 to 20 mtce per capita compared to less than 1 mtce per capita for most of the Third World. Many of the less developed countries use considerable amounts of noncommercial fuel, especially wood and agricultural wastes. It is estimated that about a billion cubic meters of wood were used annually in the 1970s for fuel throughout the world. In Africa more than 90% of fuel used was wood and in Brazil about 60%. In India wood and dung provided 58% of the fuel supply. In Sweden a lot of fuel was provided by waste from the wood products industry. The United States per capita annual energy use was about 11 mtce, but the Netherlands was 20 mtce. If the average annual world energy consumption were to increase to 3 mtce per capita and the world population were to grow to 10 billion, then the world energy consumption around the year 2030 would be 30×10^9 mtce $= 834 \times 10^{15}$ Btu $= 834$ quads. This is 3.3 times the present annual global consumption rate. One can get other projections much lower or much higher, depending on assumptions concerning prices and world politics.

Coal provided the resource for much of the increased energy use during the last century, but oil and gas have provided much of it during this century. This has, of course, created a highly troublesome problem because it is apparent that oil and gas are relatively much more limited in amount than coal. Linden and Parent (1980) state the future perspective well:

> Successful management of the transition from the abundant and underpriced supplies of oil and natural gas which sustained and subsidized the tremendous world economic expansion after World War II, to coal and other less desirable fossil fuels and then to renewable and inexhaustible energy forms based on solar and nuclear energy, is of utmost importance to the survival and eventual strengthening of the existing world order.

National use

The manner in which various nations use energy is quite different, as shown in Table 7. Note that the percentage of total energy consumption in transportation is 25% for the United States in 1978 compared with 19% for Switzerland and Canada and 14% for Japan. Intercountry comparisons of energy consumption must take into consideration such factors as the degree of industrialization, the energy inventiveness of various industries, the energy prices, the vintage of capital equipment, the climate, the population density, and the availability of resources. One interesting comparison is the number of hours one person must work to purchase a selected BASKET OF GOODS AND SERVICES. (A basket of goods and services is a selection of items including food, clothing, rent, appliances, transportation, and other services.) This figure indicates the comparative economic and social well-being of people in a country, although it may not indicate the extremes of these conditions within a population. We in the United States are very well off compared with the peoples of other countries. For example, the hours worked to earn a basket of

TABLE 7. Percentage of energy consumption by consuming sectors in 1978

Country	Industry	Residential & commercial	Transport	Nonenergy
Switzerland	29	50	19	2
United States	38	34	25	3
Canada	41	36	19	4
United Kingdom	46	37	16	2
West Germany	46	37	14	3
Sweden	47	39	13	1
Norway	53	30	14	3
Italy	53	29	16	1
Japan	59	24	14	3
Luxemburg	71	16	11	1

goods and services are 75 to 85 hours in Chicago, San Francisco, Los Angeles, and New York; 86 hours in Montreal; 98 hours in Dusseldorf; 136 hours in Tokyo, 138 hours in Paris; 150 hours in London; 189 hours in Mexico City; and 283 hours in Buenos Aires. Another measure that is sometimes used to indicate the energy use efficiency is the ratio of total energy consumption to the gross domestic product. The United States uses 1.20 mtce/$1000, whereas for Canada it is 1.13, West Germany 0.58, Japan 0.45, and Switzerland 0.28. The apparent energy frugality in Europe and Japan may be due primarily to the traditional high cost of energy. Switzerland's low ratio may also be the result of a greater concentration of light industries and of the very high net value of its goods and services.

Fossil fuel resource lifetimes

Linden and Parent (1980) have estimated the lifetimes of world fossil fuel resources at various demand growth rates, as shown in Table 8. Column A is based on proved reserves remaining of about 24,750 to 26,500 quads.

The total world reserves of fossil fuels given in Table 3 in Chapter 4 are 24,000 quads (17,700 + 3,700 + 2,600). If the annual demand growth rates are as low as 2%, these reserves will be down to a 10-year supply by the year 2024. At higher annual demand growth rates, the dates are significantly shorter. If total remaining recoverable fossil fuel resources for the world are between 161,000 and 181,000 quads, a 10-year supply

TABLE 8. Lifetimes of world fossil fuel resources at various demand growth rates[a]

Annual growth rate (%)	Date when remaining reserve to production ratio drops to 10 years[b]		
	A	B	C
4	2010	2054	2071
3	2016	2072	2095
2	2024	2103	2137

[a]From Linden and Parent (1980). Calculations are based on 1978 year-end estimates.
[b]A, Proved reserves (0.891 to 0.955 × 10^{12} tce); B, total remaining recoverable resources (5.8 to 6.5 × 10^{12} tce); C, effective doubling of B resources by use of nonfossil sources.

will remain, as shown in Column B, in the year 2103 at a 2% annual demand growth rate, in the year 2072 at 3%, and in the year 2054 at 4%. If the recoverable resources are effectively double these amounts, then, as shown in Column C, only a 10-year supply will remain in the years 2137, 2095, and 2071 for annual demand growth rates of 2, 3, and 4%, respectively. Although these numbers indicate an absolute minimum life expectancy of less than 30 years to a maximum of 150 years, it is clear that not a great deal of time remains. Unless a dramatic shift to nonfossil fuels occurs within a few decades—at most 50 years—the future could be an extremely difficult one. This projection emphasizes the necessity for the immediate development of technology to use direct solar energy, fusion energy, ocean thermal energy, or wind, tidal, and wave energy. The various forms of fossil fuel resources (crude oil, natural gas, coal, oil shale, and other bitumens) are mutually substitutable, although often at a net energy cost. On the one hand, there is a reasonable lead time still available to make the transition from fossil fuels to other forms of energy; on the other hand, time may be short, particularly when it is realized that the most difficult energy technology problems are the ones remaining—the easy solutions already having been achieved. It is not just for the next 100 years that one expects the human race to endure at a reasonable standard of living, but for thousands of years into the future.

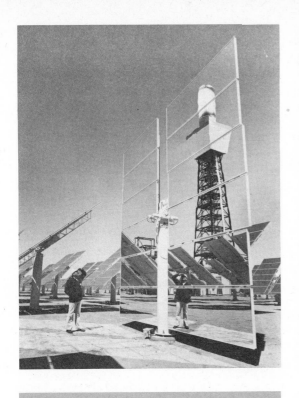

6

Solar
Energy

The solar power tower at Barstow, California reflected in one of the heliostat mirrors. Courtesy of Townsend and Bottum Corporation, Ann Arbor, Michigan.

INTRODUCTION

Solar radiation is an enormous potential source of energy for human use, and it is of infinite duration. The distribution of solar energy at the Earth's surface is uneven geographically and seasonally. Solar radiation is a dispersed source of energy, not a concentrated source like petroleum, coal, or nuclear energy. It is diurnal, except in outer space away from Earth's shadow. However, solar energy is of the utmost importance to humans, not only because it affects our weather, climate, and food supply, but also because it gives us a useful supply of thermal energy.

Solar energy may be used directly for hot water, home heating, or process heating; or it may be used indirectly to generate electricity by means of photovoltaic cells or solar power towers. Solar energy is also utilized indirectly when energy is extracted from biomass, waterfalls, wind, ocean waves, ocean tides, or ocean temperature gradients. When we include all of the various direct and indirect forms of solar energy utilization, we find that the United States is extracting more than 5% of its energy from solar sources today: biomass energy provides about 1.5 quads and hydroelectric energy about 3 quads of energy. Some estimates project solar power's share of the future energy market to be 6 to 9% by the year 2000. More optimistic estimates put it as high as 25%. Solar thermal systems can be used to pump water for irrigation, produce steam for industrial processes, generate electricity in small- and medium-size plants, and supply heat for residences or small businesses. In the United States the percentage of the total heating requirement that can be supplied by solar energy is greatest in the southwestern and least in the north central, Great Lakes, and northeastern states. At Albuquerque, New Mexico, the percentage is about 80%, and at Madison, Wisconsin, it is less than 50%. However, the total amount of energy saved in barrels of oil per year is greatest in the upper midwestern, northeastern, and Rocky Mountain states, and least in the South and Southwest, where the total heating requirement is low.

Some estimates (see Whipple, 1980) have shown that it is possible that the expanding solar energy industry could act as a substantial energy drain because the consumption of energy for new production facilities exceeds the return of energy by the entire solar energy system. This is a transient condition because ultimately there is bound to be an energy saving. Solar energy collectors, for example, substitute an initial capital investment for a lifetime of fuel savings. For a solar system there are substantial initial costs for the manufacture of equipment and its installation.

SOLAR RADIATION

The sun

The sun is a star that was formed when a massive cloud of interstellar hydrogen gas collapsed upon itself because of the gravitational attraction among its parts. As a result, the atoms of hydrogen move inward at high speed and gain sufficient kinetic energy for their nuclei (protons) to react with one another and to form into an atom of deuterium, with the spontaneous release of energy. The reason energy is released is that the two protons from the hydrogen gas have more mass than the deuterium atom the nucleus of which is made up of a proton and a neutron. One of the protons actually converts into a neutron through the process of radioactive decay, with the release of an electron and neutrino. The heat released in this reaction builds up within the sun to give an outward pressure to balance the inward gravitational pull. The reaction continues as hydrogen is converted to deuterium, a process that is still going on within the sun today. Temperatures within the sun are many millions of degrees. However, the outer gaseous extremity or photosphere of the sun as we see it from Earth radiates at a temperature of about 5760°K.

The sun is a highly turbulent mass, and its outer extremity is often interrupted with SUN-SPOTS, FLARES, and PROMINENCES. These features occur in a quasi-cyclic manner. Their numbers are associated with the so-called quiet sun and active sun, with about an 11-year periodicity. In addition there is a 22-year cycle associated with a reversal in the polarity of the magnetic field of the sun as a whole and of individual sunspot groups. There is also observational evidence of a 90-year cycle, the so-called GLEISSBERG CYCLE, which represents a modulation of the maxima associated with eight individual 11-year cycles.

In sunspots, the temperature is substantially lower (about 4000°K) than the mean photospheric temperature. During an active sun, the surface may be covered with sunspots, a condition that may result in a significant reduction in the amount of radiation emitted and in the amount of radiation reaching Earth. When solar storms occur, flares and prominences eject great masses of ionized particles into interplanetary space; when these bursts of plasma reach Earth three days later, they distort Earth's magnetic field, set up huge electric currents, and bombard the nitrogen and oxygen atoms of the upper atmosphere. The result is increased ionization of the ionosphere and shortwave radio blackout, magnetic field disturbance and compass misalignment, and auroral displays.

Changes in the weather of Earth's atmosphere do appear to have some relationship to the sunspot cycle. In particular, the 22-year periodicity in regional drought over the western United States seems to be well correlated with the sunspot cycle. It is also believed by some scientists (Agee, 1980) that the 90-year trend in the northern hemisphere surface temperature (warming since 1880, cooling since the 1940s, and a recent warming) is externally forced by the solar Gleissberg cycle.

Power from the sun

The sun radiates into space a total power of about 3.84×10^{26} W—a truly staggering amount of energy. However, Earth is at a distance of 1497×10^5 km or 93×10^6 miles from the sun, and the inverse square law shows us that the solar energy density at the Earth is only 1360 Wm^{-2}. This quantity, known as the SOLAR CONSTANT, is the irradiance of a unit area perpendicular to the sun's rays just outside the Earth's atmosphere at the mean annual distance of Earth from the sun. If one averages this flux density of radiation over the entire surface of Earth, the average value is 1360/4 = 340 Wm^{-2}. The reason for the 4 in the denominator is that solar energy is intercepted by the Earth's cross-section, taken as a disc, of area πa^2 although Earth's total surface area is that of a sphere, that is, $4\pi a^2$, where a equals the radius of Earth (6.37×10^6 m). Hence $(\pi a^2/4\pi a^2)1360 = 1360/4$. Actual values of solar flux density averaged on a daily basis vary from 200 to 1040 Wm^{-2}. The total amount of solar power intercepted by Earth is $\pi a^2 \times 1360 = \pi (6.37 \times 10^6$ m$)^2 \times 1360$ Wm^{-2} = 17.34×10^{16} W, an enormous amount of power. But this power is highly dispersed. This is about 10,000 times the power needed to supply all of the world's predicted power needs in the year 2000. No wonder people are intrigued with the possibility of capturing solar energy in space and funneling it into Earth's surface.

What really counts is how much solar power is available at any location on Earth's surface. The amount of solar power received depends on the height of the sun above the horizon. More solar power is received at the ground when the sun is directly overhead than is received when the sun is near the horizon. When the sunlight strikes the surface obliquely, as do the rays from the sun near the horizon, they are spread out over a much greater surface area than when the rays strike the surface close to perpendicular. The height of the sun above the horizon varies with time of day and time of year for any given location.

It is relatively easy to understand the geometrical relation between the Earth and the sun. Its understanding is essential to all applications of solar energy. A very brief, elementary description of the position of the sun in the sky at various geographical locations on Earth and for different times of year and day is given in the following paragraphs.

Solar altitude and azimuth

We know that the Earth is traveling around the sun in an elliptical orbit whose plane of motion is known as the ECLIPTIC. The Earth is spinning on its axis at a rate of once every 24 hours. The spin axis is inclined at an angle of 66.5° to the plane of the ecliptic, and the axis is always pointed toward the same point in the sky, a point near the North Star.

There are formulae available for calculating the position of the sun in the sky (expressed as angles of altitude and azimuth) for any geographical location and time of year and time of day. The altitude angle is the angular position of the sun above the horizontal. The azimuth angle is the angular position of the sun measured in a horizontal plane clockwise from true north. When the sun is due east of the observer, the azimuth angle is 90°; due south, 180°; due west, 270°. Charts are available for determining the position of the sun in the sky, such as the one shown in Figure 1 for latitude 40°N. A complete set of charts for many latitudes is given in List (1949) and Gates (1980). From Figure 1 we can readily see the changing position of the sun with time of year. The full circles represent angles for the sun above the horizon. The outermost circle represents the horizon and the center of the chart represents

the zenith. The heavy, curved lines represent the path of the sun across the sky at the following times of year: winter solstice, equinox, and summer solstice. The observer's position is at the center of the chart. At the time of the equinox, the sun rises due east of the observer. Sunrise for the summer solstice is far to the north of east; the azimuth of the sun at sunrise is about 58°. The early morning sun and late afternoon sun will shine against north-facing walls, slopes, and windows on June 22. However, after 8:00 a.m. and before 5:00 p.m. standard time, the sun will be south of the zenith. At noon the sun has an altitude angle of about 74°. Using charts such as these, one can design the amount of eave extension for a home so that the south-facing windows are shaded in the summer and sunlit in the winter (see Figure 2). We can work out the angles necessary to give solar panels a maximum exposure during the winter months.

At latitudes farther north than latitude 40°, we see the summer sun rise and set far north of due east and due west, and the hours of sunrise and sunset are much earlier and later in the day than they are at lower latitudes. For example, at 60°N latitude on June 22, sunrise is at 2:40 a.m. with an azimuth angle of 37° and sunset is at 9:20 p.m. standard time with an azimuth angle of 323°. However, the winter sun on December 22 rises at 9:20 a.m. with an azimuth angle of 143° and sets at 2:40 p.m. with an azimuth angle of 217°. The altitude angle of the sun at noon for this date is only 5° above the horizon.

The total amount of daily solar radiation incident on a horizontal surface outside of the Earth's atmosphere as a function of time of year and latitude is shown in Figure 3. At the equator the annual variation of the daily amount of solar radiation is quite small, while at 80° latitude it is very great. At north latitude 80° there is no sunlight from late October to late February; however, during midsummer the amount of sunlight incident on a hori-

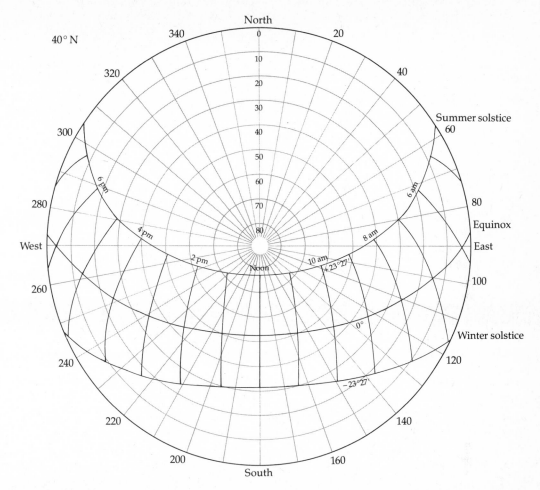

FIGURE 1. The path of the sun across the sky at 40°N latitude for December 22 (−23°27′), March 21 (0°), September 23 (0°), and June 22 (+23°27′). For any time of day, for the standard time zone on the central meridian, the azimuth and altitude of the sun can be read. The radial lines represent solar azimuth angles measured from north. The full circles are lines of solar altitude angles. The curved vertical lines are time of day.

zontal surface at 80° latitude is substantially greater than at the equator.

Atmospheric attenuation

Solar radiation traversing the Earth's atmosphere is reduced in intensity by clouds, by molecular absorption, and by scattering caused by molecules and dust in the sky. The light that is scattered by the air molecules is seen as the blue of the sky. This type of scattering is known as RAYLEIGH SCATTERING. The short wavelengths (ultraviolet and blue) are scattered much more strongly than the long wavelengths (red and infrared). The amount of scattering caused by molecules in the sky

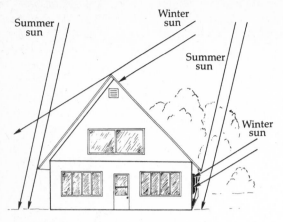

FIGURE 2. A home may have the rays of the sun pass through the south-facing windows in the winter and not in the summer.

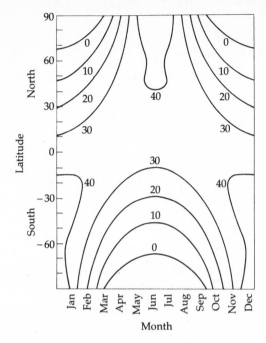

FIGURE 3. The amount per day of direct solar radiation (in MJ m^{-2} d^{-1}) on a horizontal surface outside the Earth's atmosphere as a function of the time of year.

varies inversely with the fourth power of the wavelength of the light. Double the wavelength of the light and the scattering is reduced by 16-fold. For example, the red rays from the sun are almost twice as long as are the blue rays, and therefore the blue light is scattered almost 16 times more effectively than is the red light. When the sun is on the horizon, most of the short wavelengths are scattered into the sky; the long wavelengths remain in the solar beam and the sun looks red. The scattering caused by dust and water droplets or by fog is referred to as large-particle scattering or MIE SCATTERING. Here the degree of scattering varies as the inverse 1.3 power of the wavelength for average atmospheric conditions, and in fog it is independent of the wavelength. This means that all wavelengths are scattered nearly equally. The result is a white or neutral appearance of fog and most clouds, and often a reddish appearance to a dust-laden sky.

Sunlight traversing Earth's atmosphere is not only scattered, it is also absorbed by gases in the atmosphere and by clouds. Fortunately for the benefit of life on Earth, the gases found in the atmosphere are mostly transparent to visible light. It is the visible light that provides the energy for photosynthesis by green plants. All of the wavelengths of sunlight from 300 to 2500 nm reach Earth's surface. All wavelengths shorter than 300 nm are absorbed by atmospheric gases, primarily by ozone. Some of the infrared radiation from the sun at wavelengths longer than 2500 nm (2.5 μm) reaches the Earth's surface, but much of the energy at these wavelengths is absorbed by water vapor and carbon dioxide gas in the atmosphere. These absorptions are shown in Figure 9 of Chapter 2.

When the sun is low in the sky, the solar rays traverse a long, slanted path through the atmosphere. The amount of sunlight that is scattered and absorbed is the greatest at that time, and the intensity of the sunlight received at the ground is much reduced. When the sun is high in the sky, the intensity of the sunlight received at the ground is the greatest because the amount of light that is scattered and absorbed is the least. This accounts for the great intensity of the summer sunlight when the sun is high in the sky and the low intensity of the winter sunlight when the sun is low in the sky. At higher latitudes the solar rays traverse more of the Earth's atmosphere than they do at low latitudes, and therefore the intensity of the sunlight at high latitudes is much less than at low latitudes. Details for calculating the solar intensity at any position on Earth, at any time of year and time of day, and for amounts of atmospheric attenuation and amounts of cloud cover may be found in Gates (1980), in Kreith and Kreider (1978), and in Meinel and Meinel (1976).

Daily amounts of solar radiation in the United States

Typical values of the mean daily amounts of solar radiation at the Earth's surface in July within the United States are 21 to 31 $MJm^{-2}d^{-1}$ (Figure 4). The highest values oc-

cur in the southwestern deserts and the lowest values in the Pacific Northwest. In January the mean daily amounts of solar radiation are from 4 to 12 $MJm^{-2}d^{-1}$ (Figure 5). One can see from Figures 4 and 5 that in January the isopleths (lines of constant amounts of solar radiation) run mainly east and west across the United States, and in July they run mainly north and south.

The mean daily amount of solar radiation in the United States received on a horizontal surface at the ground averaged on an annual basis is shown in Figure 6. The southwestern part of the United States receives by far the greatest amount of solar radiation and the northeastern part of the United States the least. Values shown in Figure 6 range from a low of 13 $MJ\,m^{-2}\,d^{-1}$ to a high of more than 21 $MJ\,m^{-2}\,d^{-1}$. By comparison the average amount of solar radiation received on a horizontal surface outside of the Earth's atmosphere is about 29.3 $MJ\,m^{-2}\,d^{-1}$ or 340 WM^{-2}, the number given earlier for the solar constant averaged over the total surface area of the Earth (see Bennet, 1964 for more details).

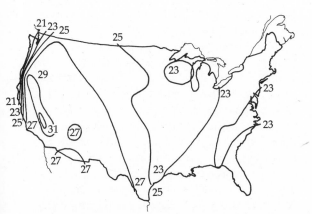

FIGURE 4. The mean daily amount of solar radiation (in $MJ\,m^{-2}\,d^{-1}$) for the month of July received on a horizontal surface throughout the continental United States.

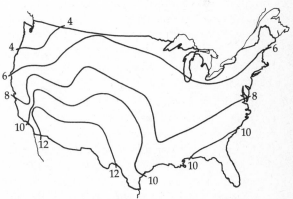

FIGURE 5. The mean daily amount of solar radiation (in $MJ\,m^{-2}\,d^{-1}$) for the month of January received on a horizontal surface throughout the continental United States.

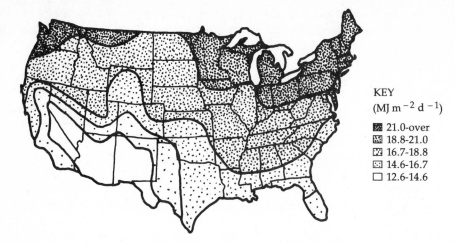

KEY
$(MJ\ m^{-2}\ d^{-1})$

▨ 21.0-over
▧ 18.8-21.0
▥ 16.7-18.8
▦ 14.6-16.7
☐ 12.6-14.6

FIGURE 6. The mean daily amount of solar radiation (in $MJ\ m^{-2}\ d^{-1}$) on an annual basis received on a horizontal surface throughout the continental United States.

SOLAR ENERGY TECHNOLOGIES

All the technologies of solar energy are considered to be using renewable energy sources. In one form or another, they are harnessing more or less continuous energy flows in the form of DIRECT SUNLIGHT or INDIRECT SUNLIGHT (WIND, BIOMASS, HYDRO, WAVES, TIDES, OCEAN CURRENTS, or OCEAN TEMPERATURE DIFFERENTIALS). Only the direct forms of solar conversion are described in this chapter; the indirect forms will be presented in other chapters. Direct forms of solar energy conversion include SOLAR HEATING AND COOLING OF BUILDINGS; solar-heated water to be used for agricultural and industrial PROCESS HEAT; SOLAR PONDS; SOLAR POWER TOWERS; PHOTOVOLTAIC CELLS; and PHOTOCHEMICAL CONVERSION.

Very detailed analyses have been made of these solar technologies in order to estimate their relative benefits to society. Their benefits include relative energy contribution, environmental values, potential price, ability to conserve oil and gas, potential for research and development breakthroughs, application to decentralized systems, and prospective use in the export market. Analyses of these technologies lead to the conclusions described in the following paragraphs. Three time periods are considered: the NEAR TERM, from the present to about 1990; the INTERMEDIATE TERM, from 1990 to about 2010; and the FAR TERM, from 2010 until about 2030.

For the near term, there are three solar technologies that are expected to contribute substantially more benefits than the others. These are solar heating of buildings, biomass, and wind. For the intermediate term, those that are likely to contribute the most are solar heating of buildings, wind, biomass, and photovoltaic cells. For the long term, three technologies with the greatest benefits are wind, solar heating of buildings, and photovoltaic cells. Solar heating is a promising energy technology, but it is costly at the present time in comparison with other energy technologies. However, this will be a less serious problem as the cost of oil increases. Solar air conditioning is much further from commercialization than is solar heating.

WIND systems appear to be highly beneficial for all time periods. Wind is attractive because of its possible energy contribution, its ability to conserve oil and gas, and its favorable economics. At least through the end of the century, wind systems will operate as fuel savers; that is, any wind system will require full backup by conventional energy systems. Using wind energy as a fuel saver is the best way to use it. Otherwise the energy generated by the wind would need to be stored in batteries, and batteries are costly to use. Large-scale wind systems (greater than 1 MWe) seem attractive; however, there are a number of disadvantages to large wind systems. Although the analysis shows that wind energy is promising, at the present time it does not appear to be developing at a very rapid rate.

BIOMASS has large benefits for the near term both because of its potential energy contribution and its ability to save oil and gas. Its near-term use will be primarily for direct combustion. Liquid or gaseous fuels derived from biomass may become competitive with other energy sources in the 1990s and beyond. The use of biomass energy may accelerate quickly because most of the components of the technology are established and incentives exist to use biomass residues as a fuel in order to avoid the disposal problems associated with these wastes.

PHOTOVOLTAIC CONVERSION devices are not likely to enter utility service on an economically competitive basis until after the year 2000. Estimated costs are prohibitive, and land use is enormous for such a system. Off-grid use by homeowners and others desiring energy independence could represent an important initial market after 1990, but significant energy displacement will not occur using photovoltaic electric generation until after 2000.

AGRICULTURAL AND INDUSTRIAL PROCESS HEAT from solar energy rates low on the benefit scale because of its generally low market potential. The available market is limited because the largest industrial energy demands,

those at very high temperatures and/or very large capacity, are virtually inaccessible to solar energy. For the near term use of solar process heat would conserve oil and gas, and near the end of the century its use would conserve coal, the anticipated fuel for process heat for the long term. One of the best potential markets is in the food, textile, and chemical industries in which large amounts of heat are needed for drying and for other processes. About 30% of all industrial process heat is used at temperatures below 300 °C; that temperature is suitable for solar heating. It is possible that by the year 2000 solar energy could displace as much as 7.5 quads of fossil fuel now used for process heat at temperatures below 300 °C.

Environmental effects

Each solar technology has a variety of special properties and environmental or ecological impacts. In contrast to fossil fuels and nuclear fuels, which are highly concentrated sources of energy, direct solar energy is highly dispersed, and some, but not all, of the indirect forms are dispersed. Wind is a dispersed form, for example, but a waterfall is a concentrated form. Biomass is dispersed, but some tidal situations are concentrated forms. Ocean waves and temperature differentials may be considered by some analysts as dispersed forms but by others as concentrated forms. The dispersed forms of solar energy require a great deal of land use, and for certain solar technologies they require very extensive physical structures in order to harness the energy. Large amounts of land used for solar energy production can preempt other necessary uses, cause ecosystem degradation, or be aesthetically displeasing. Extraction, transport, and processing of biomass may produce deleterious effects. When solar radiation is captured for energy use, there is some interference with the normal flow of this energy in the form of sunlight, wind, or water, which are central components of weather and climate, and biomass, which is the food base for human society or the primary producer for ecosystems.

Solar energy in all of its manifestations is considered a renewable energy resource. Some of the solar technologies are much more environmentally and ecologically damaging than are others. The most deleterious solar technologies are satellite solar power, new hydroelectric dams, and intensely managed biomass plantations. The least disturbing environmentally or ecologically are passive solar heating and cooling, increased electrical power generation by adding generators to existing dams, electrical generation by wind turbines, and biogasification of sewage and feedlot manures. The other solar technologies are intermediate in their environmental or ecological impacts. These technologies include photovoltaic systems, active heating and cooling systems, solar ponds, solar power towers, ocean thermal energy conversion systems, and systems using solid waste from crops or forests. The massive use of solar energy is far from benign environmentally, an assumption sometimes made by solar advocates. For example, some environmental effects can be expected to accompany the use of direct solar heating and cooling technologies (Holdren et al., 1980):

- Injuries to workers in rooftop collector maintenance
- Property or ecosystem damage from leaks of working fluid or storage medium
- Fire caused by collector overheating
- Loss of substances
- Penetration of toxics in working fluid through heat exchanger into drinking water
- Pollution from disposal of used working fluid or storage medium
- Pollution from degradation of selective collector coatings

- Hazard from falling collector glass in earthquake, fire, etc.
- Water consumption by wet cooling towers used for commercial air conditioning
- Transport of allergenic molds and fungi from storage system to room interiors
- Removal of shade trees
- Nuisance caused by collector glare
- Roof leaks from improper collector installation

A new, very well constructed, safe piece of hardware may deteriorate with age to the point of leakage or breakdown of materials. Newly fabricated materials may outgas toxic substances that are hazardous to humans and other living things. Overheating of materials, such as insulation, plastics, epoxies, and special coatings, can create much outgassing or even fires. In addition to these potential environmental effects, there are esthetic problems concerning appearance in the neighborhood, interference by a solar collector with the view by neighbors, or interference of a solar collector on one building with the collector on an adjacent building. Often there are environmental characteristics of a community that reduce the effectiveness of solar collectors. These characteristics may include shading by vegetation, orientation of buildings, density of buildings, roof configurations, and variation in the heights of buildings.

The solar technologies may impact one or more of the following important properties of ecosystems: nutrient cycling or other chemical balances; soil formation and preservation; water storage and flow; natural controls on animal and plant populations; natural controls on vectors of diseases affecting humans, animals, or plants; production of natural foods (fish, shellfish, nuts, berries); and maintenance of genetic information in the plant or animal kingdoms of the incredibly diverse biosphere from which new drugs, vaccines, crops, and other products might be obtained. More specific impacts will be described with each of the solar technologies.

Direct solar heating and cooling

Humans have always taken advantage of solar irradiation to heat their bodies or even to heat their most primitive of homes. The American Pueblo Indians built their adobe houses on ledges of south-facing cliffs such that their walls were warmed in winter and shaded in summer. Throughout the world far more clothing is dried in the sun on clotheslines or spread along the banks of streams than dried in electric or gas-heated indoor clothes driers. The majority of homes in many Japanese towns today have all of their hot water provided by solar heaters. During the 1930s and 1940s there were more than 50,000 solar hot water heaters in Florida. Despite the long history of use, solar space and water heating and solar cooling has not displaced a significant amount of fossil fuel energy in the industrialized nations of the world. However, in recent years there has been a growing interest in the technology capable of capturing as much solar energy as possible and delivering this energy for effective heating of homes or offices. It is not our purpose here to describe in detail all of the solar heating and cooling technology; rather, we shall enumerate its general features and point out that it is a materials-intensive industry.

Some type of a COLLECTION DEVICE is needed to absorb a significant fraction of the incident solar radiation, if this energy is to be used effectively. The windows or walls of a building may act as the collector, or a greenhouse may absorb the energy to be transferred to a home or building. The flat plate collector is the most common type of auxiliary device for gathering solar energy. It consists of a black metal absorber enclosed in an insulated box with a glass or plastic cover. The absorbed heat is transferred in the box to air or liquid, which is piped to its destination, usually a storage tank of rocks or water, or piped into the building, home, or pool. Flat plate collectors absorb both skylight and direct sunlight and their absorption and retention efficiencies may differ greatly summer to winter. Efficiencies can range from 70% in warm weather to 10% in very cold weather. Temperatures required are about 27°C for swimming pools, 55°C for domestic hot water, and 70°C for direct space heating. Many types of collector designs exist. Some are very simple and others are complex, but basically their designs reflect a compromise among absorption efficiency, cost, and durability.

SOLAR SPACE HEATING SYSTEMS are classified as ACTIVE or PASSIVE depending on whether the heat collected is distributed using pumps and fans or by natural radiative, conductive, and convective processes. The simplicity of many passive solar heating systems often makes them more attractive than active solar systems. Instead of installing flat plate collectors on the roof, some buildings are designed to let the sunlight pass through large south-facing windows (in the northern hemisphere) and be absorbed by the interior walls or floor. However, large temperature variations will occur with this system. The fluctuation can be reduced by decoupling thermal storage from the living space by the use of a thermal storage wall. Solar radiation entering the sun-facing windows is absorbed by a storage wall of water-filled drums or masonry (Trombe Wall). Heat is vented to the building by openings at the top and bottom of the storage wall. The wall protects the interior of the building from high temperatures during the day and transmits its stored energy to the interior at night. Sometimes heat storage systems use roof ponds comprising blackened plastic bags of water that absorb solar radiation; from these the stored heat is conducted to the interior of the house beneath the bags.

Solar ponds

A SOLAR POND captures solar energy by storing heat in a stratified pool of water irradiated by sunlight. Solar ponds have large surface areas for collecting solar energy, large amounts of thermal storage (mass of water in the pond), and a lower installed cost than most other solar systems. Water absorbs sunlight extremely well. Only about 2 to 4% of incident sunlight is reflected from water when the angle of incidence (measured from the zenith) is between 0 and 50°. For larger angles of incidence, the reflectance is 6% (at 60°) and 58% (at 80°). This means that the absorptance of sunlight is 96 to 98% when the sun is high in the sky and only drops significantly when the sun is near the horizon. A solar pond is usually only one to three meters deep and has a bottom covered with black plastic to absorb the solar radiation. More than 30% of the incident sunlight penetrates to the bottom of the pond and may raise the temperature there as high as 100°C, while the surface water is near the ambient air temperature.

Salt is added to the water used in a solar pond in order to give it thermal stability. Salts like NH_4NO_3 or KNO_3 are best because their solubilities increase with temperature, whereas salts such as Na_2SO_4 have decreasing solubility with increasing temperature. Brackish water is more dense than fresh water and therefore will concentrate in the bottom; the fresh water will float to the top of the pond. Figure 7 is a schematic drawing of a solar pond. The temperature and density profiles within a solar pond may resemble those shown in Figure 7. There is some heating by sunlight of the surface layer, as illustrated. Normally if the water near the bottom is warmer than the water above it, convective turnover will occur, with the warm water rising and the cold water falling. The salt density gradient minimizes convective mixing so that hot water is trapped in the pond bottom. Sometimes it is necessary to cover the surface of the pond with a thin clear plastic film in order to reduce evaporative cooling and losses. Because the salt concentration is greatest in the bottom layer, salt will gradually diffuse

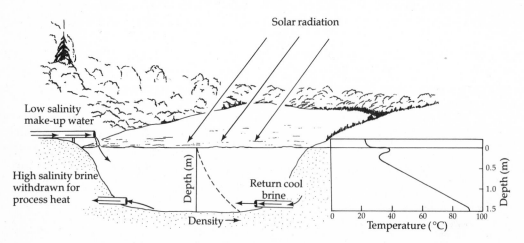

FIGURE 7. A solar pond filled with salt water. Density increases with depth. The density profile is shown as the dashed line near the center of the pond (density scale increases to the right). The temperature profile is shown in the right-hand graph. Except for a warm surface layer, temperature is highest near the bottom.

upward to the surface layer. Although this diffusion is very slow, some periodic flushing of the surface layer with low salinity water and recharge of the bottom layer with salt are required. This does create a maintenance problem and a potential environmental hazard.

In 1980 the only commercial solar pond in the United States was at Miamisburg, Ohio, where a half-acre pond provided heat for the city's olympic-size swimming pool. The pond heated the pool's water and provided 35% of the space heating needs for the pool's bathhouse. In Israel, a 150-kilowatt plant generates electric power by using hot water from a solar pond that gets saline water from the Dead Sea.

Solar ponds may be particularly useful for the generation of low-temperature process heat that may be used for grain drying, paint drying, or other commercial needs. Environ-mental hazards occur when the plastic liner in the bottom of the ponds disintegrates with age. When chemicals are used to inhibit algae or other plant growth, these may leak into the environment; or the brackish water may leak out to poison plants and other organisms. Further descriptions of solar ponds are given by Maugh (1982), Kreith and Kreiger (1978), and Meinel and Meinel (1976).

Solar power tower

In contrast to the low-technology solar pond, the SOLAR POWER TOWER is a massive, high-technology system. The power tower collects solar energy by using a large field of mirrors. It converts this solar energy into heat that generates steam, which is then used to drive a turbine and produce electricity (Figure 8). The best way to get the high temperature is with a

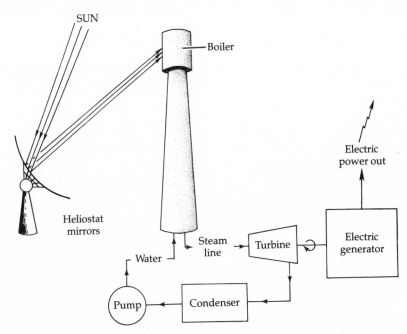

FIGURE 8. A simplified diagram of a solar power tower. Sunlight is reflected by heliostat mirrors and absorbed by blackened panels at the boiler, where it heats water to pressurized steam at 500 to 600°C. The steam turns a turbine, which rotates an electric generator.

large array of focusing mirrors, called helio-stats, that track the sun and concentrate the energy on a boiler set atop a large tower.

A 10-MWe (megawatt electric) solar power tower would have 2000 heliostats and a 100-MWe plant would have 20,000 heliostats. Each heliostat for the 100-MWe tower would be about 37 m^2 in area and would concentrate the incident solar energy 1000-fold. The total field of mirrors would occupy an area of about 3.5 km^2. The tower would be 260 m tall. In the sunny regions of the southwestern United States, where solar power towers are being tested, one can expect to have about 950 Wm^{-2} of solar energy incident on the mirrors when the sun is high in the sky. Mirror reflectance is about 90%, receiver absorptance is about 90%, dust losses are 10%, radiative and convective losses are 10%, so that the final amount of energy transferred to the working fluid is from 50 to 66% of the total incident sunlight on the mirrors. The normal overall efficiency of a solar power tower system through the turbine to the electricity delivered is about 16%.

The receiver on a 100-MWe plant is cylindrical, 17 m in diameter, and 25 m tall. The outside cylinder wall forms the absorbing surface, which is made up of 24 identical panels, each 2.2 m wide. For a water–steam receiver, each panel is composed of 170 tubes each 13 mm in diameter. The temperature of the steam generated is 500° to 600°C. The steam is used to drive a turbine. However, as in any Rankine (steam) cycle system, a considerable amount of cooling water is required through the condenser. If evaporative cooling is used, the power tower electrical generating system would use as much water as a comparable size fossil fuel plant. Not only will the steam be used to generate electricity but some of the heat will be put into storage for up to six hours to cover electrical demand during the nighttime period. Storage can be accomplished in deep geological formations, by chemical storage by means of electrolysis of water into hydrogen and oxygen, by means of other chemical reactions, or by the use of batteries. At Albuquerque, New Mexico, there is a 5-MWt (metawatt thermal) pilot facility and at Barstow, California a 10-MWe unit.

Some scientists estimate that solar plants in the southwest might supply a significant amount of our national energy requirement through electrical transmission to other regions of the country. According to some estimates a high-voltage, direct-current, electrical network could deliver power economically from the solar-rich southwestern United States to Detroit, a distance of more than 2000 km. However, the land areas required to generate significant amounts of energy would be intolerably large.

The energy required for the fabrication and construction, including the transport of materials, of a 100-MWe plant is 310,000 MWt-hours. This amount of energy could be generated in about 1.5 years; so the return on the investment seems good. Technical details concerning solar power towers may be found in Kreith and Kreider (1978) and in Metz (1977a).

Some environmental impacts will be associated with solar power towers (Holdren et al., 1980).

- Pollution from the leakage of working fluids
- Pollution from compounds used for cleaning the mirrors
- Water consumption by wet cooling towers
- Fires and the associated release of toxic substances
- Removal of agricultural land from production
- Interference with natural ecosystems
- Changes in the amount of reflected sunlight received at ground level and consequent changes in the water balance of the soil
- Thermal storage in rocks, molten salt, or liquid metals or oil

- Burns or blindness caused by concentrated light beams
- Hazardous navigational conditions around receiver towers

Some people may consider a solar tower as aesthetically displeasing, but to other people it represents an attractive technological achievement.

Photovoltaics

The use of PHOTOVOLTAIC CELLS to generate low-voltage, direct-current electricity from sunlight has been proved technically feasible during the last couple of decades. Photovoltaic cells are a standard source of power for space vehicles and for satellites. Adoption of this technology for widespread terrestrial use has not been feasible because the photovoltaic

cell used in space has been at least 1000 times too costly to be competitive with other methods of electricity generation. Costs have been significantly reduced during recent years and are still coming down.

Solar cells are produced from sliced wafers of single-crystal silicon, processed by standard methods of semiconductor fabrication, and assembled into flat plate collectors. Figure 9 shows a schematic diagram of a silicon solar cell. The key component is a semiconductor that absorbs light energy by exciting an electron (negative charge), which moves from the valence band (bound electron) to the conduction band (free electron) and leaves behind a positive hole. The electron and hole would normally recombine, but an electric field is provided within a semiconductor so that they remain separated. The electrons

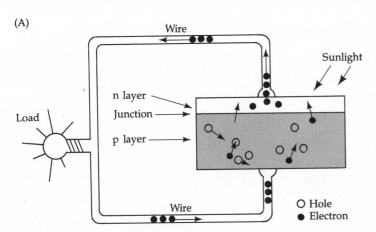

(A)

(B)

FIGURE 9. (A) Schematic diagram of a silicon solar photovoltaic cell. Pure silicon is doped with metal impurities such that one part has an excess of electrons (n layer) and the other part has an excess of holes (p layer), thereby forming a p–n junction between the two parts. When light strikes the material, an electric current flows through the circuit connecting the two halves of the cell. (B) A panel of silicon photovoltaic cells used for generating electricity.

then flow as an electric current through an external load, thereby doing useful work. Intimate contact of two materials, at least one of which is a semiconductor, produces the required electric field if the chemical potentials of electrons in the two materials are different. What is known as a p-n junction occurs between oppositely doped semiconductors, so that there are excess holes (p layer) on one side of the junction and excess electrons (n layer) on the other side. The current output of solar cells is limited by the number of carriers (electrons and holes) generated by the incident light.

Single-crystal silicon cells are the most advanced of all photovoltaic devices. Strictly speaking, silicon does not have the most desirable physical properties for a solar cell device (too small a band gap, 1.1 eV; weak optical absorption; and long diffusion distances), but they are easy to grow synthetically. Generally efficiencies have been about 6%. High-capacity, efficient processes for less costly silicon production and purification are being explored, and cells of 15% efficiency have recently been fabricated on slices of cut polycrystalline silicon ingots. New processes have provided single-crystal silicon ribbons directly from a silicon melt. These are shown to have efficiencies of about 15%. Other semiconductor materials used are gallium arsenide, cadmium sulfide, copper sulfide, amorphous silicon, indium phosphide, and copper indium selenide.

One key to the future use of photovoltaic cells is the cost of the extremely pure semiconductor grade of silicon needed for the cells. Raw silicon is melted and remelted so many times during purification and crystal growth that it is among the most energy-intensive commercial materials in the world. Furthermore, another problem is waste. About 80% of the purified silicon is left as scrap in the crucibles used to grow large cylinders of the material in crystalline form or it ends up as

sawdust after the cutting of the cylinders into discs. New methods of growing and cutting are reducing these wastes.

When photovoltaics are used for central power plants, the land requirements are considerable and are similar to solar power tower land requirements. The semiconductor materials used in photovoltaics are hazardous to workers. Such hazards include dust from silicon handling and poisons from cadmium sulfide, gallium arsenide, and other metallics. The electrical transformers that must be used to boost the voltages from photovoltaic cells contain toxic substances. Batteries used for electrical storage contain metals such as lead, cadmium, and nickel. These metals are toxic and hazardous throughout their manufacture, handling, and use.

Many of the environmental impacts associated with photovoltaics occur at the time of their manufacture or during mining and processing of the raw materials. For example, the production of cadmium sulfide photovoltaic cells is accomplished by spraying of cadmium sulfide on glass or by vacuum deposition on copper foil. Particulate releases of cadmium and cadmium sulfide occur, and cadmium oxide fumes also are released. Cadmium compounds are bioaccumulated and are highly toxic to humans and animals. Despite these risks, it is generally true that cadmium emissions are maintained at very low levels during the manufacturing process. Photovoltaic cells made from silicon, gallium arsenide, copper sulfide, and indium phosphide have toxic materials that are released during grinding and cutting.

When large arrays of photovoltaic cells are used in a central power station for the generation of electricity, substantial environmental impacts are associated with construction activities. These impacts are similar to those that might occur during solar power tower construction. These central power station photovoltaic plants will most likely be located in desert areas because deserts are regions receiving large amounts of sunshine. Typical site construction activities include grading of the site, construction of access roads, transport of components, and assembly and placement of the photovoltaic arrays. Construction activities will temporarily degrade the air quality of the region because of fugitive dust and vehicle exhaust.

Deserts have a crustal surface made up of organic compounds generated by fungi and bacteria in the soil. These compounds help to bind sand particles together and to minimize erosion. The destruction of this crustal surface releases dust as well as fungi and bacteria into the atmosphere. Some of these organisms can cause serious lung disease in humans.

Site construction may destroy large numbers of burrowing desert species of animals. These animals seek shelter underground during hot daytime hours. Disturbance of their habitat through dirt moving, equipment operation, and vehicle noise will occur. Sessile species will be destroyed and mobile species forced to emigrate to adjacent areas. Animals moving into territories claimed by other populations will often be killed or driven away. The removal of vegetation sets up the potential for increased erosion by wind or water. Local water quality will undergo temporary degradation because of increased runoff attributable to vegetation loss, to the presence of oil and grease released from construction equipment, and to toxic substances released from either herbicides or pesticides used at the site.

Orbiting photovoltaic systems

Harvesting solar energy from outer space for use on Earth sounds like the ultimate utopian Aladdin's lamp. Large arrays of silicon or gallium arsenide photovoltaic cells would be mounted on a satellite in geosynchronous Earth orbit such that it remains stationary above one location at the ground. The electrical energy generated by these photovoltaic cells would be converted to microwaves and beamed to large receiving antennas located at the Earth's surface, where the energy would be rectified and put into utility electrical grids.

The microwave beam would operate on a frequency of 2.45 gigahertz with a power density of 230 Wm^{-2} at the center of the receiving antenna. The antenna surface area would be about 150 km^2. At the edge of the antenna the power density would be 10 Wm^{-2} and the side lobes in the radiation pattern would contain about 1 Wm^{-2} at a distance of several kilometers to one side. The hazards to human health and to plants and animals of this microwave radiation are not well understood. Birds and insects will fly through the beam, but the spreading of the radiation field beyond and around the central receiver may irradiate large numbers of plants, animals, and humans.

The quantity of materials that must be placed in orbit is very large; on the order of 50×10^6 kg. To do so will require more than 200 large rockets. The exhaust fumes from so many large vehicles will have serious consequences to the atmosphere in terms of changes in photochemistry, ozone concentration, radiation balance, and precipitation. The potential for accidents among so many rocket launches is not insignificant. For installation and maintenance of the photovoltaic and microwave equipment in space, workers will need to shuttle between Earth and satellites. This incurs hazards to the workers, including weightlessness and bombardment by cosmic rays. Weightlessness can cause a loss of calcium from bones and teeth, and cosmic radiation can cause genetic damage. The microwave beam can produce ionospheric disturbances and can interfere with radio communications, or it may influence a variety of molecular reactions in the atmosphere, including those affecting the ozone layer and even climate.

7

Biomass

Energy cane, shown at the left of this picture, has much more biomass than ordinary sugar cane, shown on the right, growing in a field in Puerto Rico. Courtesy of Melvin Calvin, Lawrence Berkeley Laboratory, University of California.

INTRODUCTION

BIOMASS sources of energy include wood, grass, agricultural crops, food processing wastes, animal wastes, oil-bearing plants, and kelp from ocean farms. There are three ways in which humans extract the energy contained in organic matter. The most efficient way is by direct combustion of plant materials—in a stove, a fireplace, or an electric power plant furnace. Another extraction method uses gases, such as methane, which are derived from the transformation of animal wastes into BIOGAS, burned in a furnace. Biogas can be readily transported through pipelines or it can be compressed and transported in tanks. A third extraction method involves conversion of biomass into liquid fuels, such as ETHANOL or METHANOL, which are burned in an engine. Liquid fuels are easily transported in tanks.

There are several major difficulties associated with utilizing BIOMASS for solar energy conversion. They are (1) the relatively small percentage (0.1 to 0.9%) of incident light energy that is converted into biomass by plants during an annual cycle; (2) the dispersed form and low concentration of biomass per unit area of land and water; (3) the scarcity of highly productive land beyond that already under cultivation; (4) the high moisture content (15 to 95%) of plant material, which makes it heavy to transport and energy consumptive to dry; and finally (5) the environmental and ecological effects associated with erosion, nutrient supply, dust, and habitat disruption, as well as the effluents released during processing. Nevertheless, despite some of these disadvantages, biomass promises to be a significant fuel source for certain applications.

Most of the lesser developed societies of the world use biomass as their principal energy supply. So also did the United States during its pioneer stage. In 1850 more than 90% of the United States energy supply came from biomass sources. Unfortunately, for many people in the world today, local supplies of biomass are becoming acutely short. Ironically, this is occurring at a time when the United States is considering means to increase the biomass fraction of its total energy supply.

Sources of biomass in the United States

There are currently about 304 million hectares (750 million acres) of forested land in the United States, and of that, 101 million hectares are used commercially. The commercial forest land has a net productivity of about 2.5 metric tons per hectare per year. Improved varieties of hardwoods can yield up to 10 mt $ha^{-1} yr^{-1}$.

At the present time, wood with a total energy equivalent of about 5 quads/yr is being harvested from U.S. forests. Of this amount, about 1.7 quads/yr ends up in finished forest products, such as lumber, furniture, paper, and boxes. About 1.2 to 1.3 quads/yr of wood is burned by the forest products industry. Approximately 0.2 to 0.4 quads/yr are burned in home stoves and fireplaces and about 2 quads/yr are returned to the forest soil by decomposition or are being burned at the logging site.

The amount of energy that could be supplied from biomass in the United States is very substantial. Optimistic estimates by the U.S. Office of Technology Assessment (Gibbons, 1980), indicate that 12 to 17 quads/yr could come from biomass energy by the year 2000. Whether this is achieved or not depends on variables such as availability of croplands, improvement of crop and forest yields, development of efficient processes of conversion to alcohol, proper resource management, and appropriate pricing levels. If energy consumption in the United States by the year 2000 is about 106 quads/yr, then as much as

16% of this might be provided from biomass. This biomass energy may be derived from the following components: up to 10 quads/yr from wood; 0 to 5 quads/yr from grasses and legume herbage; 1 quad/yr from crop residues; 0.3 quad/yr of biogas from manure; and 0.2 quad/yr of ethanol from grain. The energy to be derived from aquatic plants, such as kelp, and from oil-bearing arid land plants is estimated to be insignificant before the year 2000. Eventually municipal solid waste may turn out to be a significant source of biomass energy.

Achieving 10 quads/yr of energy from wood would require more intensive forest management than at the present time. Approximately 70 to 75% of potential wood growth is in the eastern United States, where there is a mixture of softwoods and hardwoods; in the western United States, there are mainly softwoods. Much of the forest land in the eastern United States has been harvested several times; the highest value trees have been removed and the poorer quality timber has been left for future growth. On these lands more intensive forest management will include clear-cutting and replanting with trees of higher commercial value. Many stands of trees will need to be thinned, and brush and poorer quality timber, including dead or diseased trees, will need to be removed in order to improve forest production. The collection and burning of waste wood helps to eliminate breeding places for insects in the forest. Eventually fuelwood harvests would come from thinning of forests, from logging residues, from the removal of dead, dying, and diseased trees, and from the use of fast-growth plantation stands.

The Commonwealth of Puerto Rico grows a great deal of sugarcane, but it is also completely dependent on imported oil for its energy needs. This dependence on imported oil, which makes up 99% of Puerto Rico's energy use, is both costly and risky. The U.S. Depart-

ment of Agriculture in Puerto Rico is developing a form of sugarcane known as ENERGY CANE. Energy cane contains much more solid biomass than does sugarcane for the same crop area. It is possible to use its stalks and leaves to fuel electric power-generating plants. Stalks and leaves from cane plants are called *bagasse*. The energy content of bagasse is 8440 kJ/kg (4000 Btu/lb) at 51% moisture and 10,550 kJ/kg (5000 Btu/lb) at 40% moisture. High-yield energy cane may produce from 25 to 75 mt ha^{-1} yr^{-1}. Puerto Rico may be generating a considerable amount of electrical energy within a couple of decades by burning energy cane bagasse.

Use of biomass in other nations

Several nations have begun ambitious programs of biomass energy development, notably Brazil, Sweden, and China. Brazil derives 27% of its total energy supply from wood. Brazil currently runs 330,000 automobiles on a water and alcohol mixture, representing 10% of its oil supply. Brazil, however, has plentiful crops of sugarcane and cassava from which the alcohol is derived.

Sweden's interest in wood is an intriguing case. Sweden imports about three-quarters of its total energy needs, most of it in the form of increasingly expensive crude oil from the Middle East. The Swedes consume more imported oil per capita than any other nation. Sweden's 1978 energy mix included oil, 72%; hydroelectric, 12%; lyes and wood waste, 8%; coal and coke, 4%; and nuclear, 4%. Seventy-five percent of its electric power generation comes from hydroelectric plants and 25% from nuclear plants. Sweden has six additional nuclear power plants that will be coming on-line soon and will deliver an additional 9440 MWe. However, all of them are slated to be phased out in 20 or 30 years. Sweden hopes to have 100 large-scale wind turbines operating by the year 1990.

Willow (*Salix*), poplar (*Populus*), birch (*Betula*), and alders (*Alnus*) grow extensively in Sweden. The wood of these trees has a thermal value of 21,000 kJ/kg (9950 Btu/lb) at a 3% moisture content. Some poplar is rated at about 10% less fuel value. According to one report, about 4.5 million tons of tree remnants or waste wood, corresponding to 1.9 million tons of oil, are left to rot each year in Swedish forests. Sweden intends to boost its use of wood for fuel by intensively harvesting waste wood in the forests and by planting so-called energy forests of fast-growing species.

The National Board for Energy Source Development in Sweden says, "Forest energy (including logging wastes) tops the list of our energy options. By 1990, 20% of our energy could come from forests." They go on to show that 3 million hectares, or nearly 7% of Sweden's total land area, could be used for energy plantations. They are cultivating 11 different fast-growing clones of poplars, willows, birches, and alders. Certain species of willow can convert over 5% of the incident solar energy into biomass during the growing season, but the annual average would be about 1%. Some clones grow four meters per year. Rotation periods of one to three years appear feasible. They anticipate having 100,000 hectares sprouting high-yielding energy forests by 1990. The harvested trees will be taken to processing centers where they will be converted to powder or chips for use in power plants, in industrial boilers, and eventually in home furnaces. (See Lonnroth et al., 1980 for further discussion of Sweden's energy options.)

WOOD

Wood has a number of advantages over other sources as a domestic fuel source. Wood can be grown locally, and it is widely available and renewable. Wood may be burned directly, it may be converted to charcoal and then burned, it may be gasified, or it may be converted to alcohol. The major drawbacks to wood are its bulk, its relatively low energy content per pound compared with oil or gas, and its environmental disadvantages.

Wood, in terms of organic matter content, is about 50% CELLULOSE, 20% HEMICELLULOSE, and 30% LIGNIN. The following definitions are important:

- 1 dry kilogram (0% moisture) of wood weighs 1 kg and has an energy content of about 19,000 kJ/kg.
- 1 dry kilogram (50% moisture) of wood weighs 2 kg (1 kg of wood and 1 kg of moisture) and has an energy content of 16,900 kJ/kg.
- 1 green kilogram (50% moisture) of wood weighs 1 kg, is equal to 0.5 dry kilogram (50% moisture), and has an energy content of 8450 kJ/kg.

By comparison, bituminous coal has an energy content of 24,400 to 32,500 kJ/kg and crude oil has an energy content of 38,000 kJ/kg. Most wood has a moisture content of 50% as shown by the numbers in Table 1. The moisture in the wood reduces the net energy content for two reasons. If the wood is weighed green at 1 kilogram, then its weight is only equivalent to 0.5 dry kilogram (50% moisture). Therefore, 20,000 kJ/dry kg = 10,000 kJ/green kg. In addition, approximately 2100 kJ/kg is used to evaporate the moisture from the wood when it is burned. Thus the effective net energy available is only about 7900 kJ/kg. The values given in the last column of Table 1 are the energy content of wood per green kilogram of mass. These values range from a low of 7634 kJ/green kg to a high of 11,764 kJ/green kg.

Forest productivity yields about 2.5 mt ha^{-1} yr^{-1} on a commercial basis today in the United States, but 10 mt ha$^-$ yr^{-1} is entirely possible for some regions of the world where soils and climate are favorable for growth. It is suggested that productivity of plantation for-

TABLE 1. Average heat content, moisture content, and effective heat content of wood in northern Vermont[a]

Species	Heat content (kJ/dry kg)	Moisture content (%)	Heat content (kJ/green kg)
White and Red Pine	20,090	55	7,634
Balsam fir	20,611	53	8,331
Spruce	19,957	46	9,599
Hemlock	20,657	51	8,818
Red oak	19,377	43	9,946
Yellow birch	20,247	49	9,074
Paper birch	21,943	49	9,937
Sugar maple	19,047	44	9,541
Red maple	19,233	44	9,646
Beech	19,477	35	11,764
Ash	19,964	45	9,830
Aspen	20,085	48	9,216

[a]From Sherwood and Meadows (1978).

ests in the tropics using highly selected species of *Eucalyptus* may give very high yields. Abelson (1982) reports that some experiments in Brazil with selected *Eucalyptus* clones having a seven-year growth cycle were initially yielding 23 mt/ha; and with further selection they improved to 33 mt/ha and finally to 40 mt/ha. The five best clones of trees produced as high as 61 mt/ha. Because the growth cycle was seven years, annual yield would be 8.7 mt/ha.

Two wood-fired utility plants are in operation in the United States. Burlington Electric Company in Vermont operates an 8-MWe plant using wood chips from forest residues and noncommercial dead or culled trees in addition to mill waste. The Electric Board in Eugene, Oregon produces heat and process energy for a local business district, hospital, college campus, cannery, and greenhouse from a 33.8-MWe plant. For eight months of the year, this plant also uses its steam to generate electricity. This plant consumes 275,000 metric tons of wood per year and displaces about 180,000 barrels of oil. At a $10 differential between the cost of oil and wood, the utility saves about $1.8 million per year in fuel costs.

In late 1982, the Dow Corning Corporation of Midland, Michigan began operating a 22.4 MWe cogeneration electric power plant using wood as a fuel. The power plant consumes up to 180,000 dry tons of wood per year. Cogeneration is achieved by using the higher-grade spent steam for process heat and the lower-grade spent steam for heating the building. The plant manufactures various silicone products. Approximately 50% of the wood comes from forest land within a 125 km radius of Midland. The other 50% of the wood comes from a variety of sources including dead or dying trees in municipalities, from land cleared for agricultural uses, from new construction projects, from wood scraps being discarded by industries, and from sawmill residues.

Home stoves

The use of wood for home stoves has increased dramatically during the last few years. More than a million households in the United States are now using wood stoves or furnaces as a primary source of heat. In some parts of the United States the availability of inexpensive, easy-to-gather wood is running low and costs for purchasing wood for the privilege of cutting it are rising steeply.

The environmental effects of burning wood in the home can be serious. The Environmental Protection Agency has found six chemicals in wood smoke that are known to cause cancer in laboratory animals. Most of these chemicals are polycyclic organic compounds of the same type as those in cigarette smoke. Other pollutants are carbon monoxide, soot, and small particles.

Ironically, it is the very action of using airtight stoves and dampening them down in order to produce a slow burn of the wood that produces the most dangerous pollution. This action lowers the temperature of burning, and the reduced temperature does not break down the organic compounds as efficiently as does a higher temperature. Hence, a partial pyrolysis of the wood occurs, and the organic compounds mix with the air coming out of the chimney. Wood stoves need to be redesigned for more complete burning or need to have a secondary chamber to burn off gases and organic matter.

Fly ash, the small particles in wood smoke, can produce a blue haze in the atmosphere. Many towns in Vermont and New Hampshire, located in mountain valleys with temperature inversions, are finding wood smoke pollution a serious health problem. In Denver, Colorado on evenings and weekends as much as 25% of the fine particles suspended in the air seem to be associated with wood burning.

Gasification

Gasification of wood by partial oxidation at high temperature produces a crude gas consisting primarily of H_2, CO, CO_2, and about 46% nitrogen when air is used to oxidize the wood. The nitrogen can be removed by condensation with liquid air at very low temperatures, but this uses energy. Reliable, high-efficiency, air-blown gasifiers using wood as a fuel could provide a means to convert oil- or gas-fired boilers to wood while allowing the return to oil or gas if wood were in short supply. Several different processes are available for converting wood to gas. The cost of converting oil- or gas-fired boilers to a biomass gasifier is cheaper than the cost of converting to the direct combustion of wood.

ENERGY FROM RESIDUES AND FORESTS

Agriculture and forestry RESIDUES include corn stalks, sugarcane bagasse, husks, wheat stalks, straw, waste paper, and wood. These materials are described as containing lignocellulosic compounds such as cellulose, hemicellulose, and lignin. The composition of these residues is 30 to 50% cellulose, 18 to 35% hemicellulose, and 15 to 25% lignin. Cellulose and hemicellulose are polysaccharides, or sugar polymers, and can be hydrolized to fermentable sugars. Lignin has a three-dimensional phenolic structure that is highly resistant to microbial attack or to chemical hydrolysis. In fact, it may be most practical to use intact lignin in plastics, adhesives, and in other ways that make use of its particular properties. (See Bungay, 1982 for a detailed description of biomass refining.)

The net energy derived from harvesting biomass for fuel may not be understood properly unless all factors are taken into consideration. Piementel et al. (1981) present a thorough analysis of this issue with regard to crop and forest residues. Of the plant biomass produced annually in the United States, approximately one-third is harvested and converted to agricultural and forest products. Residues remaining on sites after harvest amount to 17% of the total annual biomass production. In the United States the residues remaining in the field are estimated to be about 540 million metric tons (Mmt) by dry weight and have a gross heat energy content of about 12% of the fuel consumed annually by the United States. Of this 540 Mmt of residues, about 430 Mmt are from crop harvesting and 110 Mmt remain after trees are harvested for lumber and pulpwood.

Net energy from corn residues

Corn stalks left in the field after the corn crop has been harvested are referred to as STOVER. A corn crop yields about 5500 kg of grain per hectare and about 5500 kg of corn stover per hectare, with about a 50% moisture content. It is assumed that when collecting corn stover for energy use it is necessary to leave about 2000 kg/ha on the land to protect the soil from erosion and to maintain the necessary soil organic matter. Piementel et al. (1981) report the energy content of corn stover to be 13,800 kJ/kg. The amount of corn stover available per hectare is 3500 kg/ha. The energy available per hectare from corn stover is then 48.3×10^6 kJ/ha. When the corn stover is burned directly and 55% of its energy is converted to useful heat, the heat content of corn stover is effectively only 7600 kJ/kg and the energy available per hectare is only 26.6×10^6 kJ/ha.

It is important to consider the net energy available from corn stover. Energy used for collecting the stover, energy used to transport it to the furnace, energy losses associated with soil erosion, and energy losses for replacing nutrients are significant. For example, it is estimated that corn yields are reduced by about

4% (220 kg/ha) for each 2.5 cm of topsoil lost. Each ton of topsoil lost carries away about 5 kg of nitrogen and 1 kg of phosphorous. This loss can be partially offset by increased application of nutrients requiring the expenditure of additional fossil fuel energy.

The collection of corn residues requires an energy expenditure of about 181 kJ/kg. The transport of the residues from the field to the furnace requires an energy expenditure of $5.48 \text{ kJ kg}^{-1} \text{ km}^{-1}$. If the average round trip distance for the vehicles is 40 km, then the energy expended for transportation will be 219.2 kJ/kg. The energy expenditure necessary to offset the effects of soil loss on productivity is estimated to be 2143 kJ/kg. The energy required to replace nutrients lost by the removal of corn residues is estimated to be 703 kJ/kg. All of these expenditures of energy come to a total of 3240 kJ/kg. The energy expenditures may, in fact, be greater than this, as shown by Piementel et al. (1981), for land with greater amounts of soil erosion than for the numbers I have used here.

The total energy content of corn stover was given earlier to be 13,800 kJ/kg. The net energy available from corn stover may now be calculated to be $13,800 - 3,240 = 10,560$ kJ/kg. If it is burned directly, then the 7600 kJ/kg thought to be available is now in effect a net energy of only 4360 kJ/kg.

Electric power from corn residues

The possibility of using agricultural residues or forest residues to fuel an electric power-generating plant is of much interest. The conversion of biomass into electrical power has the following advantages: electricity can be more easily transported to the consumer than the raw biomass material; electricity is a high quality form of energy; electric power plants can be located close to the source of residues; and crop and forest residues burn more cleanly than coal.

Electric power plants are rated according

to their maximum capacity C for power production. An ordinary-size coal burning power plant would have a capacity of 600 MWe, and a very large plant would be rated at 1000 MWe. However, power plants do not operate 100% of the time. The plant factor P is the percentage of capacity at which the plant operates when averaged over a year. A typical value for P is 70%. A power plant also has an efficiency of operation E, which is the number of kilowatt-hours of electrical energy produced per kilowatt-hour of fuel input to the plant times 100. A typical value for E is 38%. One can combine all the characteristics of the power plant with the energy content of the fuel used (B, in kJ/kg) to obtain the total annual fuel requirement (AFR, in kilograms):

$$\text{AFR} = \left(\frac{P}{100}\right)C\,(\text{MWe})\,10^3\left(\frac{\text{kW}}{\text{MW}}\right)3600\left(\frac{\text{sec}}{\text{hr}}\right)$$
$$\times\,24\left(\frac{\text{hr}}{\text{d}}\right)365\left(\frac{\text{d}}{\text{yr}}\right)\left(\frac{100}{E}\right)\frac{1}{B}\left(\frac{\text{kg}}{\text{kJ}}\right) \quad (1)$$

Combining all of the conversion factors in Equation (1), we can rewrite it in abbreviated form:

$$\text{AFR} = 31.54 \times 10^9\,\frac{PC}{EB}\left(\frac{\text{kg}}{\text{yr}}\right) \quad (2)$$

As an example, we shall calculate the amount of corn residue required per year to fuel a 100-MWe power plant that has a plant factor $P = 70\%$ and an efficiency $E = 38\%$ and that burns corn stover of net energy content $B = 10,560$ kJ/kg.

Substituting these numbers into equation 2, we get

AFR = $(31.54 \times 10^9)\,[(70 \times 100)/(38 \times 10,560)]$
   = 550×10^6 kg/yr
   = 550,000 mt/yr

Now it is easy to determine how much land must be available in corn production in order to fuel a 100-MWe power plant. We can get 3500 kg/ha of corn stover. Hence, the area necessary to produce 550 million kilograms of corn residue per year is

$$\text{area} = \left(\frac{550 \times 10^6\,(\text{kg/yr})}{3500\,(\text{kg/ha})}\right) = 157,140\,\text{ha/yr}$$

This is an area of dimensions 39.6 × 39.6 km or 23.8 × 23.8 miles. This is a tremendous area, and it must contain nothing else but corn. Furthermore, a 100-MWe power plant is not particularly large. It would supply power for a population of about 75,000 to 80,000 people. (See Chapter 12 for further information concerning electric power-generating plants.)

If no energy expenditures are considered and the potential fuel value of corn residues is taken to be 13,800 kJ/kg, then the annual amount of corn stover required to fuel a 100-MWe plant would be 420.9 × 10^6 kg/yr. The land requirement would now be 120,200 ha/yr or a piece of land of dimensions 34.6 km × 34.6 km. This means that oil, gas, coal, or nuclear energy supplied all the subsidies to make up for the energy consumed by the collection and transport of the corn stover and the energy used to supply nutrients to replace losses by soil erosion and by the removal of the residues.

Electric power from forest residues

There is a considerable amount of interest in the possibility of supplying forest residues as the fuel for an electric power-generating plant. The same procedure used above for estimating the amount of corn stover required to supply a 100-MWe power plant with fuel may be used here for forest residues. First one needs to estimate the net energy content of the forest residues as delivered at the power plant.

Forest residues containing 50% moisture have a potential energy content of 14,000 kJ/kg. The collection of forest residues requires an energy expenditure of about 559 kJ/kg. This is considerably greater than for corn stover because of the additional energy required for chipping the slash. The transport of the chips from the forest to the furnace requires an energy expenditure of 219 kJ/kg, the same as for the corn stover. The energy expenditure necessary to offset the erosion of soil is 85 kJ/kg, a quantity much smaller than for corn stover, providing good collection practices are observed. The expenditure to replace nutrient loss that goes out with the wood chips is 661 kJ/kg. The total energy expenditure necessary for harvesting forest residues is estimated to be 1524 kJ/kg. The net energy available from the wood chips would be 12,476 kJ/kg.

The wood chips will be supplied to a 100-MWe electric power-generating plant having a plant factor $P = 70\%$ and an efficiency $E = 38\%$. Using equation 2 we can calculate the annual quantity of forest residue needed to fuel the power plant as

$$AFR = 31.54 \times 10^9 [(70 \times 100)/(38 \times 12,476)]$$
$$= 466 \times 10^6 \text{ kg/yr}$$

Because the energy content of forest residue is greater than the energy content of corn stover, the total quantity of wood chips needed to fuel a 100-MWe power plant is less than that using corn residue.

The forest residues may be harvested at about 24,700 kg/ha. In order to harvest 466×10^6 kg/yr, an area of 189 km^2 is needed on an annual basis for a 100-MWe power plant. This is an area 13.7 km $\times$ 13.7 km. The rate at which a forest may regrow will determine how soon a harvested forest may be harvested again. In the southern United States it is possible to have a rotation period as short as 7 to 10 years, but in northern regions the rotation period is more like 30 years. The total area required to supply forest residues in sufficient quantities to operate a 100-MWe power plant would be from a minimum of 1320 km^2 in the southern United States to a maximum of 5661 km^2 or more in the northern United States. These areas represent pieces of land of dimensions 36.3 km $\times$ 36.3 km and 75.2 km $\times$ 75.2 km, respectively.

Electric power from whole forests

If instead of using only residues, the entire standing biomass of a forest is used to fuel a 100-MWe electric power-generating plant, it may be possible to reduce the quantity of land required. According to Whittaker (1975), a mature, climax forest has a standing biomass above ground of 170 to as much as 585 metric tons per hectare, depending on the type of forest. If one uses a value of 250 mt/ha, for example, with an average net energy content of 12,476 kJ/kg, the annual forested area required to fuel a 100-MWe plant would be 1864 ha/yr or 18.6 km^2/yr. This is an area equivalent to 4.3 km $\times$ 4.3 km each year. However, it takes a very long time to grow a climax forest—a time often exceeding 100 years. On this basis these numbers would not seem very good.

Earlier we mentioned that experiments in Brazil were producing highly productive clones of *Eucalyptus* with as high as 61 mt/ha with a seven-year rotation period for a productivity of 8.7 mt ha^{-1} yr^{-1}. Using this number we calculate that the annual area required is 63,218 ha/yr or 632 km^2/yr. Each year an area equivalent to 25.1 km $\times$ 25.1 km of plantation forest must be harvested. It would seem that this would be the most optimistic minimum amount of forest required on a sustaining annual basis to fuel a 100-MWe power plant. We also mentioned earlier that on a commercial basis the United States forests only yield about 2.5 mt ha^{-1} yr^{-1}. The land requirement now would be 186,500 ha/yr or 1864 km^2/yr. This would be an area equivalent to 43.2 km $\times$ 43.2 km—a very large forest.

As mentioned earlier, a 100-MWe power plant is not particularly large and would only supply power to about 75,000 people. It is highly improbable that forests will be used productively to fuel electric power-generating plants of even moderate size (100 MWe), let alone of large size (500 to 1000 MWe).

Environmental and ecological effects

Soil erosion. Soil sediments washing off the land carry with them nutrients and pesticides, which have strong ecological impact on the flora and fauna of streams and lakes. The added nutrients may increase aquatic productivity, thereby resulting in eutrophication, and suspended sediments may reduce light levels in aquatic ecosystems and reduce productivity. Pesticides may be toxic to some aquatic organisms. Nearly all of the effects on ecosystems of increased soil erosion are deleterious and must be avoided or strongly mitigated if crop and forest residues are to be used for energy production.

Corn stover must be collected by vehicles moving through the fields. Whether or not all of the corn stover remaining in the field is removed or only part of it depends on the method of cultivation used. If conventional tillage is used for corn production on land with a slope between 6 and 12%, soil erosion will be about 45 mt ha^{-1} yr^{-1}. When the best conservation practices are instigated, it is possible to reduce soil erosion to as low as 4 mt ha^{-1} yr^{-1} on land with a slope less than 5%.

It is not easy to estimate the effects of soil erosion on crop productivity because of the many variables involved. Evidence suggests that corn yields are reduced by an average of about 4% for each 2.5 cm of topsoil lost from a base of 30 cm or less. Various methods are used to prevent or reduce soil erosion. CONSERVATION TILLAGE may be used. It includes contour planting, crop rotation, cover crop planting, and no-tillage cultivation. Growing rye and vetch between corn crops may reduce soil erosion by as much as 43%. In addition, because vetch is a legume with the ability to fix nitrogen, it provides 112 kg of nitrogen per hectare. The use of cover crops during the winter is of great importance if all crop residues are removed.

NO-TILLAGE cultivation also may be used.

The procedure here is to plant a new crop directly in the crop residues left after harvesting. A winter cover crop may be planted and then chemically killed in preparation for planting the new crop in the spring. Piementel et al. (1981) write,

> Corn residues left on the surface of the land in no-tillage culture generally reduce soil erosion to about one-third that in conventionally tillaged corn grown continuously on a 4- to 5-degree slope; combining no-tillage culture with a corn-corn-oats-hay rotation would reduce soil erosion to about one-ninth that in conventional tillage.

There are environmental and ecological problems associated with the chemicals used to kill winter cover crops or to poison meadowland in no-tillage cultivation.

Residue removal. Crop and forest residues left on the land are vital to the well-being of agricultural and forest ecosystems. Their presence maintains soil fertility, organic matter production, and soil structure. In addition they help to control erosion, sedimentation, and flooding. Residues left in place protect the soil against the effects of wind and precipitation. When residues are removed, soil erosion increases markedly, and with increased erosion there is reduced soil fertility because nutrients are carried away with the sediment and with the water flow.

ALCOHOL FUELS

Alcohols have long been used for fuel and particularly for lighting. About 1830, alcohol replaced the use of fish and whale oils as a lighting source; then it was displaced by kerosene. During both world wars, alcohol was used as a substitute for gasoline in automobile engines in Europe. Now, as oil scarcities loom and gasoline prices rise, alcohol is seen as a useful additive to gasoline; the mixture is known as *gasohol*. Usually a mixture of 10% alcohol and 90% gasoline is used, a blend that permits its use in automobiles without a modification of the engine compression ratio or its timing.

Alcohol can be used for home heating as a clean, safe fuel and also in power plants for generating electricity. In a recent set of power boiler demonstrations, methyl alcohol was tested against No. 5 fuel oil and natural gas. With methyl alcohol no particulates were released from the stack, the amount of NO_x was less, the CO concentration was less, no sulfur compounds were emitted, the amounts of aldehydes, acids, and unburned hydrocarbons were negligible, and essentially no soot deposit occurred in the stack.

What are they?

ALCOHOLS are a group of organic compounds made up of hydrocarbons containing hydrogen, carbon, and oxygen. METHANOL (CH_3OH, wood or methyl alcohol), is the simplest alcohol compound; it is a colorless, odorless liquid and the alcohol most easily made. However, it is more corrosive than ethanol and toxic when ingested. ETHANOL (C_2H_5OH, grain or ethyl alcohol), is used in alcoholic beverages and has many industrial applications. All alcohols can be made from natural gas, from oil shale, from coal, and from any relatively dry plant material such as grain, wood, grasses, and herbage. Alcohols can be stored and shipped in tanks and pipelines. The concentration of alcohol in any liquid is indicated by its proof. PROOF is a number that is twice the percentage by volume of alcohol present. Industrial ethanol is usually 190 proof (95% alcohol) and ethanol put into gasoline is 200 proof (100% alcohol). Rum, whiskey, or gin may be marked 80 proof (40% alcohol).

Methanol

The methanol molecule may be thought of as two molecules of hydrogen gas (H_2) made into a liquid by adding one molecule of carbon monoxide (CO). The chemical reaction is $CO + 2H_2 \rightarrow CH_3OH$. Since 1929 a process has been in use for synthesizing methanol from the action of hydrogen gas on carbon monoxide gas at high temperature and pressure in the presence of a catalyst. Methanol freezes at $-97.8\,°C$, boils at $64.6\,°C$, and has an ignition temperature of $467\,°C$ and an octane rating of 106. Gasoline has an ignition temperature of $222\,°C$ and an octane rating from 90 to 100. The heat of combustion of methanol is 17,700 kJ/liter (63,500 Btu/gal), about half of the value for gasoline. Methanol burns with a clean blue flame.

In the production of methanol from biomass, wood or plant herbage is put into an oxygen-blown gasification chamber and burned to make carbon monoxide. The carbon monoxide is mixed with hydrogen and the synthetic process just described is followed to produce methanol. One metric ton of wood yields about 120 gallons of methanol, and one metric ton of grass is estimated to give about 100 gallons. Methanol derived from wood is more expensive than methanol obtained from coal, but probably will be comparable in price to the more expensive synthetic fuels from shale. It is a simpler process to produce methanol from wood than it is to produce ethanol from wood. Therefore, the large-scale commercialization of methanol production from wood may occur before large-scale ethanol production.

Methanol is useful for many different fuel applications and is especially suitable for use in fuel cells for generating electricity. Methanol can be added to gasoline in quantities up to 15%; and, according to Reed and Lerner (1973), there will be improved economy, a lower exhaust temperature, lower emissions, and improved engine performance in comparison with gasoline only. However, Wigg (1974) contests this conclusion and suggests that methanol gives little in the way of improved efficiency or reduced emissions when used in the newer model cars where carburation is more lean because of emission controls. Methanol is the only fuel used in the Indianapolis 500-mile auto race because methanol is nonexplosive. Methanol fires can be quenched with water, but gasoline fires cannot be. This is the reason why alcohol stoves are used on boats in place of gasoline cooking stoves.

Many deaths have ensued from drinking methyl alcohol. It has a specific action on the optic nerve and many cases of blindness have resulted from drinking the liquid or from repeatedly inhaling its vapor.

Ethanol

Ethanol is a somewhat more complex molecule than methanol and is more expensive to produce. In the United States today, it costs about three times the price of methanol. It is colorless and has a characteristic odor. Ethanol freezes at $-117.3°C$, boils at $78.5°C$, and has a density of 0.79. Ethanol has many of the same properties as methanol, but it is less corrosive and much less toxic. Its fuel content is 23,500 kJ/liter (84,300 Btu/gal), about 33% greater than that for methanol, but much less than gasoline at 34,840 kJ/liter (125,000 Btu/gal). Ethanol can be produced by fermentation of biomass whereas methanol cannot be. All of these advantages, except cost, make ethanol preferred over methanol as a mixture in gasoline. Ethanol can be produced from the same dry plant material as methanol and also from grains, sugar crops, and fermentable wastes. Ethanol is produced by the fermentation of sugars, such as maltose and dextrose, by the catalytic action of yeast enzymes that act on organic compounds. The sugars in grain are most easily fermentable to ethanol, but in principle most carbohydrates also can be fermented. Starch is a carbohydrate that does not undergo fermentation, but it can be converted to dextrose by the en-

zymes that are in malt, a derivative of barley. Before fermentation is started, grain is cooked to form a starch mass. The mixture is then cooled and malt is added to supply the enzymes to hydrolyze the starch to fermentable sugars. When molasses from sugarcane or sugar beets is used as the initial material, it contains enough sugar to be fermented directly by yeast.

The chemical reaction of fermentation to form ethanol from dextrose is

$$C_6H_{12}O_6 \xrightarrow{\text{catalyst}} 2C_2H_5OH + 2CO_2$$

Two pounds of fermentable sugar are required for each pound of ethanol produced because an equal amount of carbon dioxide is also produced. Very large scale ethanol production could release substantial quantities of carbon dioxide gas to the atmosphere. Furthermore, when ethanol is burned, it breaks down to carbon dioxide and water vapor, further adding to the carbon dioxide concentration of the atmosphere.

During fermentation of a brew of starch and enzymes, many complex chemical reactions occur and many compounds are formed, including a number of alcohols, other hydrocarbons, and acids. Although these impurities are in low concentrations, they must be separated from the ethanol by distillation, a process requiring heat. How this energy is obtained determines whether or not the net energy gained from ethanol is positive or negative and by how much. This will be discussed later.

It is most desirable to produce ethanol from feedstock other than grain, for example, from agricultural and forestry residues. However, it is not only a question of availability but also of collectability and conversion to alcohol. There are also competing uses because some of the residues must be left in the field to retain soil fertility and tilth and to reduce erosion. In some areas the cropland erosion is so great that no residues should be removed. It is possible that of the estimated 400 million tons of crop residues, 80 million tons could be removed for alcohol production. This quantity could potentially produce as much as 9.7 billion gallons of alcohol, about the same amount that could be produced from 50% of our corn crop.

Plant residues contain hemicellulose, cellulose, and lignin. To break these materials down to fermentable compounds (e.g., sugars), they must be separated and hydrolyzed. The processes involved are complex, requiring treatment of the residues with a variety of acids and enzymes. Once separated, cellulose can be fermented to alcohol by standard processes. Lignin is a major component and has a particularly high energy content, but its complex structure makes it difficult to degrade. Research is showing encouraging progress so that fermentation of these woody plant compounds to ethanol looks feasible. Considerable effort is being made to develop new yeasts through gene selection that will speed up the conversion and fermentation process to yield ethanol. Recently, microorganisms have been discovered that directly convert the entire lignocellulose complex into ethanol and may render obsolete the need to separate the lignin.

A metric ton of fermentable sugar will give 137 gallons of ethanol. A metric ton of grain will yield 93 gallons of ethanol, and a metric ton of wood or plant residues will produce an estimated 70 to 120 gallons of ethanol. Another way of stating the yield is that one bushel of corn will yield 2.57 gallons of ethanol and 16.3 pounds of carbon dioxide gas.

The situation in the United States

Currently the only fuel alcohol being produced from biomass in the United States is ethanol, from grain and from some processing wastes. By 1980 the total U.S. capacity to distill grain into alcohol was between 100 and 200×10^6 gal/yr (0.1 to 0.2% of total U.S. gasoline consumption). When ethanol production reaches levels above 2×10^9 gal/yr, competition between food and energy uses for American grain will begin to drive up grain prices. If this occurs and the consumer has to pay higher food prices, ethanol will become the most expensive of all synthetic fuels. A lot depends on how rapidly technology develops for producing ethanol from wood or herbage.

How much grain would be required to produce the ethanol needed to make a 10% alcohol–90% gasoline blend for the nation's automobile consumption? We consume about 110×10^9 gallons of gasoline annually. If all the gasoline we consume were changed to gasohol, it would require 11×10^9 gallons of alcohol and that would use 4.3×10^9 bushels of corn or about 60% of the nation's corn crop. This would require a drastic change in our livestock industry and eliminate most exports of corn. To what extent are we willing to expand our ethanol production using only grain? It is here that the potential for converting wood cellulose to alcohol becomes of great importance.

When using crops for fuel, the greatest problem concerns conflicting use for human food production. With the world's population increasing, the chances are that all possible crop production will need to be used for human food consumption.

Net energy gain

An important concern with ethanol production from grains and sugar crops, more so than for methanol from wood and herbage, is the net energy balance of the process. One can get any number of answers to this concern depending on the assumptions made. Most recent studies seem to indicate that with currently available technology the net energy output of alcohol used for gasohol balances, or is somewhat greater than, the input energy for growing, transporting, and processing the corn into alcohol. But this is a complex issue. It is possible that the need for liquid fuels will outweigh considerations of net energy balance.

Several people have estimated the net energy gain or loss associated with alcohol production. Hopkinson and Day (1980) have evaluated the net energy gain in alcohol made from sugarcane grown in Louisiana. They, as well as others, conclude that the net energy produced depends on the type of fuel used to run the industrial processing plant producing the ethanol. The net energy gain must take into account energy expended for all of the agricultural activities associated with the sugarcane agriculture, including fertilizers, transportation, machinery, planting, and harvesting; process energy, including distillation, fermentation, cooking, purification, evaporation, drying, and machinery; and the energy content of the bagasse and the ethanol. If all available bagasse (waste herbage left over during the extraction of sugar from cane—mainly leaves, stems, and roots) were converted to steam, the net energy gain would be 1.8/1.0. However, this energy gain is based on the fact that all of the bagasse generated is used to generate steam and the steam is used for process heat at the plant and elsewhere. Heat is needed for the fermentation and distillation processes. However, about 27% of the steam generated is excess and probably would not be used productively. When the amount of bagasse used for steam is only that required in the production process, the net energy yield is 1.5/1.0. When 50% bagasse and 50% fossil fuel is used for steam production (currently the situation in Louisiana plants), then the yield is 1.2/1.0. When only fossil fuel is used, the net energy gain is negative at 0.9/1.0. In Brazil, where gasohol has been used for a number of years, the situation is different because fuel, machinery, and nitrogen inputs to the production process represent only 83% of the total inputs, compared with 90% in Louisiana. In Brazil, labor costs are very low. Agricultural yield in Brazil is similar to that in Louisiana (54 versus 53 metric tons per hectare), but the net energy gain is 2.4/1.0. Chambers et al. (1979) concluded that in terms of petroleum only, gasohol is an unambiguous energy producer because most energy inputs can be supplied by coal; but, in terms of total nonrenewable energy, gasohol is close to the energy break-even point.

Environmental impacts

Most biomass production has associated with it a series of potentially serious environmental and ecological impacts. A considerable amount of dust is put into the air by agricultural activities such as harrowing and plowing. Dust coats the leaves of plants and reduces the amount of light reaching the interior of the leaf and in turn reduces photosynthesis. Dust can affect the local climate by interfering with the amount of sunlight incident at the ground or by preventing the escape of radiant heat to outer space.

Agriculture, even with the arid land plants, makes use of water. Pesticides, herbicides, and fertilizers used on fields will get into groundwater and affect areas downstream. Non-point-source pollution by agricultural wastes is one of the most serious contamination problems in the United States today. Soil erosion and sedimentation of streams, lakes, and reservoirs is a widespread problem. The conversion of biomass into useful products such as alcohol or methane gas leaves a sludge residue that must be properly disposed. The conversion of natural ecosystems into intensely managed ecosystems completely changes the community of organisms present and always results in a lack of diversity along with a decrease of stability. Even natural areas surrounded by highly managed land will be enormously changed in terms of the populations of plants or animals within them. Monocultures, whether crops or trees, will tend to attract exotic species of plants and animals. Even the noise generated by agricultural practices will seriously disturb natural habitats nearby. (See Holdren et al., 1980 for a detailed description of the environmental effects of renewable energy resources.)

Soil erosion. Increased use of land for energy plantations or for any other agricultural uses will most likely cause increased soil erosion. In the contiguous United States, we are losing from agricultural land an estimated 2.7×10^9 metric tons of topsoil per year to streams, and an additional 1×10^9 metric tons is eroded by the wind. Erosion rates in cultivated soils average more than 20 mt/ha annually. Since agricultural practices were first begun in the United States, about one-half of the original topsoil has been lost from one-third of the cropland. Sedimentation fills reservoirs, silts spawning grounds for fish, fills harbors, and clogs navigation channels. Sedimentation increases flood damage, spoils the recreational value of streams and lakes, raises the cost of water treatment for urban use, complicates hydroelectric power generation or cooling tower use, and obstructs drainage and irrigation ditches.

Sustained soil loss can damage land productivity, although it often takes a long time to be noticeable. For example, a net loss of 25 mt ha^{-1} yr^{-1} leads to a loss of only 2.5 cm of topsoil in 15 years. The loss of productive potential during this time may not be great on some lands because of their depth of topsoil. Even a significant loss of soil may go unnoticed by being masked in the short term by productivity increases resulting from improvements in other farming practices such as more intensive use of agricultural chemicals. Sheet and rill erosion alone on intensively managed croplands averages 18 mt ha^{-1} yr^{-1} in the corn belt. Continuation of this loss rate indefi-

nitely will cause a decline in farmland productivity. New intensive crop production for ethanol is likely to cause more severe erosion than ordinary food production, for several reasons. First, the lands likely to be shifted into ethanol production appear to be about 20% more erosive than land presently in intensive crop production—perhaps it is steeper or the soil is less binding. Second, if the land is less productive than existing cropland, erosion rates per unit of production will go even higher because of less plant cover per hectare. Third, the increased use of lightly managed ecosystems, such as forested land, will increase erosion. In each instance, the removal of most of the biomass for ethanol or methanol production would increase potential soil erosion.

If grasses (perennial, close-grown crops) are used for alcohol production, erosion problems may be reduced because perennials provide more soil erosion protection than do annuals, and close-grown crops more than raw crops (such as corn or wheat). The use of grasses for ethanol or methanol production will have few of the erosion problems associated with the production of grain for conversion into alcohol. Grasses have extensive root systems that bind the soil. Only the aboveground parts of grasses are harvested so that the root system containing most of the nutrients remains intact. (See Batie, 1983 for a detailed analysis of the soil erosion problem in the United States.)

Air pollution. Distillation facilities for producing ethanol from grains and sugar crops probably will require an energy expenditure of about 14,000 kJ/liter (50,000 Btu/gal) of ethanol produced to provide electricity and to power the distilling, drying, and other operations. A distillery producing 50 million gallons per year will consume slightly more fuel than a 30-MWe power plant. A distillery industry producing 10 billion gallons per year will consume an amount of energy equivalent to that consumed by a 6,000- to 7,000-MWe power plant. The power for a distillery is usually generated on the site from coal or biomass residues. Particulate emissions will be a major source of pollution. The use of high-sulfur coal as a fuel, quite likely in the Midwest, could lead to high local concentrations of SO_2.

In contrast to an ethanol distillation plant, very little of the energy required for a methanol production process is supplied by external combustion sources; most of the energy is obtained from the heat generated during gasification of the feedstock (wood or lignocellulose). The gasification process will generate a variety of compounds such as hydrogen sulfide and cyanide, carbonyl sulfide, and tars and oils containing a multitude of oxygenated organic compounds (such as organic acids, aldehydes, ketones), aromatic derivatives of benzenes (such as phenols), and particulate matter.

The concentrations of toxic inorganic and organic compounds in the gas stream from the gasifier will make raw gas leakage a substantial occupational hazard unless careful controls are maintained. However, cleanup of the gas stream will be necessary because contaminants will foul the catalysts that are used in the final step in the production of methanol.

Water pollution. Water effluents from ethanol plants will require careful controls. The untreated effluent from the initial distillation step in ethanol production, called STILLAGE, has a very high biological oxygen demand and must be kept out of surface waters. It contains many organic compounds, including sugars, starch, and acids, as well as malts and yeasts. The stillage from corn and other grains is a valuable feed by-product for cattle and it will be recovered for that purpose. Stillage from other ethanol crops is less valuable and yet must be kept out of aquatic ecosystems.

In methanol production, pollutants in the effluent include tars and oils, oxygenated hydrocarbons, and small amounts of phenols or other benzene derivatives.

Nutrient losses. Whenever biomass is removed from an area, there is a removal of considerable quantities of nutrients contained in its organic matter. Standard agricultural practices require the addition of fertilizer each season to compensate for the nutrient loss sustained with crop removal and leaching from the exposed soil following harvesting. With forest harvesting, it has not been standard procedure to restore the soil nutrient status by adding fertilizer. However, to sustain a highly productive forest yield, it will be necessary to add nutrients after harvesting.

When whole-tree chipping operations are used, all parts of a tree are harvested, including the boles, branches, tops, and leaves. The small branches, tops, and leaves make up the slash that is usually left on the site following harvesting of boles. The slash contains a disproportionate amount of the nutrient content of the tree. There are few long-term studies of the nutrient losses from forested ecosystems. Best known is the work of Likens et al. (1977) for the Hubbard Brook Forest in New Hampshire. Boyle et al. (1973) report on the nutrient losses from whole-tree harvesting in Michigan. The loss of nutrients following harvesting may be extremely different on sites of different soil composition, bedrock, and vegetation type. At Hubbard Brook, heavy nutrient loss by leaching follows cutting of the forest because the soils are relatively thin and underlaid by impervious bedrock, whereas in northern Michigan, on sandy soils, there is little nutrient loss following cutting of aspen according to Richardson and Lund (1975). Aspen propagates mainly from root sprouts. When the parent tree is injured, such as by cutting, a hormone is released and dozens of root sprouts will receive a signal to grow. The result is a rapid regrowth of large numbers of small aspens accompanied by a rapid uptake of nutrients by the young trees and very little leaching of soil nutrients.

Wildlife impacts. Harvesting a forest sets back succession and changes the habitat and diversity of species. Grasses, herbs, and shrubs may enter the cutover area, as well as many young trees from sprouts or nearby seed sources. The animals dependent on the more mature forest, such as woodpeckers, will retreat from the area, and the more open young ecosystem will become attractive to grouse and deer. When the slash is left in place, it not only provides cover for rabbits and other small animals, but returns nutrients to the soil. As trees develop from either root sprouts or seeds, the young forest begins to grow rapidly and the diversity of wildlife present increases markedly. At this stage there is a lot of ground cover in the form of grasses, herbs, shrubs and young trees, and even ferns, such as bracken. Many birds nest in such rich habitats and mammals find there an abundant food supply, whether from vegetation or from a variety of organisms, such as insects, constituting the food web. Small rodents and hares may occupy the area and larger mammals such as foxes may prey on them.

It is only as a forest becomes more mature that the diversity of organisms within it begins to diminish. If the forest is harvested at some intermediate stage, the cycle of succession and the greater wildlife diversity of early and intermediate stages may be promoted.

ARID LAND PLANTS

Oil and rubber can be derived from plantations of plants grown under semiarid conditions. Oils derived from the arid land plants are useful as edible products, cosmetics, lubricants, and waxes, or as chemical feedstocks

and fuels. Various species of gourds (*Cucurbita*) are used as well as jojoba (*Simmondsia chinensis*), gopher plant (*Euphorbia lathyris*), and guayule (*Parthenium argentatum*). Buffalo gourd (*Cucurbita foetidissima*) is a perennial, reproduces asexually, grows as a weed in regions of low rainfall, and produces a large crop of seeds rich in oil and protein. The roots, which may weigh up to 50 kilograms, are rich in starch. This starch can be hydrolyzed chemically or enzymatically to produce glucose. The glucose may be used as a sweetener in foods and beverages. The vines have a high protein content and therefore a high forage value for livestock. The oil of the seed can be extracted by a solvent process or by mechanical pressing. The remaining seed meal may be fed to animals. Seed yields of up to 3000 kg/ha with an estimated 16% hydrocarbon content can produce about 3.5 barrels of crude oil per hectare. Hence, the buffalo gourd is not a particularly promising producer of crude oils.

Jojoba

The JOJOBA plant grows naturally as a shrub of the Sonoran Desert of the United States and Mexico. Jojoba seeds contain about 50% oil by weight. Hydrogenating jojoba oil produces a hard, colorless wax resembling beeswax in its chemical structure and properties. The oil can be used as a substitute for sperm whale oil. Jojoba oil has potential uses as a fuel, as a chemical feedstock, and as a replacement for vegetable oils in foods, cosmetics, and hair oil. The oil can also be a source of long-chain alcohols, antifoaming agents, and lubricants. The hydrogenated oil, a wax, can be used for waxing floors and automobiles, for waxing fruits, for impregnating paper containers, for manufacturing carbon paper, and for making candles that have slow-burning qualities (see Foster, 1980 for more information).

Jojoba is mainly harvested from the wild; it

is found widespread throughout 260,000 km^2 of the Sonoran Desert. Cultivated plantations of jojoba cover more than 2500 ha but do not contribute significantly to the annual harvest of seed. Ecological effects of harvesting may include the disruption of its natural reproductive cycle and the consequences of reducing a food resource for seed-eating rodents. For example, of 219 seedlings growing initially, only 3 seedlings survived after 4 years. Moreover, conditions favorable to seedling survival may occur only once in 10 years or more. Years that are favorable for seed germination may not be the best years for seedling survival. A jojoba shrub must produce thousands of seeds during decades just to replace itself. The impact of seed harvesting on natural propagation could be significant and serious. Bailey's pocket mice (*Perognathus baileyi*) use jojoba seeds for food; they store as much as 800 grams of seed in their burrows. They can eat these seeds because they possess a unique ability to detoxify the cyanic compound, simmondsin, in the seeds. In turn, the Bailey's pocket mice play an active role in the dispersal and propagation of jojoba. A reduction in pocket mice numbers also will affect interspecies competition and will affect animals higher in the food chain that prey on the mice. As research on jojoba progresses, more cultivated plantations of the plant will be established, and eventually it may become an important agricultural product of arid lands.

Gopher plant

The GOPHER PLANT is one of several arid land plants being studied for its hydrocarbon content. To be agronomically practical on a long-term basis, a plant cultivated to produce hydrocarbons should not compete for land used for food or fiber production. The gopher plant grown in California has yielded 25 barrels of crude oil per hectare. This was achieved from plants grown from wild seed

that had not been subject to genetic selection or agronomic development. Considering the agronomic success of increased rubber yield achieved with the rubber tree (*Hevea brasiliensis*; another member of the family Euphorbiaceae, to which the gopher plant belongs), it seems reasonable that a great increase in latex yield could be achieved for the gopher plant. Perhaps a gain of 250% could be accomplished, to produce some 65 barrels of crude oil per hectare.

Of 120 million hectares of land in the arid southwest of the United States, where gopher plants could be grown, it is estimated that 8 to 12 million hectares might actually be used to grow this crop without irrigation or with only minor amounts of irrigation. Today 34 million hectares are used in the United States for corn production and 32 million hectares for wheat. At 65 barrels of oil per hectare, 12 million hectares would yield more than 10% of the current U.S. petroleum demand.

Calvin (1978) and Nielsen et al. (1977) estimated the costs of growing gopher plants to be about $10 per barrel, assuming an annual yield of 25 barrels/hectare (modest). Removing the oil from the plant involves drying, grinding, and multiple solvent extractions. These processes cost another $10 per barrel. Thus, the total cost of producing crude oil is estimated to be $20 per barrel. World crude oil prices now exceed $25 per barrel. A great deal of research is now being done to improve yields and to get more efficient methods of oil extraction. In the next few years, we may see a considerable advance in the technology of using gopher plants as a source of oil.

Guayule

GUAYULE is a rubber-producing shrub native to the Chihuahuan Desert in southwestern Texas and northern Mexico. Rubber produced from guayule has properties nearly identical with rubber from *Hevea brasiliensis*. Guayule was a commercial source of rubber in the early 1900s and was used on a limited basis during World War II. Increased petroleum prices have increased the price of synthetic rubber, and natural rubber prices have kept pace with synthetic rubber prices. Therefore, there is now renewed interest in guayule as a source of rubber. One of the primary difficulties to commercial development is the fact that it is propagated by nursery-grown seedlings. Processing costs are high, but harvesting and cultivating costs are normal.

BIOLOGICAL CONVERSION OF WASTES

The biological conversion of organic wastes (municipal sludge and animal manure) to fuel can be accomplished by bacterial digestion in the absence of oxygen. A pound of dry organic solids can produce 3 to 5 cubic feet of gas with a 55% methane content. Some estimates suggest that as much as 10 to 13 cubic feet of gas may be derived from a pound of organic matter. The residual sludge that remains after the digestion may also be dried and burned. Methane production plants have been built both for municipal sewage plants and for processing animal manure at some feedlot operations. The annual production of animal solid waste alone is 1.3 billion metric tons in the United States; a quantity that is seven times greater than all solid urban waste produced.

The conversion of biological wastes into fuel is contemplated by many people throughout the world, but in China's Sichuan province a massive program is already underway. There 5 million converters (digesters) had been constructed as of August 1979. In certain rural areas of the province, nine-tenths of all the families had one built for their household. Somewhat larger digesters are operated communally. Each digester is an airtight chamber in which the fermentation of animal dung, human excrement, and crop residue yields a clean-burning gas that is one-half to three-fourths methane. A 10-cubic meter digester can yield as much as 2 cubic meters of biogas per day, a quantity said to be sufficient for a family of five to cook its meals and boil 15 liters of water.

As time passes the parasitic organisms in the fermenting wastes tend to die off. Hydrocarbons are being liberated as gas, but nitrogen, phosphorus, and other plant nutrients remain in the solid materials. Twice a year, at planting time, the sediment in the digester is removed and mixed into the soil as a fertilizer that is more hygienic than the unprocessed excrement was.

The China example is being emulated in India and elsewhere. We can do the same with feedlots in this country. A feedlot producing 100,000 cattle annually produces 150,000 tons of dry organic waste. On the basis of 5 cubic feet of gas per pound, this could be converted by anerobic digestion into at least 1.5 billion cubic feet of methane per year, enough to supply the natural gas needs of 15,000 people at present rates of use.

Not only can animal wastes be put into a digester, including human wastes from sewage plants, but solid waste consisting of garbage and paper products can be added. They will tend to dilute the effect of toxic substances in sewage, such as acids and chromate used in sewage treatment processes. The toxic substances inhibit the digestive activity of microorganisms. Sewage, including animal manure, is rich in nitrogen, phosphate, and other nutrients, whereas solid waste is composed of paper and wood, rich in cellulosic compounds containing carbon, but poor in nutrients. Water is necessary in the fermentation process. Wastewater from the digester will be loaded with nutrients and must be properly disposed. The water and sludge may be used on crops and forests.

Several large plants for converting manure

to methane have been constructed in the United States. There is one at the community of Guymon in the Oklahoma panhandle. This plant transforms 600 tons of cattle manure a day into 1.6 million cubic feet of methane. This does not appear to be particularly efficient because it works out to be 1.33 ft^3 of gas per pound of manure. If all the animal manure produced in the United States today could be fed into digesters, it would yield over 4 trillion cubic feet of methane per year, the equivalent of 18% of the country's current natural gas consumption. This is certainly not economically feasible at the present time, but eventually we shall probably see many more large-scale digesters installed at major feedlots throughout the country.

8

Coal

A coal seam in the Black Thunder Mine, Powder River Basin, Wyoming. Courtesy of Arco Coal Company, Denver, Colorado.

INTRODUCTION

Coal, that black product of millions of years of organic decay and accumulation, exists in massive amounts in the world, but its distribution is extremely uneven. The largest coal reserves are located within the boundaries of China, the Soviet Union, and the United States. Coal is found in stratified beds or seams that are continuous over extensive areas. The global distribution of coal is shown in Figure 1 and distribution in the United States in Figure 2.

MINABLE COAL is defined as 50% of the coal actually present in a bed. The U.S. Geo-logical Survey has estimated the global reserve of minable coal to be 7.6 trillion metric tons. Our estimate, from Schurr et al. (1979), is 5.0 trillion metric tons. Included in this figure is coal in beds as thin as 36 cm (14 inches) and beds extending to depths of 1.2 km (4000 feet) or, in a few cases, to 1.8 km (6000 ft). The United States has an estimated 1.64×10^{12} tons of coal, which potentially is more energy than all the remaining petroleum, natural gas, oil shale, and tar sands combined. However, coal is dirty, and there are many disagreeable environmental and social problems connected with mining it and burning it.

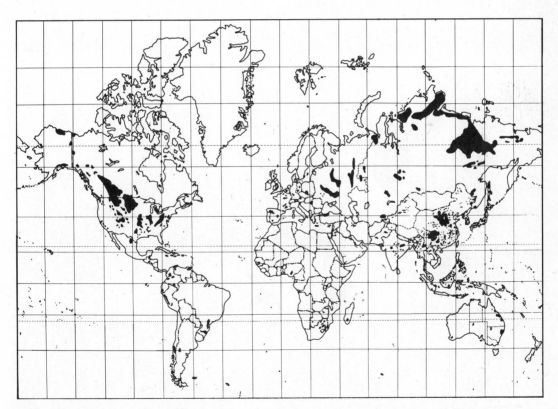

FIGURE 1. Distribution of known world reserves of coal (in black). (After Skinner, 1969.)

Coal consumption in the United States

The U.S. government has made a commitment to promote greatly increased use of coal. Techniques exist to convert coal into synthetic natural gas, into gasoline, and into other liquid fuels and to clean up the emissions from coal-fired boilers. Synthetic fuels derived from coal appear to cost twice as much as oil or gas, and they release much more carbon dioxide to the atmosphere. The problem of coal conversion seems to be more of a political and economic issue than it is a technical one because of uncertain governmental policies concerning environmental and safety regulations, international oil policies, and government subsidies.

Coal, which at one time provided most of our domestic energy, today provides about 19% of our total energy needs. We are exporting more than $1 billion worth of coal annually or 43×10^6 metric tons each year. The United States now uses coal for only about 49% of the electric power generated from fossil fuels. Coal production in the United States is expected to increase from 667×10^6 mt/yr in 1980 to 1.2×10^9 mt/yr in 1990 and 2.0×10^9 mt/yr by 2000.

Future coal consumption is very sensitive to the electricity growth rate. This growth rate has been decreasing, particularly compared with estimates made only a few years ago. Each percentage point change in the compound growth rate would change utility coal demand by about 136×10^6 mt/yr in less than a decade. Electricity growth depends on the health of the economy, the success of conservation efforts, the rate of home building, and consumer life-style choices. The market for coal depends on its price and competitiveness with other fuels. The cost of producing coal depends on a host of factors: severance taxes; the cost of reclaiming strip-mined land and the extent to which reclamation is required; mining equipment technology and productivity; union wage demands; mine

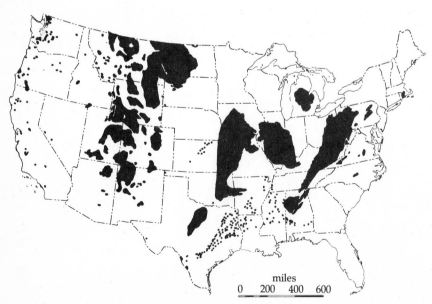

FIGURE 2. Distribution of known coal reserves (in black) in the United States. (After Skinner, 1969.)

safety costs; black lung disease payment requirements; and coal leasing policies for federal lands. Other possible constraints include the availability of investment capital, government subsidies, and strikes. The cost of using coal is dependent on the availability and cost of transportation. Coal consumption by utilities is highly sensitive to the price of oil. The unknown extent to which nuclear power plants are brought on-line makes the future of coal production uncertain. Coal use will depend on progress in synfuel technology for coal gasification and liquefaction. Future coal production will also depend on export demand (see Perry, 1983).

PROPERTIES OF COAL

Coal is composed of a highly complex, heterogeneous group of substances, and it possesses a wide range of chemical and physical properties. Because coal is formed by the deposition, accumulation, and modification of plant matter, its final properties depend on the nature of these processes and subsequent events during metamorphosis into coal.

Coal classification

Coal is classified into a series, depending on the carbon content, ranging from lignite at the low end of the scale through subbitumi-

nous and bituminous, to anthracite at the high end. This scale is based not only on carbon content but also on volatility, caloric value, and agglomerating characteristics (see Table 1).

Lignite, also called brown coal, has up to 50% water, a low carbon content, and low heat value (partly because of the water content). Large deposits exist in North and South Dakota and Montana. Smaller deposits are in Texas, Arkansas, Louisiana, Mississippi, and Alabama. Subbituminous coal comes from the Rocky Mountain states. Bituminous coal is the most widely distributed and most abundant of all forms of coal. Because of its high value, it is much used by electric utilities. Deposits are found in eastern, midwestern, Rocky Mountain, and far western states. Anthracite, the so-called hard coal, has the lowest moisture content and the highest heat value. The largest deposits of anthracite coal are in eastern Pennsylvania.

Trace elements in coal

The trace elements of heavy metals in coal are a serious air pollution threat, as well as being potential contaminants in water. Trace elements found in coal include arsenic (1 to 16 ppm), barium (31 to 380 ppm), beryllium (0.6 to 2.4 ppm), boron (15 to 85 ppm), lead (4.0 to 6.0 ppm, except 0.6 ppm in Powder River

TABLE 1. Classification of coals

Class	Moisture (%)	Carbon (%)	Volatile matter (%)	Caloric values	
				kJ/kg	Btu/lb[a]
Anthracite	2	86–98	2–14	32,564–37,216	14,000–16,000
Bituminous	2–15	50–86	14–50	24,423–32,564	10,500–14,000
Subbituminous	20–30	40–50	—	19,306–24,423	8,300–10,500
Lignite	30–50	40	—	14,654–19,306	6,300– 8,300
Peat	70–95	—	—	11,630	5,000

[a] 1 tce = 7×10^9 cal = 29.3×10^9 J = 27.8×10^6 Btu (assumes a heating value of 12,600 Btu/lb).

Basin and 33 ppm in Illinois), mercury (0.05 to 0.20 ppm), molybdenum (2.0 to 9.8 ppm), vanadium (12 to 30 ppm), and zinc (15 to 140 ppm). In addition, coal contains concentrations of uranium (1.3 to 2.5 ppm) and thorium (3.3 to 5.4 ppm), both radioactive elements. Cadmium, antimony, and selenium may also be found in coal in concentrations greater than average in the Earth's crust and are hazardous to humans. All of the above elements can be toxic to plants and animals.

Properties of coal for utilities

Characteristics of coal that are of particular importance to electric utilities are heat, sulfur, ash, moisture, and trace-element contents. A high value for heat content and a low value for sulfur, ash, moisture, and trace elements are desired.

Eastern coal and western coal generally have sulfur contents that range from 2.0 to 3.3% and 0.4 to 0.7%, respectively. Some coal is as low as 0.2% and as high as 7.0% sulfur. Sulfur content is important because of the serious health and environmental, as well as ecological, effects of sulfur dioxide, which is produced by combustion of sulfur-containing coal. These deleterious effects have led to regulatory restrictions on sulfur dioxide emissions that require use of low-sulfur coal or removal of sulfur from coal before combustion. In 1969 the U.S. power industry discharged 7×10^6 metric tons of sulfur in the form of SO_2. In 1980, the industry emitted approximately 20×10^6 metric tons of SO_2.

USES OF COAL

Direct burning

Direct burning of coal is the most efficient process for extracting its heat content, but it causes the most environmental problems. During the nineteenth and early twentieth centuries much coal was burned for individual home heating and by industry in America and Europe. The virtually uncontrolled burning resulted in great volumes of soot and oxides of sulfur disgorged into the air of large cities. London's smog, a combination of smoke and fog, became notorious as a death-dealing, lung-choking cloud. Pittsburgh, Pennsylvania and St. Louis, Missouri were blackened cities and were known to be extremely unhealthy places to live. The changeover from coal burning to oil and gas use cleaned up all these cities. Today, we know that if we are to use coal it is mandatory to clean up the effluent. Burning coal at high temperatures in large power plants generates enormous amounts of carbon dioxide, oxides of sulfur, oxides of nitrogen, and fly ash. The technology exists for reducing most of these emissions.

Coal was once burned in large chunks, but this method is wasteful and less efficient than burning pulverized coal in large boilers. Coal is crushed into a fine powder and is blown with a small amount of air into a furnace. This is the method most frequently used in power plants. However, another process known as fluidized-bed combustion is even more efficient and environmentally acceptable. In this process pulverized coal is mixed with crushed limestone, both of which are suspended in a blast of air in the furnace and ignited. Combustion under these circumstances occurs at a lower temperature and therefore does not yield the large volume of nitrogen oxides and melted ash that conventional boilers do. Furthermore, about 90% of the sulfur in the coal combines with the limestone to produce calcium sulfate, which is then disposed of as a solid waste. This method greatly reduces the volume of sulfur dioxide released into the atmosphere. The direct burning of coal will continue to be done on a large scale and its percentage use in electric utility boilers will probably increase as oil and gas become scarcer and more expensive.

Gasification

GASIFICATION of coal is an old art. The first gaseous fuel to be obtained by the destructive distillation of bituminous coal was COAL GAS, a mixture of hydrogen and methane. It has an energy content of 18,640–22,370 kJ/m^3 (500–600 Btu/ft^3). Coal gas is produced by placing coal into a series of retorts and heating it. The heat causes the coal to undergo complex changes that result in the release of a large number of compounds. Volatile products escape from the heated coal and are condensed to form a coal tar. Ammonia (NH$_3$) also is produced because of the presence of nitrogen in the air in the furnace. However, some vapors do not condense at this point in the process. They pass on to another condenser and then through a purifier that contains lime or iron oxide to remove any sulfur compounds. A ton of coal produces at least 28.3 m^3 (1000 ft^3) of gas. Left behind in the retorts is coke, also a very useful material, particularly for the smelting of iron ore.

WATER GAS, a mixture of carbon monoxide and hydrogen, is produced by passing superheated steam over very hot anthracite coal or coke. The reaction of carbon with steam is given by

$$C + H_2O \rightarrow CO + H_2$$

Water gas burns with a pale blue, nonluminous flame. Its heat value is 9320–12,120 kJ/m^3 (250–325 Btu/ft^3). It is extremely poisonous and has no odor. To give it an odor and to improve its illumination quality, it must be enriched with hydrocarbons, particularly with methane, acetylene, and ethylene.

In the 1920s more than 150 companies worldwide manufactured coal gasification equipment. Use of coal gas and water gas was common at one time in many American communities. At first oil and then natural gas replaced coal gas in most applications, and the technology available for coal gasification became obsolete. The future of coal gasification seems uncertain in the United States even though it is being used in Europe and elsewhere. Coal gasification processes available are very inefficient and costly. The complexity of a coal gasification plant can be illustrated by listing its components. The core of a pilot plant is the gasifier vessel, typically about 2 meters in diameter. With its support structure, it is about 40 meters high. In addition there are coal hoppers, coal pretreatment systems, additional reactor vessels, intricate valves, gas cleanup and sulfur recovery systems, process vessels for converting the raw synthesis gas from the gasifier into methane, power stations, oxygen and steam plants, and miles of piping. This complexity is necessary because a solid, impure fuel is processed at high pressure and temperature.

Why convert coal to gas at all? Some analysts question whether it makes economic sense to convert coal to gas. This is mainly an argument over energy distribution systems because the main use of coal outside of utilities has been discontinued (home and business distribution systems), and what remains is an electric grid and the gas pipeline network. At the present time, the gas pipeline system transports about three times as much energy as the electric grid, not counting the gas used to generate electricity. But as U.S. gas production ultimately declines, the pipelines must either be used to transport gas made from coal or be abandoned.

One type of gas that might be economically produced from coal is POWER GAS, a gas of low heating value (5590 kJ/m^3 or 150 Btu/ft^3) compared with synthetic natural gas, which has a much higher heating value (37,280 kJ/m^3 or 1000 Btu/ft^3).

When the carbon monoxide produced by the water gas process is combined with hydrogen, methane is produced:

$$CO + 3H_2 \rightarrow CH_4 + H_2O$$

Power gas is composed of carbon monoxide,

hydrogen, and methane. However, because of its low heating value, transporting it is not economical. Therefore, it must be converted to more useful forms of energy near the site of production. One way of doing this is to generate electricity by using a combined cycle system of power gas and steam turbines. Low-Btu power gas is much easier to generate than high-Btu synthetic natural gas. Sulfur present would be formed into hydrogen sulfide and an absorption system would remove this and other sulfur compounds. The processes for removing hydrogen sulfide from power gas are expected to be so highly effective that sulfur emissions from the power plant will be almost completely eliminated. As a method for removing the sulfur from dirty fossil fuels, gasification appears to be superior to the major alternative of removing sulfur dioxide from the stack gases of a conventional, coal-fired power plant.

Coal can be gasified by burning it *in situ* deep in the ground. UNDERGROUND COAL GASIFICATION, commonly called UCG, is an old process that the Russians began developing in the 1930s. They have been operating two full-scale electric generating plants fueled by gas from UCG since the 1950s. Although some exploratory research with UCG in eastern U.S. coal was done at a site in Alabama in the 1940s, the poor properties of eastern coal for this purpose caused the project to be abandoned. More recently, intensive research has begun on UCG in western coal beds, particularly in Wyoming. The method developed by the Laramie Energy Research Center is called REVERSE COMBUSTION. Two wells, spaced between 20 and 50 m apart, are drilled to the base of the coal seam. Lighted charcoal is dropped down one well to ignite the coal, and air is injected into the second well. Western coal is sufficiently permeable to allow the air to diffuse toward the first well and draw the flame toward the second well. The two wells are linked near the

base of the coal seam by a channel as much as 1 m in diameter (see Figure 3). The fire expands and eventually consumes all the coal between the two wells. The rate of air injection controls the quality of partial combustion. Low-energy gas is emitted from the first well. It takes from 10 to 12 days for the burning to link two wells, 20 m apart, during which time gas is being produced. If necessary, explosive fracturing can be used in the coal seam to improve the movement of air.

The underground coal gasification (UCG) process puts a considerable amount of nitrogen from the air into the gas. Its energy content is from 3727 to 6335 kJ/m^3 (100 to 170 Btu/ft^3). At such a low energy content, it does not pay to transport this gas over a very great distance. The use most often suggested for this gas is for electric power generation near the site of the coal. Experiments are being conducted on increasing the energy content of the gas by injection of pure oxygen into the

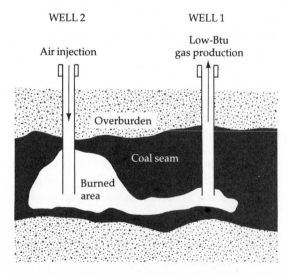

FIGURE 3. Coal gasification system. Two wells penetrate the coal seam; the coal seam is ignited, air is injected down well number 2 and low-Btu gas emerges from well number 1.

coal seam. The UCG process has an overall efficiency of 60 to 65% because about 80 to 85% of the coal is consumed and about 80 to 85% of its energy is recoverable in the gas. Underground mining, by comparison, has an overall efficiency of about 50%. The UCG process yields between 4.5 and 5.3 Btu of energy for each Btu expended. A large-scale coal gasification system might increase this ratio to nearly 8.0, a value characteristic of underground mining.

Liquefaction

Coal contains hydrogen molecules and carbon molecules in a ratio less than one. Liquids may be formed from coal if the hydrogen to carbon ratios are 2.0 or greater. To do this one must either remove carbon or add hydrogen to the coal molecules. The first method involves pyrolysis and the second hydrogenation.

Liquefaction is the most difficult coal conversion technique. Not only is it technically demanding, requiring complex rearrangement of coal's chemistry, but there are economic barriers as well. However, liquid fuels are highly useful and in much demand. New methods for liquifying coal under conditions much less extreme than those of gasification hold considerable technical promise. Coal is liquified by exposing finely powdered coal, dispersed in oil, to hydrogen gas under high pressure and in contact with a catalyst at a temperature of about 500°C.' Under these conditions various products are formed, including lubricating oils, kerosene, and gasoline. Coal also can be destructively distilled by heating so that its volatile components are given off and can be condensed as a liquid. The net result is to add hydrogen or remove carbon, in the process shortening the length of the hydrocarbon molecular chains. Hydrogen is generated by gasifying some of the coal, and this is a substantial part of the cost of liquefaction. Hydrogen is also used to remove sulfur from coal by converting it to H_2S. Heavy, oil-boiler fuel is cheaper to make than to refine into gasoline because this second step requires about twice as much hydrogen.

The only operating coal-to-liquid plant in the world in 1976 was in South Africa. It produced synthetic gasoline, pipeline gas, ammonia, and other products. The coal is gasified to produce synthetic gas (carbon monoxide and hydrogen) and then, using the Fischer-Tropsch process originally developed in Germany, is passed over a catalyst and partially converted to a mixture of hydrocarbons. The process is inefficient, and in the United States, where coal is more expensive than in South Africa, there is no commercial interest in the process. The South African plant produces 40,000 barrels of gasoline a day. It is estimated that a coal liquefaction plant in the United States will require 30,000 tons of coal a day to produce about 50,000 barrels of oil.

ENVIRONMENTAL AND ECOLOGICAL IMPACTS FROM MINING OPERATIONS

At nearly all stages of coal production and use, there are serious environmental and ecological impacts. These are summarized in Table 2.

The most straightforward way to consider the environmental and ecological impacts of mining activities is to take them up sequentially in the order in which they might occur. The first activity is exploration, and it is followed by site clearing, construction, excavation, removal, storage, transport, and use. In addition, enormous quantities of tailings and waste rock are associated with some mineral mining, such as mining for oil shale and uranium. These products are not usually associated with coal mining.

Environmental impacts

Surface clearing. Surface mining activities require the removal of large amounts of vegetation whereas underground mines have much less impact on vegetation; but they do create some disturbance. For example, during the period 1976 through 1985 in the Powder River Basin of Wyoming there will be lost an estimated 500,000 tons of range forage, 3400 game animals, 2500 game birds, 170 predators, and a multitude of small mammals, birds, and reptiles.

Construction of facilities. Construction of buildings, parking areas, conveyors, haul roads, and other facilities usually does not create extensive impact on wildlife beyond the initial clearing. However, the influx of large numbers of workers does have an impact through recreation, hunting, poaching, and inadvertent collision of wildlife with vehicles. Considerable noise may also be associated with the construction activities and may disturb some animals in the immediate vicinity. Dust can be an environmental problem if the surface is not kept moistened during construction.

Excavation. Excavation to reach the coal bed has major ecosystem impact. After surface mining activities, the level of the reclaimed surface will usually be substantially below that of the original surface, thus creating new land forms around the edge of the site and a changed topography for the area. Many large mining operations, such as for uranium or oil shale, may never be properly reclaimed and may remain as open pits or as extensive dumps for the disposal of wastes and overburden. The procedure used for strip-mining coal is to remove a broad strip of topsoil and stockpile it nearby. Then the overburden on top of the coal bed also is removed and stockpiled. After the coal is excavated, the overburden is spread out and the topsoil is replaced on top of the overburden. Obviously the new surface will be lower than the original surface was.

TABLE 2. Environmental and ecological impacts associated with coal

Mining

Land disturbance whether underground or strip-mining
Dust released into the atmosphere from moving masses of dirt or from coal-handling
Acid drainage from mine water
Habitat disturbance on surface and necessity to restore and reclaim the land
Noise of machines, trucks, etc.
Underground mining hazardous to health and safety

Transport

Railroad right-of-way and associated problems
Shipping by barge or boat and associated problems
Slurry pipeline and associated problems

Stockpiling

Acid drainage into aquatic ecosystems
Fugitive dust into the atmosphere
Loss of primary productivity and small animal habitat

Burning

Carbon dioxide released to the atmosphere
Acid precipitation from SO_x, NO_x, etc.
    On aquatic ecosystems
    On terrestrial ecosystems
    On limestone buildings, statuary, etc.
Gaseous pollutants (SO_x and NO_x)
    Effects on vegetation
    Effects on animals
    Effects on humans
Heavy metals and radioactive elements released to the environment
Fly ash released to atmosphere
    Deposited on vegetation
    Inhaled by animals and humans
    Blackening of urban areas and of vegetation; industrial melanism occurs.

Land subsidence. Underground mining causes subsidence of the land, a phenomenon often encountered in Appalachia where whole houses sometimes suddenly disappear into the ground. It is estimated that for every two acres of coal mined in central Appalachia, 10.5 acres become vulnerable to subsidence, thereby adversely affecting the use of the land for building or agriculture. It is estimated that over 2 million acres of land have subsided in the United States because of coal mining. The Department of the Interior has estimated that it will cost $7 billion to stabilize the land that has caved in and crumbled because of subsidence.

Coal removal. Removal of coal from the mine does not cause significant ecosystem impact except where it may interrupt or otherwise disturb the flow of surface water and local aquifers. Dust and noise may be associated with trucks, conveyors, or other machinery used to remove the coal.

Coal storage. The open storage of coal may contribute to the deterioration of air or water quality and indirectly to ecosystem and wildlife disturbance. Acid drainage is an example and will be discussed in detail later in this chapter.

Coal transport. The transport of coal may require extensive haul roads, very long railroad lines, or slurry pipelines, which consume land. Transport activities contribute to increased roadkills of wildlife, interruption of migration routes, habitat losses, and dust and noise along the transportation corridors.

Personnel transport. The movement of large numbers of people between work and home causes increased disturbance of wildlife, increased roadkills, and interference with migration routes. When there are dirt roads, dust is put into the atmosphere by vehicles using the roads.

Community growth. The sizeable work force for the mine causes a large increase in the population of neighboring communities or the establishment of new communities. These population increases bring with them many impacts on wildlife including loss of habitat, increased hunting and poaching, interference with migration routes, and disturbance from reclamation.

Reclamation. Reclamation of the mining site, while desirable and necessary, will produce considerable ecosystem changes involving topography, vegetation composition and diversity, location and flow of water, and wildlife diversity and abundance.

Mining hazards. Mining, particularly deep mining underground, is a very dangerous profession—one of the most dangerous in the United States. Since 1907, more than 90,000 coal miners have died in mining accidents in the United States. Coal mining deaths exceed 260 per year and another 4000 deaths per year are attributable to cancer or black lung disease. If one evaluates this on the basis of the amount of energy generated, the production of coal results in one death per billion kilowatt-hours generated, a rate that is more than 30-fold greater than the next most dangerous mining (i.e., uranium) and about 150 times higher than the rates associated with oil and gas production.

Ecosystem impacts

There is no doubt that mining activity has a considerable impact on the ecosystem. Components of the impact may be temporary or transient or they may be long-lasting or permanent. Because ecosystems comprise abiotic factors and biological communities, the impact on both will be considered.

Airborne materials. Dust is a major problem associated with mining activities, much more so for surface mining than for underground mining. Large particulates may settle out on vegetation nearby and produce changes in productivity. Fine particulates may remain airborne for days, weeks, or months and may contribute to the load of global atmospheric dust. Direct effects of these airborne particulates on vegetation and wildlife is believed to be minimal.

Groundwater and surface waters. Considerable amounts of water are often required for coal mining activities, for example, for dust control and for revegetation. It has been estimated that in the Powder River Basin coal development will require about 55 acre-feet of water (43,560 ft^3) per million tons of coal extracted.

Mining activities affect surface and groundwaters in many ways. They interrupt and divert the flow, consume some of it, and contaminate it with pollutants. The diversion of surface waters and shallow aquifers can wipe out some aquatic ecosystems or greatly disturb others, particularly in the west where water is a scarce resource. Degradation of water quality by the increased siltation of toxic substances into local streams and impoundments may adversely affect wildlife such as waterfowl, fish, and aquatic invertebrates; in addition, mammals, birds, and amphibians are dependent on the water supply. Acid mine drainage is characterized by low pH, low alkalinity, high sulfate, high iron content, and increased sedimentation. Acid mine drainage definitely affects adversely the biota of streams. In general, the disturbance causes a reduction in diversity, evenness, and number of taxa present—the reduction being proportional to the stress imposed. Thus, benthic (bottom) community characteristics are often used in assessing the impacts of mining on streams. The effects of acid mine drainage may be noticeable for many years after the drainage has been removed.

Changes in the flow of streams and the character of ponds may affect directly all wildlife using these waters, particularly during nesting, brooding, migration, or spawning. In addition, riparian areas are often used as corridors for wildlife movement from one area to another, and disruption can create barriers to these movements. Natural streams usually have stable banks and contain a mixture of spawning areas, pools, and riffles. When a stream is relocated, it should have these features if degradation of the habitat is to be avoided.

When stream water is partially diverted, many problems are created that may not occur during the normal, natural life of a stream. Low water levels, lower than on natural cycles, deny spawning areas to fish or modify food supplies or disrupt the food web in parts of the stream. Shallow or slow-flowing water is more susceptible to freezing—with adverse effects on fish population—than is deep or fast-moving water.

There can be positive effects of mining on ground and surface waters, if water from underground aquifers is used and released in good quality and at a steady rate. The additional surface water flow, particularly into impoundments, can create long-term benefits to wildlife.

Soils. Erosion of soil materials may increase considerably when soil is removed, stockpiled, and replaced improperly. Handling of the soil can make it more erodible because of the destruction of soil aggregation. Also when soils are left exposed for long periods of time following removal of vegetation, leaching occurs and nutrients are lost. Thus, the soil becomes less amenable to revegetation. Soil compaction during handling can lead to poor soil structure because compacted soil has reduced porosity and moisture infiltration. As a result, the soil becomes less suitable to revegetation and more susceptible to erosion. Disturbance of soils also affects microbial activity, disrupts the activity of small mammals and invertebrates that live in the soil, and, in general, reduces productivity. Microbial activity facilitates nitrogen fixation, and invertebrate activity generally aerates the soil and promotes the flow of materials.

Large, long-term storage piles of soil can be avoided by the direct transfer of salvaged soil to areas ready for revegetation. It is also best to disturb and mine only a limited area each year. Large stockpiles of topsoil will suffocate microorganisms, small mammals, and invertebrates, each of which will need to reestablish in the soil once it has been replaced in the mined area. Postmining vegetation may differ greatly from premining vegetation on the site. For example, according to Streeter et al. (1979), "in northeastern Wyoming uniform spreading of topsoil will probably preclude reestablishment of such communities as sandhill grasslands, and scoria grasslands or savanna which require a coarse substrate." Thus, care is required in the handling and relocation of soil to assure a variable thickness and composition reasonably comparable to what was there prior to disturbance.

Topography. The topography of a site is extremely important in terms of exposure to solar radiation, water drainage characteristics, interaction with the flow of air, evaporation rates, and temperature. Low pockets of ground may contain cold air, and south-facing slopes may be warmer in winter than north-facing slopes. Organisms living in and on the soil are often extremely sensitive to such factors and select microhabitats to suit their needs. Mule deer (*Odocoileus hemionus*) and elk (*Cervus canadensis*) find the south-facing slopes in Wyoming to have less snow cover and to be warmer during the winter; therefore, they use them for resting and feeding areas. Reptiles may also be more abundant on south-facing slopes than on north-facing slopes at certain times in spring and autumn.

Topographic features such as caves, rough breaks, cliff faces, hummocks, hills, valleys, and canyons are important wildlife habitats. Many of these features provide windbreaks against very severe wind chill in the winter, particularly on the windswept plains of Wyoming. Raptors will occupy cliff faces because these are often the only remote, protected areas available for nesting and rearing young without human interference. Mining may create topography attractive to raptors and other wildlife by leaving high walls and other protective features.

Vegetation and animals. As the result of a mining operation, the vegetation coverage may be completely destroyed, removed, and replaced with a new vegetation type, or removed and replaced with the original plant community. Any change of the original vegetation will have a variety of effects on wildlife. These effects include destruction of food source; destruction of cover; displacement of mobile species; death of nonmobile species; increased competition for resources on adjacent areas; modified nutrient and energy flow to adjacent areas; and increased wind and water erosion resulting in increased turbidity and altered chemical composition of water supplies.

ACID WATER FROM MINES

The drainage of acid water from underground mines is a serious problem. During the mining process, iron sulfides in coal are exposed to air and water and oxidized to form sulfate and sulfuric acid. Water leaching through the mine carries this acid into waterways, where it plays havoc with aquatic organisms and ecosystems. As the pH of a stream becomes lower, spawning success by fish decreases and finally stops. A detailed discussion of this topic is given in Chapter

10. More than 11,000 km of streams, most of them in Appalachia, are badly contaminated with the discharge of acid water from coal mines. Techniques for partially controlling acid mine drainage include sealing off old mines, pumping them out, and neutralizing acid water with lime.

As coal is mined and removed, the iron disulfides occurring as marcasite or pyrite in the unmined coal and exposed strata become oxidized and decomposed. During this decomposition a complex series of chemical reactions occurs, forming compounds that readily dissolve in water to produce acid and associated hydrous iron complexes. Acid mine drainage often has a pH of about 2.0, with an acidity of from 4 to 20 mg/liter of hydrogen ions; it contains an abundance of Fe, often 50 to 500 mg/liter of ferrous iron; and it has a high sulfate content in the range of 500 to 10,000 mg/liter. Ferrous iron is generated by the oxidation of iron pyrite; and the ferrous iron is further oxidized to ferric iron. It is the ferrous ($Fe(OH)_2$) and ferric ($Fe(OH)_3$) iron hydroxides that impart the red color to acid mine drainage and the precipitated iron hydroxides that give the yellow or YELLOWBOY appearance to the streams.

Water quality

Water quality is often described in terms of alkalinity or acidity. ALKALINITY is defined as the capacity of a sample of water containing a compound, with or without hydrolysis, to neutralize a strong acid to pH 4.5. Carbonate (CO_3^{2-}) and bicarbonate (HCO_3^-) are the predominant components of natural alkalinity and result from the action of carbon dioxide dissolved in water (carbonic acid) upon basic materials in the strata (calcium carbonate). (See Chapter 10 for the definition of pH.)

Many people consider low pH and acidity synonymous; however, in fact they represent different things. The free acidity of water is the capacity of compounds in the water to neutralize a base and does not include a neutralizing capacity attributable to dissolved carbon dioxide (carbonic acid). The pH represents the concentration of hydrogen ions in a solution. A water sample may have no free acidity; but, because of a high dissolved carbon dioxide content, it may have a low pH. Iron bacteria, which catalyze oxidation reactions, thrive in water of low pH.

The chemical composition of water in a stream or lake depends in part on the surface structure, vegetation, rock formation, and soil types it flows over or through in the watershed and on the chemistry of the precipitation. Hence, acid rainfall or acid mine drainage into limestone (calcium carbonate) formations may affect the chemistry of the runoff only slightly whereas the same inputs into basic rocks such as granites, slates, or schist may have a larger effect on the water chemistry. In eastern Kentucky, where many coal mines are located, the majority of streams are slightly alkaline, with pH values from 7.0 to 7.9; all of the streams with pH values less than 6.0 are strongly affected by acid mine drainage. Drainage from two levels, separated vertically by 100 feet, from a Kentucky coal mine, went into two distinct watercourses. One of the streams had a pH of 8.1 and the other 2.8, although they were each receiving about the same acid wastes.

The manganese and sulfur concentrations in a stream seem to be sensitive indicators of acid pollution. Streams with a low pH release considerable quantities of iron, aluminum, calcium, manganese, and magnesium. For eastern Kentucky, Striffler (1973) reported 0 to 14 ppm for Fe, with 140 stations exceeding maximum permissible levels of 0.3 ppm; 0 to 45 ppm for Al; 5 to 400 ppm for Ca; 0 to 15 ppm for Mn; and 0.9 to 82 ppm for Mg.

Ecological impacts

The ecological consequences of acid mine drainage are many and complex (see Warner, 1973). Waters containing acid mine drainage are heavily loaded with cations containing Al, Fe, Ca, Mg, and Mn. Roaring Creek in West Virginia, where Warner made observations, is severely polluted by mine drainage. The critical pH level, below which the organism diversity decreased abruptly, was 4.2. Stream tributaries with pH 2.8 to 3.8 were inhabited by highly acid-tolerant biota consisting of 3 to 12 kinds of bottom-dwelling invertebrates and 10 to 19 species of algae. Among the acid-tolerant animals were alderflies, midges, *Ditiscidae* beetles, and a caddisfly. Physical features of the stream, and not water chemistry, determined the exact number of species at any particular place, although the water chemistry did affect the range of species numbers. Streambeds having heavy coatings of iron precipitates were inhabited by fewer species than other reaches with comparable pH and clean streambeds. Headwater streams and tributaries that were not severely polluted and had pH values greater than 4.5 were inhabited by 25 or more species of bottom-dwelling invertebrates and 27 or more species of algae. Although acid-tolerant forms of these organisms were present in the nonpolluted parts of the stream, they were never in dominant numbers. In waters of higher pH there were many species of caddisflies, blackflies, crayfish, and algae. This observation was found to be true over and over again in other streams of West Virginia and Pennsylvania.

What is it about these ions in the water that so affects plants and animals as to reduce species diversity? High concentrations of hydrogen and sulfate ions may be directly toxic. The osmoregulatory mechanisms of organisms may be upset by high concentrations of mineral salts. Sulfate and hydroxyl salts of Fe, Al, Zn, Pb, and Cu may be of sufficient concentration to be toxic. The metallic cations are often directly toxic, and there are synergistic effects. When the metals present are oxidized, the oxygen concentrations may become too low for some organisms. Low pH may cause acidemia in fish because of high carbon dioxide tensions in their blood—with ensuing death. (For further information, see Cairns, 1979; Cairns et al., 1979; Herricks and Shanholtz, 1976; and Herricks et al., 1975.)

REVEGETATION OF COAL SPOILS

The extensive amounts of COAL SPOILS in the United States and elsewhere represent badly disturbed ecosystems and aesthetically repulsive landscapes. Revegetating these spoil heaps is difficult because of their acidity and because they are often environments of high summer temperatures and drying winds. Only certain plant species are tolerant to these harsh conditions. However there are some tree species that will grow on most types of coal spoil and a few shrubs and herbaceous species are known to be adapted to the most extreme sites. The species of plants that may be used most successfully for revegetation depends very much on local conditions of climate and soils. Therefore, to describe successful attempts at revegetation, it is necessary to do so by specific regions. Revegetation of strip-mined sites in Wyoming and restoration of lignite sites in Texas will be described last.

Temperature of mine spoils

Surface temperatures of strip-mined spoils can often be exceedingly high, making revegetation more difficult. Coal spoils usually are dark, thereby absorbing solar radiation effectively and making for high surface temperatures. As a result darker spoils revegetate more slowly than lighter ones and many remain barren for many years (Lee et al., 1975). Deely and Borden (1973) observed that maximum surface temperatures consistently reached 50° to 55°C on dry, light-colored spoils and 65° to 70°C on dark-colored spoils during the summer months in Pennsylvania. In humid regions spoil moisture content is not a limiting factor once plant roots are well established; however, new seedlings can find the environment of dark-colored spoils to be particularly harsh and inhibiting. According to Lee et al. (1975), the higher surface temperatures of the darker spoils can be reduced significantly by painting the surface with whitewash to increase its surface reflectivity.

Revegetation in Pennsylvania

Tree species with the widest range of adaptability are black locust (*Robinia pseudoacacia*) and European black alder (*Alnus glutinosa*) in Pennsylvania (Miles et al., 1973). Spoil sites are classified according to acidity, slope, and stoniness. The four groups of sites classified on the basis of acidity and the ability of plants to become established with various pH ranges are

1. 75% of area above pH 5.5 and less than 25% below pH 4.0. Predominantly limestone overburden. Black locust (*Robinia pseudoacacia*), timothy grass (*Phleum pratense*), and sweet clover (*Melilotus* spp.).
2. Less than 75% of area above pH 5.5, but at least 75% above pH 4.5. Not more than 25% below pH 4.0. Acid shales and sandstone overburden predominate. Aspen (*Populus* spp.) and poverty oatgrass (*Danthonia spicata*).
3. 50% or more of area below pH 4.5, but not more than 50% is below pH 4.0. Pyritic material predominates. Plants, such as poverty oatgrass, occur only in scattered areas.
4. More than 50% of spoil area below pH 4.0. High in pyritic material. Characterized by bareness and erosive and acidic qualities.

The two categories of sites classified on the basis of slope are (1) sites with no slope to slopes under 25% and (2) sites with slopes greater than 25%. Only the first category is useful for hayland or pasture.

Land use and the operation of mechanical equipment are the characteristics considered in the stoniness classification. There are two categories: (1) nonstony (will not interfere with seedbed preparation or limit mechanical tree planters) and (2) very stony (will restrict seedbed preparation but will not restrict tree and shrub planting by hand).

Japanese larch (*Larix leptolepsis*) and European larch (*Larix decidua*) are adapted to acidity groups 1 and 2 and provide good protective ground cover. Red oak (*Quercus borealis*) will grow on spoil acidity groups 1 through 3. Cottonwood (*Populus deltoides*) will not grow well on any spoil sites. Autumn olive (*Elaeagnus umbellata*) does well on most spoil sites, partly because it possesses symbiotic organisms that fix nitrogen. Bristly locust (*Robinia fertilis*) is an excellent plant for erosion control on most spoil sites; and being a legume, it fixes nitrogen. Tall oatgrass (*Arrhenatherum elatius*) does well on all sites. Chinese lespedeza (*Lespedeza cuneata*), also known as *Sericea lespedeza*, is a good ground cover on most of the spoil sites. The plant species that work successfully for revegetating coal spoil sites in Pennsylvania do not necessarily do well elsewhere.

Revegetation in Kentucky

Surface mining for coal in the mountains of Appalachia creates spoil banks with steep outer slopes, which must be revegetated to reduce erosion and to restore the aesthetic qualities of the area. Black locust (*Robinia pseudoacacia*) is a fast-growing tree species that grows well on a wide range of site conditions, including very acid spoil banks. Because it is a legume, it is able to fix nitrogen readily. When black locust was seeded alone and fertilized, it grew much better than when seeded with a ground cover. A herbaceous cover of spoil sites is usually desired to create soil stabilization, aesthetics, and forage for wildlife. Four cover mixtures were used. Their dominant species were (1) Korean lespedeza (*Lespedeza stipulacea*), (2) weeping lovegrass (*Eragrostis curvula*) and Kentucky-31 tall fescue (*Festuca orundinacea*), (3) Kentucky-31 tall fescue alone, and (4) Italian ryegrass (*Lolium multiflorum*). Fertilizer greatly increased the herbaceous cover on seeded plots. The different species competed with one another to varying degrees because of the time and habit of their growth. According to Vogel and Berg (1973),

> Italian ryegrass grows rapidly early in the spring and monopolizes most of the available moisture and plant nutrients before slower growing species, including black locust, get started. Tall fescue also starts growth early, but does not develop quite so rapidly as Italian ryegrass. Weeping lovegrass and annual lespedeza are warm-season species, that germinate later in the spring and make much of their growth during the late spring and summer after black locust has germinated and started growing.

Revegetation in Indiana

By 1968 38,143 ha in southern Indiana had been disturbed by surface mining of bituminous coal. The Indiana coal-mining industry believes they were the first in America to have an organized revegetation program for surface-mined lands. Records begin in 1918. Analysis of the success of these efforts led Medvick (1973) to conclude that the following species mixed together in equal proportions are most successful for acid sites: red oak (*Quercus rubra*), sweet gum (*Liquidambar styra-ciflua*), sycamore (*Platanus occidentalis*), river birch (*Betula nigra*), and European alder (*Alnus glutinosa*). For sites with a pH greater than 5.4, Medvick recommended 20% each of red oak, yellow poplar (*Liriodendron tulipifera*), sweet gum, and black walnut (*Juglans nigra*); 2.5% Chinese chestnut (*Castanea mollissima*), and 17.5% white ash (*Fraxinus americana*) and red maple (*Acer rubrum*) or silver maple (*Acer saccharinum*). In addition every eighth row should be planted in pine (for aesthetics), and pine should be planted near roadways. White pine (*Pinus strobus*), pitch pine (*Pinus rigida*), jack pine (*Pinus banksiana*), and Virginia pine (*Pinus virginiana*) are recommended.

Revegetation in Iowa

Most of the strip-mining for coal in Iowa has been in the southcentral and southeastern parts of the state. The Iowa coal is mostly located in bedrock described as Pennsylvanian shale. It is covered with a loess that was deposited during the Wisconsin glaciation and with a till. The region was covered originally with tall grass prairie, but after European settlement much of the area was converted to cropland, pasture, and woodland.

The spoil banks from strip mining in Iowa are generally very steep and have narrow draws between them. The draws are often badly eroded. The spoil surfaces are often dry and have a high surface temperature when not protected by vegetation.

Glenn-Lewin (1979) reports a study of the natural revegetation of six spoil sites in Iowa. The dominant species were not always the same on each site, but generally the following species were in greatest numbers: cottonwood (*Populus deltoides*), American elm (*Ulmus americana*), sumac (*Rhus glabra*), box elder (*Acer negundo*), and sweet clover (*Melilotus* spp.). Most sites were not well covered with vegetation, and even the oldest site only had one-third of its surface covered with plants.

Revegetation in Kansas

In Kansas there are nearly 20,250 ha of coal spoil needing reclamation. More than 70% of the spoil is nontoxic, with a pH above 4.0. Available phosphorus and potassium are in concentrations of 83 and 81 ppm (amounts adequate for most plants); exchangeable Ca was high, and organic matter was about 1.7% (Geyer, 1973). For the revegetation of strip-mine spoils in the Kansas prairie–forest ecotone, sycamore (*Platanus occidentalis*) and black locust (*Robinia pseudoacacia*) did best. It is interesting to note that sycamore did poorly in Pennsylvania. Also used in Kansas were eastern red cedar (*Juniperus virginiana*), bur oak (*Quercus macrocarpa*), loblolly pine (*Pinus taeda*), and shortleaf pine (*Pinus echinata*). In contrast to Pennsylvania, the cottonwood (*Populus deltoides*) did well in Kansas. Forage and small grain crops are also being planted successfully on Kansas coal spoil sites (see Geyer, 1973 for more information).

Revegetation in Germany

Revegetation experiments of coal spoil sites are going on in many parts of the world. Particularly detailed ecological information is derived from long-term observations in the Ruhr. To quote Bauer (1973),

> The main ecotype groups on the excavated mines are gravel slopes, gravel plains, and reclaimed fields, coal-dust areas, and depressions and lakes. On the first three sites, and particularly on the coal-dust areas, edaphic, hydrologic, and microclimatic conditions are often extreme. Nevertheless there is a spontaneous, rapid colonization by various plants and animals. Then tree and shrub seedlings become established. The landscape changes appearance from desert-like to grass-steppe and then more gradually to savanna-like. Even later the tree canopies close, environmental conditions change, and the initial in-

vaders largely disappear. Only then does competition begin to play a dominant role in the determination of the final composition of the plant and animal association.

Pioneer species of vegetation in the Ruhr included *Calamagrostis epigejos, Oenothera biennis, Poa annua*, and *Bromus sterilis*. The pioneer species are mainly annuals, which supply material for humus when they die. Seeds of woody vegetation are carried in by wind or water and include species of *Salix* (willow), *Betula* (birch), *Populus* (aspen), and *Robinia* (locust). After about eight years the slopes are covered with a young forest. The black locust (*Robinia pseudoacacia*) shows an allelopathy toward some other plants because it contains poisons in roots, bark, seeds, and leaves. Many insects, birds, and other animals increase in numbers with the increasing number of plant species.

Knabe (1973) describes tree growth on deep-mine refuse piles in the hard coal region of the Ruhr, where the piles either loom as ugly eyesores or improve the landscape as picturesque wooded hills. Refuse heaps in the Ruhr are of all shapes, forms, and compositions. They are composed of a mixture of coal refuse, topsoil, overburden, fuel ash, and domestic and industrial wastes. Textures range from stony to silt and clay. Their forms vary from steeply conical, irregular heaps, table mountains, and dams to level mounds of various ages. Some refuse piles are bare, some are covered with grass, and some are covered with trees. After a few decades, substantial amounts of airborne dust accumulate on the vegetated sites, where it adds organic matter and nutrients. On bare sites the dust blows away. A dense network of roots form in the dust layer and this promotes soil development and tree growth.

Burned shale sites are usually naturally colonized by grasses (mainly *Agrostis tenuis*) or by European white birch (*Betula pendula*) and willow (*Salix caprea*). Unburned gray shales may remain unvegetated longer than burned shales. Naturally seeded birches and willows are found in gaps between artificial plantings of poplars. As stated elsewhere, black locust (*Robinia pseudoacacia*) appears to inhibit the growth of young birch underneath them, but there is a shrub layer of *Sambucus racemosa*, *Holcus mollis*, *Poa annua*, *Urtica dioica*, and *Athyrium filix femina* as well as *Agrostis tenuis* and *Epilobium angustifolium*.

The studies of coal spoil sites in the Ruhr show that they can be successfully reforested. Stands of hardwoods have perpetuated themselves and promoted soil development. Earthworms are found only under tree stands, particularly where there has been considerable accumulation of a dust layer. Black alder (*Alnus glutinosa*), speckled alder (*Alnus incana*), black locust (*Robinia pseudoacacia*), and European white birch (*Betula pendula*) are best for revegetation, with variations from site to site. A variety of other trees can be introduced for long-term stability of the forest cover. These include red oak (*Quercus rubra*), sycamore maple (*Acer pseudoplatanus*), ash (*Fraxinus excelsior*), cherry (*Prunus avium*), and linden (*Tilia platyphylla*). Other species may also be used along borders at the bases of refuse piles. Conifers generally will not grow successfully in the Ruhr Valley because of heavy air pollution.

Microorganisms, millipedes, woodlice, insects, and other organisms that may migrate to spoil sites are as important as reforestation. Good plant growth is assured if there is fertile soil, and that depends on the soil fauna (including microorganisms) responsible for its formation. Nutrients, originating from minerals or from fertilizers, are assimilated by plants and returned to the soil as leaf litter. It is the breakdown of this litter by a whole array of fauna and microorganisms that release the nutrients for uptake by other plants. Usually leaf litter is eaten and reduced to small pieces by primary decomposers (arthropods and other invertebrates) and then decomposed by microorganisms into humus. A soil that is rich in fauna, and particularly in microorganisms, will utilize nitrogen effectively and keep it recirculating. If the leaf litter does not break down and humic acid forms, the soil will become acidic and deficient in the necessary organisms. Therefore, it is of great importance to avoid acid soils in order to keep nitrogen active in the ecosystem. Fertilizer must be used to start the soil culture. Then vetch, lupine, alder, and other plants with root nodules containing nitrogen-fixing bacteria must initiate the process until such time as a rich soil fauna develops. The soil fauna will need to migrate in from neighboring areas. If this does not happen (for a variety of reasons), soil containing the proper organisms may need to be brought into the site.

WESTERN COAL

The western coal reserves in the United States are vast and are mainly located in New Mexico, Colorado, Wyoming, Montana, and North Dakota. These reserves have two advantages: they are relatively close to the surface and they are of low sulfur content. Because the energy content of western subbituminous coal (20,000 kJ/kg) is less than that of eastern bituminous coal (28,000 kJ/kg), it takes 140 kg of western coal to equal 100 kg of eastern coal. Thus, more land must be disturbed in the west than in the east for the same amount of energy. In terms of total bulk, 70% of our coal reserves are west of the Mississippi; but, in terms of the total heat content in the beds, about 55% is east of the Mississippi.

Vast reserves of low-sulfur, subbituminous coal underlie the Powder River Basin in northeastern Wyoming. Much of the coal is found in thick seams that lie close to the surface. In Campbell County, Wyoming alone more than 20 billion tons of coal are available for mining. Several mines have already been opened in Campbell County and others are scheduled to begin production soon. Each new mine is expected to produce from 5 to 25 million tons of coal annually.

The Black Thunder mine

The best way to describe the activities associated with mining western coal is to use a specific mining operation as an example. Atlantic Richfield's Black Thunder mine is a large mine characteristic of many in the Powder River Basin of Wyoming where it occupies 2640 ha (6524 acres) on a federal coal lease about 45 miles south-southeast of Gillette. The coal is in a sequence of rocks of late Cretaceous to early Eocene age. The most minable coal seam outcrops on the eastern edge of the Black Thunder site and dips about one degree to the west-southwest into the center of the Powder River Basin. The coal seam averages 22 m thick. The overburden of clay shales and sandstones ranges from 0 to 65 m thick, the average being 42 m. The coal is a low-sulfur (0.4%), low-ash (5%), subbituminous coal. Trace element (heavy metal) concentrations are generally low in the bulk of the coal seam, but they are much higher in the uppermost and lowermost portions of the coal seam and in the overburden and floor.

It is expected that the Black Thunder mine will eventually produce about 20 million metric tons of coal each year. The energy content is 19,770 kJ/kg (8500 Btu/lb). The first step in the mining operation is to remove the topsoil and stockpile it nearby for future use. Stockpiled topsoil is revegetated to minimize erosion until it is eventually replaced on top of the recontoured overburden. Next the tightly compacted overburden is blasted loose, dug up by electric shovels three stories high, and loaded into huge trucks, each carrying 170 tons. The overburden removed is stockpiled. As sufficient pit room becomes available, the overburden is used as a backfill in the previously mined sections. Then the exposed coal seam is blasted loose, dug out with giant shovels, and loaded into large trucks. Only two pits or trenches of about 32 ha each are mined at a time in order to minimize the amount of land disturbed. As the minefront advances, overburden and topsoil are replaced behind it in reverse order of removal and then contoured and replanted.

The trucks carry away the coal to the primary crusher, which is underground, thereby eliminating the need for truck ramps. Here the coal is crushed to sizes less than 20 cm. The pieces are conveyed on belts to the secondary crusher, where the pieces are broken down to less than 5-cm diameters. The crushed coal is sampled for quality and sent either to storage or loaded on trains. All of the conveyor belts and storage bins are enclosed to minimize dust release. As the coal is removed from the storage bins, it is mixed to obtain a desired quality and sent to the trains.

Coal transport. UNIT TRAINS of 100 cars pass slowly under the 50 meter-high loader where 100 tons of coal are gravity fed into each car. Coal samples are taken and analyzed to determine the exact quality of the load and each car is weighed before and after loading. All information is put into a computer for record-keeping and billing. Most of the Black Thunder coal goes to utilities in Texas, Oklahoma, Nebraska, and Wisconsin.

Coal SLURRY PIPELINES are being developed to connect the coal fields of Wyoming with the centers of consumption in the Midwest and South. Coal is ground fine and mixed with water to form a slurry. The slurry is pumped through a pipeline to its destination, where water and coal are separated. Slurry pipelines now in existence are relatively short. One in Ohio is 180 km long and carries 1,300,000 mt of coal annually; the pipeline in

Arizona is 455 km long and carries 4,800,000 tons annually. However, pipelines as long as 2300 km are planned to carry 25,000,000 mt of coal each year from the Powder River Basin. Water for the coal slurry is pumped from the Madison Formation at 1000 m beneath the surface. This aquifer is recharged by water flowing from the Black Hills to the east and from the Big Horn Mountains on the west. Recharge rates are estimated at 150,000 acre-feet annually. The coal slurry pipeline to serve the ARCO property will withdraw about 15,000 acre-feet annually. The Madison Formation is estimated to contain a total of 1 billion acre-feet of water under the entire Powder River Basin. Its depth is from near the surface at its eastern- and western-most limits to as deep as 5000 m. Water drawn from most of these depths is far too expensive for agricultural use. The water to be used at the Black Thunder mine would cost about $400 per acre-foot, whereas irrigation water cost about $10 to $15 per acre-foot.

Ecosystems at Black Thunder. "As the Black Thunder site is mined, the soil-plant-animal associations now present on the site will be destroyed. Although the site will be reclaimed, the ecosystems developed on the site are likely to be different than those present now." Thus states the report entitled "Final Environmental Assessment, Black Thunder Mine Site, Campbell County, Wyoming, October 1976."

In their natural undisturbed state, most of the soils of the Black Thunder site are good grassland soils with adequate fertility in the upper horizons and less than adequate fertility in the lower horizons. In a few areas salt concentrations may accumulate to toxic levels at the surface. The natural vegetation of the site is very stable and erosion is minimal.

Three major vegetation communities exist at Black Thunder. The largest and most diverse vegetation community is made up of big sagebrush (*Artemisia tridentata*), blue grama grass (*Bouteloua gracilis*), needle and thread grass (*Stipia comata*), and plains prickly pear (*Opuntia polycantha*). This vegetation occurs on undulating or rolling uplands predominated by sandy loam soils. A second community type occurs on undulating uplands of fine-textured loam soils and consists of big sagebrush and bue grama grass without the *Stipa* present. The third major vegetation community is made up of big sagebrush, blue gama grass, and western wheat grass (*Agropyron smithii*). This vegetation occurs on most of the steeper sloping, shallow soil areas. It is much utilized by deer for browse. Additional vegetation inhabiting Black Thunder include blue-bunch wheat grass (*Agropyron spicatum*), slender spikerush (*Eleocharis acicularis*), foxtail barley (*Hardeum jubatum*), and prairie junegrass (*Koeleria cristata*). The blue grama grass dominates much of the vegetation of the Black Thunder site and reduces the natural value of this grassland to wildlife, which prefer more palatable species such as the bunchgrasses.

On the Black Thunder site (1) secondary succession following cultivation reestablishes native vegetation after 40 years; (2) abundance and composition of natural species is directly related to the available seed sources; and (3) on disturbed oil well sites, the most abundant pioneer species is prostrate knotweed (*Polygonum oviculare*) and common Russian thistle (*Salsola kali*).

Generally microfungi constitute 75 to 90% of the microflora of the grassland soil and are among the most active decomposers. The mycorrhizae (fungus-root), which are common in shortgrass prairie, enhance nutrient uptake and as a result improve primary productivity by as much as 400 to 500%. The roots of young plants become infected with soil fungi to form mycorrhizae, which consist of the mycelia (threads) of fungi that live in mutualistic association with the living roots of plants. Nitrogen-fixing bacteria interact with root tissue of legumes to form structures that increase the ability of the plant to extract minerals from the soil. In return the bacteria are supplied with some of the photosynthate generated by the plant.

Species of the fungus *Aspergillus* are the most common of the fungi occupying grassland soils. Comparison of disturbed with undisturbed soil sites showed the following effects on microfungi: (1) a 50% reduction of microfungal density; (2) a change in species composition from dominance by *Aspergillus* to dominance by *Chrysosporium*; (3) similar mycorrhizal densities in both disturbed and undisturbed sites, indicating that mycorrhizae can reestablish in prairie soils following disturbance; and (4) slightly increased rates of decomposition in disturbed soils. This may be due to the increased presence of *C. pannorum*, a species more capable of decomposing cellulose than are the *Aspergillus* species.

These soils had low densities of microarthropods, which were mostly in the upper 5 cm of soil. Undisturbed areas averaged 500 to 1500 microarthropods per square meter. The most abundant orders present were mites and collembolans. Because both orders are strongly influenced by temperature and moisture, there was a great deal of seasonal difference. Microarthropods in grassland soils have little direct influence on energy and nutrient cycling, but may have indirect effects.

Nematodes were more abundant and more significant than the microarthropods. Densities in wetter sites were 2 to 3 million/ m^2 and in disturbed sites as low as 0.5 million/m^2. However, one sample from a disturbed oil well site gave 36 million/m^2. Disturbed sites generally had fewer trophic groups than undisturbed sites. Soil pore size, amount of water, amount of aeration, and temperature mainly determine the abundance of nematodes. The saprophytic (bacterial feeding) nematodes can be important in regulating nutrient cycling, by either stimulating or inhibiting bacterial turnover. The plant parasitic nematodes may consume a significant proportion of the primary production, at times greater than the vertebrate grazers. Interactions between the plant parasites and their hosts are quite complex.

A high proportion of a grassland biomass production goes below ground into the root–soil system. Studies elsewhere have shown that approximately 79% of the net primary productivity is translocated to plant roots in an ungrazed prairie. Respiration by primary and secondary consumers accounts for only 1.2% of this net primary productivity on ungrazed prairie and 3.2% on grazed prairie. Respiration by saprophagic soil organisms, those that feed on decaying organic matter, accounts for 73% of the net primary productivity on ungrazed prairie and 87% on grazed prairie. These numbers can only be considered as approximate because many fundamental processes of prairie ecosystems are poorly understood.

Vertebrate species are abundant at the Black Thunder site. During 1974, 95 species of terrestrial vertebrates were observed: 21 mammal species, 68 bird species, 4 reptile species, and 2 amphibian species. The most abundant group of mammals was the rodents, of which there were 11 species. There were four species of rabbit and hare (Leporidae); three species of mustelids or polecats (Mustelidae); two species of deer (Cervidae); and the pronghorn antelope (*Antilocapra americana*), the only living representative in the world of the family Antilocarpridae. Rodents included the deer mouse (*Peromyscus maniculatus*), the harvest mouse (*Reithrodontomys magalotis*), the grasshopper mouse (*Onychomys leucogaster*), the pocket mouse (*Perognathus fasciatus*), the kangaroo rat (*Dipodomys ordii*), and the thirteen-lined ground squirrel (*Spermophilus tridecemlineatus*).

Jack rabbits, the white-tailed (*Lepus townsendii*) and the black-tailed (*Lepus californicus*), both occurred on the site, particularly where there was good sagebrush cover. Cottonwood rabbits (*Sylvilagus nutallii grangeri* and *S. audubonii baileyi*) were in the area. Birds on the site included the bald eagle (*Haliaeetus leucocephalus*) and the golden eagle (*Aquila chrysaëtos*), 6 species of hawks, and 2 species of owls. Meadowlarks (*Sturnella neglecta*), red-winged blackbirds (*Agelaius phoeniceus*), lark buntings (*Calamospiza melanocorys*), Brewer's sparrow (*Spizella breweri*), and vesper sparrows (*Pooecetes gramineus*) were present. The most abundant bird species was the horned lark (*Eremophila alpestris*). Sage grouse (*Centrocercus urophasianus*) nested on the site. The most common hawks were ferruginous hawks (*Buteo regalis*), red-tailed hawks (*B. jamaicensis*), and marsh hawks (*Circus cyaneus*).

Birds of prey should receive special attention from managers of mining operations because they are at the top of the food web and control through their predation a variety of animals of significance to agriculture and ecosystems. They control many herbivores, such as mice, that might otherwise wreak havoc on a revegetation project. In fact, at the Black Thunder site of ARCO Coal, ecologists have put up dozens of kestrel boxes as nesting sites. The idea is to saturate the area with American kestrels (*Falco sparverius*) so they will prey on the field mice.

Ecologists working for ARCO's Coal Creek Mine, Powder River Basin, Wyoming have pioneered a technique to induce golden eagles to transfer their nesting sites from potential coal mining areas to noncoal lands, or from unmined to reclaimed land. They were able to successfully transfer an eaglet from its natural nest to an artificial platform about 175 m away. The platform was atop a pole about 13 m high. Eagles are extremely attentive parents and when the eaglet was moved, during the adults' absence from the nest, the adults later located the eaglet and attended it. This operation was continued successfully; the eaglet was transferred to another platform and then another, where it was 1375 m from the original nest. The adults returned to the eaglet each time with food and finally the eaglet fledged. The secret to this transfer operation is to know well the habits and movements of the eagles and then place the platforms along a route within the adult's existing territory. The reason for all of this effort is that the mining permit severely limits mining activities in the general vicinity of eagle nests.

Revegetation of strip-mined sites. Prior to mining, the topsoil will be stripped from the overburden to an average depth of maximum root concentration. Topsoil removed during the first two years of mining will be stockpiled until backfilling begins. Then the topsoil will be spread over the reclaimed spoil bed of overburden material that has been removed and replaced. Following this initial phase, overburden and topsoil removal and replacement will be a continuous operation. This procedure will avoid the problem of long-term storage and deterioration of topsoil. Approximately 80 to 100 acres of surface will be rehabilitated each year when production reaches 10 million tons of coal per year. The surface will be topographically blended with the surrounding terrain. Drainage systems will be reestablished and upland areas will be graded such that no slopes are greater than 4:1. The goal is to control erosion and to rebuild a stable biotic community capable of maintaining itself. Productivity on the restored site should be equal to or greater than it was before disturbance.

A number of procedures are recommended for mitigating the impacts of surface mining on soil microfauna. Soil should not be stored any longer than necessary; when soil is spread on a mined area, microrelief in the form of rocks, shrubs, and varying topography must be provided. Microclimate extremes that adversely affect soil organisms are ameliorated by microrelief. This will provide a more suitable habitat for the return of a diversity of soil microfauna and invertebrates and be attractive to vertebrate fauna as well. Heavy mulching improves soil structure and promotes the return of soil organisms. Mulching may encourage decomposers and may provide adverse conditions for parasites and some plant pathogens. The addition of glucose to the soil increases soil aggregation because of increased fungal and bacterial growth.

Revegetation efforts should use a variety of plant species that are native to the Black Thunder area. To produce some shading of the soil, shrubs such as big sagebrush should be reestablished. Native plant species may be more resistant to nematodes than are exotic species. A routine of alternate drying and wetting of the soil may speed up nutrient cycling. This practice often increases mineralization of organic matter, thereby resulting in increased inorganic nitrogen concentrations in the soil. Microbial populations also may be enhanced by the process. Alternate wetting and drying may also help control plant parasitic nematodes. Some use of fertilizers may be necessary to increase revegetation success. Phosphorus may be added to the soil; nitrogen fertilizers should be used only if the site is irrigated. Initially there should be an excess of nitrates in reclaimed topsoil. If this is the case, the use of ammonium nitrate fertilizer will have no effect on initial plant growth unless irrigation is used.

Nursery areas near the mining site should be maintained to serve as sources of seeds of native plants. The seed mixture used for revegetation may also include short-lived, introduced grasses and legumes to help stabilize reclaimed soils. Late autumn seeding is preferred to spring seeding, but either may be used. Late autumn seeding has several advantages: (1) soils tend to be drier and easier to work; (2) more time is permitted for the seeds to stratify in the seedbed prior to germination; and (3) winter conditions are often necessary to break dormancy and weaken the hard seed coats of native species. Rains come in the spring in the Powder River Basin of Wyoming.

The stockpiled topsoil should be seeded with a grass mixture containing approximately 40% crested wheat grass (*Agropyron cristatum*), 20% thickspike wheat grass (*Agropyron dasystachyum*), 20% western wheat grass (*Agropyron smithii*), and 20% streambank wheat grass (*Agropyron riparium*). These species have different establishment characteristics and moisture tolerances. The crested wheat grass establishes most quickly, but does not have rhizomes. It is a grass introduced from Eurasia, and it spreads readily by reseeding. The native wildlife do not like it, nor do the cattle except when it is young or in seed. The western wheat grass establishes more slowly, but develops rhizomes that help to hold the soil. It is preferred by the wildlife.

Texas lignite

Enormous reserves of LIGNITE (brown coal) are found in Texas, Arkansas, Louisiana, and Mississippi. Since 1971 the mining of lignite in Texas has grown from 2 to over 20 million tons per year. Lignite was first mined for fuel here in the 1880s; production grew to an average of 1 million tons per year between 1915 and 1930. In the 1930s, Texas electric utilities discovered the ready availability of cheap natural gas. So they turned away from lignite. However, during the next decade or two the situation changed dramatically and by 1954 the Texas Utilities Company began generating electricity from lignite, electricity the Alcoa Corporation used to process aluminum from bauxite.

Today the main consumers of Texas lignite are three subsidiaries of Texas Utilities: Dallas Power and Light Company, Texas Electric Service Company, and Texas Power and Light Company. In 1976 the Texas Utilities Company produced 31.5% of its energy from lignite; this figure has recently now changed to greater than 50%. Currently, Texas Utilities is paying about 50¢ per million Btu for lignite. To fuel the same plant with natural gas would cost about $2.00 per million Btu or more. It is estimated that by 1985 total lignite power-generating capacity in Texas alone will be 14,500 MWe.

Lignite is very difficult to handle, yet relatively easy to mine. The terrain is flat or gently rolling and easy to restore. No blasting is needed to break up the overburden or the lignite. Draglines remove the overburden and power shovels take out most of the lignite, although front-end loaders and backhoes are also used. Lignite is hauled to crushers, crushed to 15-cm pieces, then to 2-cm pieces, and finally pulverized for burning. Lignite cannot be stockpiled because it is too fragile. Only a 2- or 3-day supply can be kept at a time. Lignite is dusty and can blow away in the strong Texas winds. When heated by sunlight, it ignites. Too much handling makes it too fine, and excessive cold freezes it into chunks. When rained on it becomes sodden and must be blended with drier lignite—too much rain can almost shut down an operation.

Texas lignite varies in energy content from 14,600 to 19,300 kJ/kg (6300 to 8300 Btu/lb). It has a sulfur content of 0.6 to 1.2%, an ash content of 12 to 15%, and a moisture content of 30 to 50%. Seams of lignite range in thickness from one to five meters and as many as seven or eight seams may exist, one above the other. The ratio of overburden removed to lignite mined was at first about 5:1, but now is 15:1. Texas Utilities alone has 750 million recoverable metric tons of reserves available.

Exxon owns about 600 million metric tons and hopes to use this lignite for gasification. They have shipped 16,500 metric tons of lignite to the Sasol Lurgi coal gasification plant in South Africa for a test of its suitability for gasification. Exxon feels that gasification of lignite is the best use from an economic standpoint. Lignite can be ignited underground to produce a low Btu gas that can be released through a pipe tapping the underground reservoir.

Land restoration. The Texas Utilities Company follows a procedure very similar to other strip-mining operations in the West. The soft brown coal is mined from a series of long trenches, each 30 to 40 m wide. The overburden from one trench is placed as fill in an adjacent trench. Texas Utilities found that when the topsoil is mixed with the subsoil, the productivity is greater than when the topsoil is returned to the surface layer; this procedure is not used or desirable elsewhere. After the fill is bulldozed approximately to its original contour, the surface is tilled, fertilized and planted with winter wheat, coastal Bermuda grass, alfalfa, and crimsom clover. Sometimes only coastal Bermuda grass and red clover are used. These fields have been highly productive. Experiments are being conducted on other revegetation crops, particularly row crops and trees.

PEAT

PEAT is a geologically young coal, a precursor to the formation of lignite. It is essentially an accumulation of plant remains in various stages of decomposition. It is thousands of years old, rather than millions of years old as coal is. It has not undergone burial to great depths or metamorphosis under high pressure and temperature. Peat is usually formed in bogs, which essentially are shallow depressions where water saturates or covers dead vegetation. The bog itself (i.e., the surface of the underlying material) is alive with healthy vegetation that forms a characteristic plant community. Such vegetation may include a number of *Sphagnum* moss species, leather leaf (*Chamaedaphne calyculata*), Labrador tea (*Ledum groenlandicum*), pitcher plant (*Sarracenia purpurea*), sun dew (*Drosera* sp.), black spruce (*Picea mariana*), and blueberry (*Vaccinium oxycoccus*). This vegetation will often form a dense mat composed of a fine network of roots. As bog plants die and decay, they accumulate in the water and build slowly into an organic muck that eventually consolidates into peat. The water blocks the action of aerobic bacteria, thus retarding decay of plant debris and allowing the cellulosic matter to be retained. Bogs are very beautiful, fascinating places and are of great benefit for filtering and cleansing water flowing through them into underground aquifers. Bogs are important habitats for a large variety of animals, and some species are indigenous to bogs.

World use of peat

Peat amounts to only about 1.1% of the world's fossil fuel resources and is not considered of particular importance in meeting the world energy needs. It is, however, of considerable local importance for meeting the energy needs of some groups of people. Peat has been used for centuries as a fuel, particularly by people in northern Europe. Ireland has long burned peat and is currently using about 3 million metric tons per year, accounting for approximately one-third of its total energy supply. The Netherlands, Germany, and Denmark all used peat as fuel on a large scale until the exhaustion of their reserves forced them to use coal, gas, and oil. The Soviet Union burns an estimated 70 million metric tons or more per year, mostly to produce electricity from some 70 or more power plants. Peat as fuel provides about 2% of its total energy supply. The city of Leningrad, lo-

cated near many peat bogs, gets about 17% of its energy needs supplied by peat. Finland imports much oil and gas and is now using increasing amounts of peat for fuel. Sweden abandoned the use of peat many years ago, but is now starting to use peat for electric power plant fuel. Greece has been considering running a power plant with peat.

Location of peat

Peat bogs are found throughout the world at all latitudes and on all continents, although most of the larger peat bogs are in a relatively narrow belt of the temperate zone where the climate has been favorable for peat formation for at least the last several thousand years. There is an estimated 20 to 40 million hectares of peat in Alaska, but much of it is frozen and very inaccessible. Outside of Alaska, the world's total peat reserves are in the Soviet Union, 60%; Canada and Finland, 20%; and the contiguous states of the United States, 5%. Minnesota is estimated to have about 3 million hectares of peat bog, covering about one-seventh of the state's total land area. Other large amounts occur in Wisconsin, Michigan, Florida, New York, and Maine.

Uses of peat

Peat and peat bogs have many uses. In the first place, a peat bog can be developed for growing vegetables, wild rice, cranberries, blueberries, forage grasses, and certain kinds of timber such as black spruce, which can be used for Christmas trees or for making paper pulp. One peat bog in northern Minnesota has been producing vegetables for more than 40 years. Near Minneapolis, bogs have been used for growing special vegetables and bluegrass turf. Peat bogs can be used for growing cattails, sedges, reeds, and grasses that can be burned for fuel.

Sphagnum moss from bogs was used for surgical dressings for wounded soldiers in

World War I and was long used as a packing material for shipping—before the use of Styrofoam. Peat can also be applied as a soil conditioner for improving the physical and chemical properties of soils. This product is the well-known PEAT MOSS purchased by gardeners. A number of products are derived from peat including single-cell proteins used as cattle feed, humic acids that are used as viscosity control agents for oil well drilling and water pollution control, and various waxes and steroids.

In the Netherlands, Germany, and Poland, activated carbon for use in filters has been derived from peat. In Scotland, peat is used in the distilling of Scotch whiskey.

Peat as fuel

Peat has advantages and disadvantages as a fuel. It is relatively easy to mine because it is located on the surface. It has a low sulfur content and hence, when burned, will not contribute much sulfur dioxide to the atmosphere. The heating value of air-dried peat is less than that of wood and lignite, and about half that of some coal (see Table 1). Even this heating value can only be realized if the peat is air- or sun-dried because peat is about 70 to 95% water. This water must be largely removed before peat can be burned. When fossil fuel is used to evaporate the water, the net heating value of the peat is quite low.

Other disadvantages to the use of peat are its location in remote areas distant from major users; its bulkiness, which makes it costly to transport; and the mining process, which is disruptive to the landscape, destructive of wildlife habitats, and potentially disturbing to the water table.

Peat may be burned directly in stoves or furnaces for heating homes and small commercial establishments or it may be burned to generate electricity. It is also possible to gasify peat to produce a high Btu gas that can be transported through pipelines.

If peat were burned in the furnace of an electric power plant, it would require enormous tracts of peatland to supply sufficient energy for a small power plant. The tracts of land would be much greater than would be those required if forests were used to supply wood for electric power generation.

Harvesting peat

The method of harvesting peat in Europe is known as the MILLED PEAT PROCESS. A bog is drained, stripped of surface vegetation, and leveled so that machines can till the top half-inch or so of soil to facilitate drying by the sun. Another machine is used to suck up or scoop up the dried peat. The milled peat is either burned directly or compressed into briquettes for home heating.

Much of the harvesting can only be done during the warmer months of the year. The entire operation is labor intensive for harvesting, transporting, handling, and processing; hence, the final product tends to be expensive.

Environmental impacts

Peat harvesting causes a major change in the landscape. The land surface may be lowered by 3 to 7 meters, depending on the depth of the peat. All vegetation must be removed from a large area, and the entire area drained. The drainage process affects the water quality of adjacent streams or lakes, the local and regional water tables, and leads to the potential for serious flooding at times of heavy precipitation. Proponents of peat mining claim that proper restoration of the mined area through revegetation can mitigate many of the expected impacts. Even the value of a peat bog for wildlife habitat may be restored, although it will not be at all like its original form because the plant community will be completely different after restoration.

9
Carbon Dioxide and Climate Change

A change of climate resulting from an increase in the atmospheric carbon dioxide concentration may bring increased drought to the interior of continents (top) and increased precipitation to many other regions (bottom). Courtesy of the National Center for Atmospheric Research and Art Bumpas.

THE GREENHOUSE EFFECT

A greenhouse stays warm because sunlight can come in through the transparent glass, but longwave (infrared) radiation cannot be transmitted out because the glass is opaque to infrared radiation. The radiation trapped inside warms the interior surfaces of ground, plants, and other materials. A greenhouse is also effective because it does not let convection (air movement) remove heat to the external environment, except as desired through controlled vents (see Figure 1A).

In the atmosphere, certain gases (notably carbon dioxide and water vapor) act like the glass of the greenhouse. They allow sunlight to pass through to the Earth's surface, where it is absorbed, but they interfere with the passage of longwave radiation outward. This process, known as the GREENHOUSE EFFECT, balances the incoming and outgoing radiation at a level that warms the Earth's surface. Figures 1B and 2 illustrate the exchange between sun, Earth, and space.

Carbon dioxide gas is odorless, colorless, and, at low concentrations, nontoxic. Carbon dioxide is transparent to shortwave radiation. The ultraviolet, visible, and near infrared parts of the solar spectrum—with wavelengths from 300 to 2500 nm—are included in the term SHORTWAVE RADIATION. Earth's surface radiates to outer space a broad continuum of wavelengths—ranging from 4000 to 20,000 nm (4 to 20 μm). Radiation within these wavelengths is known as LONGWAVE RADIATION. Carbon dioxide has an extremely strong, broad absorption band centered at a wavelength of 14 μm, but extending from 12 μm to beyond 20 μm. As shown in Figure 9 of Chapter 2 this overlaps with some of the wavelengths of radiation emitted from Earth's surface. Therefore, carbon dioxide absorbs this radiation. Water vapor also has strong absorption regions in the infrared, mainly in a broad band centered at 6 μm and in another extending beyond 13 μm. The carbon dioxide and water vapor molecules reradiate the energy they have absorbed, sending

(A)

(B)

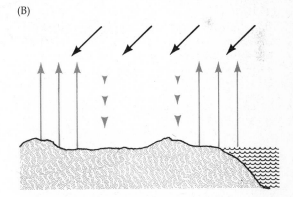

FIGURE 1. (A) A greenhouse absorbs solar radiation but does not allow infrared radiation (heat) to escape. The process is known as the GREENHOUSE EFFECT. The long arrows represent shortwave solar radiation; the short arrows (shaded) represent longwave infrared radiation emitted by the greenhouse glass and ground. (B) The exchange of radiation between the sun, the Earth's surface, and outer space. Shortwave solar radiation (long downward arrows) warms the surface, and longwave infrared heat (shaded upward arrows) escapes to space, thereby cooling the surface. Carbon dioxide and water vapor in the atmosphere absorb some of this infrared heat and reemit some of it back to the ground (shaded short arrows).

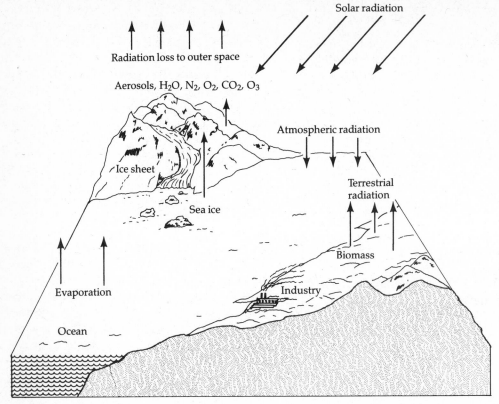

FIGURE 2. The exchange of energy between the sun, the Earth, and outer space involves many processes. Solar radiation brings energy to the atmosphere and to the ground surface. The atmosphere, containing aerosols and the gases H_2O, N_2, O_2, CO_2, and O_3, reradiates energy to the ground as well as to outer space. The ground surface and sea surface radiate heat to outer space. Energy is removed from the Earth's surface by the evaporation of moisture from the ocean and from the land.

half of it in the direction of the ground and half of it in an outward direction. The process is actually a complex cascade of radiant energy absorbed and reradiated until the outer limit of Earth's atmosphere is reached. However, each time an absorption and emission takes place, some energy gets returned to the ground. The net result is a warming of Earth's surface. However, while the ground and lower atmosphere is warming by this process, the stratosphere is cooling because the stratosphere radiates freely to outer space

through the thin layer of atmosphere above it.

The Earth's surface and atmosphere maintain a balance between the incoming flux of solar radiation and the outgoing infrared radiation. As a result of this radiation balance the mean surface temperature of the Earth is about 15°C. Disturb this balance and the temperature may go up or down. Increased cloud cover can reflect sunlight to space, reduce the amount of sunlight reaching the ground, and cause cooling. Increased amounts of carbon dioxide and water vapor (not in the form of clouds) can cause warming of the global climate. Climatologists are carefully considering this phenomenon in their calculations of climate change by using large-scale, numerical, computerized climate models. They find that a doubling of the atmospheric concentration of carbon dioxide could produce an increase of the global mean temperature of the Earth of about $3\,°C \pm 1\,°C$. It turns out that this much of an increase in the mean surface temperature is very significant and is in fact greater than any naturally occurring changes in temperature during the past 10,000 years.

There is strong evidence that the concentration of carbon dioxide is inexorably increasing in the atmosphere as the result of human activities, such as burning fossil fuels and cutting forests. Earth's climate may be warming, and extremely serious consequences may result: glaciers may melt, sea levels may rise, agricultural production may be affected, and ecosystems may undergo substantial changes.

ATMOSPHERIC CARBON DIOXIDE CONCENTRATION

Earth's atmosphere and biosphere evolved together over time. The molecular composition of the atmosphere is the direct result of gas exchange among the atmosphere, biosphere, lithosphere, and hydrosphere. Green plants, through photosynthesis and respiration, have had significant influence on the atmospheric concentrations of carbon dioxide, oxygen, and water vapor.

Preindustrial amount

Primordial Earth contained carbon in the igneous rocks (estimated at 1.2×10^{18} metric tons of carbon). From 1 to 5% of this carbon was released to the atmosphere by venting of CO_2, CO, and COS through fumeroles, hot springs, and volcanic eruptions. Much of the volatile carbon released to the atmosphere has been dissolved in the oceans and precipitated in sedimentary limestone. Uplifting of Earth's crust has produced a recycling of some of this carbon in the form of CO_2 released to the atmosphere. Periods of unusual volcanic activity may have vented substantial amounts of CO_2, thereby causing a buildup of this gas in the atmosphere until equilibrium was reestablished with the oceans and sediments.

Prior to 1860 and the industrial age, an approximate equilibrium existed between sources and sinks for carbon, which included the atmosphere, soils, vegetation, animals, oceans, and sediments. Botanists Brown and Escomb in 1905 reported 91 measurements of atmospheric carbon dioxide concentrations taken at Kew, England between 1893 and 1901. Whether or not the air they sampled was contaminated by industrial pollution depended on the direction of the wind—whether it was away from London or toward London. Taking out the apparently contaminated records, Keeling (1978) reported the most probable CO_2 value in 1900 to be 290 ppm. This is not necessarily the preindustrial carbon content of the atmosphere because much industrialization, much coal and wood burning, and much forest destruction had already taken place by 1900. Stuiver (1978a, 1978b), after analyzing the carbon content of tree rings of trees growing during the eighteenth and nineteenth centuries, concludes that the preindustrial atmosphere contained a carbon dioxide concentration between about 268 ppm and 274 ppm by volume. The lack of adequate CO_2 measurements 125 years ago leaves considerable uncertainty as to the exact concentration of carbon dioxide in uncontaminated air prior to the industrialization of society. However, it is most probable that the carbon dioxide concentration was about 270 ppm, equivalent to 572 Gt of carbon in the whole atmosphere.

The recent record

Very carefully calibrated measurements of the carbon dioxide concentration of the atmosphere were begun at Mauna Loa, Hawaii in 1957 and simultaneously at the South Pole, Antarctica. The atmospheric concentration at that time was 315 ppm. Because in 1983, 25 years later, it was 340 ppm, there was an average annual increase of 1 ppm per year. The record of measurements taken at Mauna Loa is shown in Figure 3. It is clear from this record that for about one decade, from 1958 to 1968, the annual rate of increase was substantially less than during the period 1968 to 1981. Although the annual rate of increase of atmospheric CO_2 may change from year to year, there is absolutely no doubt about the fact that it is continuing to increase. The increase since 1900 has been about $(340-290)/$290 $\times$ 100 = 17% and since 1860 may have been as much as $(340-270)/270 \times 100 =$ 26%. Other measurement stations in the world have verified the amounts recorded since 1957 at Mauna Loa; they also show the steady upward trend in the concentration of atmospheric carbon dioxide year after year.

The atmospheric carbon dioxide record in Figure 3 shows an annual variation in the CO_2 concentration, with a minimum in late summer and a maximum in late winter. This is believed to result from photosynthesis by green plants and respiration by all organisms. During the growing season plants will absorb carbon dioxide from the atmosphere and assimilate it into carbohydrates. Respiration is the process by which all plants and animals derive chemical energy by oxidizing organic compounds and releasing carbon dioxide to the atmosphere. Respiration by organisms continues at all times of year, summer or winter, although it is much reduced during the cold of winter. Respiration rates always

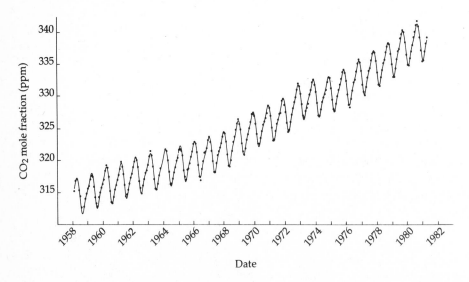

FIGURE 3. The atmospheric carbon dioxide concentration measured at Mauna Loa, Hawaii from 1958 to 1981. The annual amplitude is caused by the annual cycle of ecosystem assimilation and respiration—largely by vegetation in the northern hemisphere.

increase with increasing temperature. Photosynthesis by plants growing at high latitudes is vigorous during sunlit, warm summers, but is minimal or zero during cold winters of low light intensities. The net exchange of carbon dioxide between organisms and the atmosphere is the difference between the amount assimilated through photosynthesis and that released by respiration. During the summer growing season, more CO_2 is assimilated than is released by respiration. During winter, more CO_2 is respired into the atmosphere than is drawn out of the atmosphere by photosynthesis. Therefore, the annual cycle of the CO_2 concentration coincides with the seasonal influence of photosynthesis and respiration. A phase shift occurs between the timing of these events as they occur in the northern hemisphere, taken as a whole, and the arrival of this changing flux of CO_2 in the well-mixed atmosphere above Mauna Loa, Hawaii. Therefore, the minimum in the Mauna Loa record is in late summer or early autumn and the maximum is in late winter or early spring.

Sources of carbon

The carbon cycle and the reservoirs of global carbon are shown in Figures 8 and 9 in Chapter 1. The carbon reservoirs of the world are immense. Approximately 32,000 Gt of carbon are in ocean sediments, 12,000 Gt in fossil fuel reservoirs, and 1760 Gt in terrestrial plants and animals. The atmosphere in 1982 contained about 721 Gt of carbon at a concentration of carbon dioxide of 340 ppm. The exchange rates of carbon among these reservoirs are complex and not well understood.

When scientists fully realized what was happening to atmospheric carbon dioxide, they asked why. Studies of the global carbon cycle revealed the source of this carbon dioxide to be the burning of fossil fuels, the man-ufacture of cement, and the cutting of forests. Fossil fuel use, whereby carbon in oil, gas, and coal is oxidized by burning, is contributing about 5.3 Gt/yr and cement production contributes a small fraction of this, about 0.5 Gt/yr.

Figure 4 shows the annual CO_2 emissions to the atmosphere from fossil fuel burning and cement manufacturing expressed in amounts of carbon. There was a steady growth in the rate of release of carbon dioxide from 1950 until 1973 of about 4.58% per year, but the rate of growth then dropped by half to about 2.25% per year. In 1950 the rate of release of carbon to the atmosphere from fossil fuel burning and cement production was about 1.59 Gt/yr; but in 1960 it was about 2.57 Gt/yr, and in 1980 5.2 Gt/yr. The atmosphere seems to be retaining about 55% of the CO_2 being put into it, the remainder going into a sink, most likely the oceans.

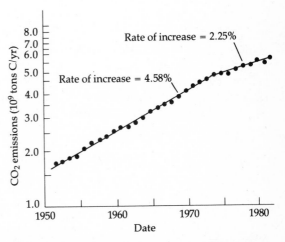

FIGURE 4. The annual worldwide emissions of carbon dioxide to the atmosphere from fossil fuel burning and cement manufacturing for the period 1950 through 1980. The slope of the line is the rate of increase in emissions. From 1950 to 1973, the rate of increase was 4.58%, and from 1973 to 1983, 2.25%.

The most controversial source of carbon is the net amount released by the destruction of forests. Stuiver (1978a) has suggested that during the last century the rate of CO_2 production from the destruction of forests greatly exceeded the rate of production from fossil fuel use. The forests were used for lumber and fuel, but many simply burned through carelessness. When forests were lumbered, there was much waste and only about 17% of the timber ended up as a useful product. When the wood left in the forest decayed or burned, CO_2 was released. But during this century the use of fossil fuel has increased rapidly and now exceeds the contribution by biomass to the atmospheric CO_2 concentration.

Woodwell et al. (1978) suggest that the destruction of the world's forests is releasing as much as 5 Gt/yr, whereas Bolin (1977) has estimated 1 Gt/yr. Others have suggested that the forests of the world are a net sink for carbon because many of them are in vigorous to moderate growth stages.

A growing successional forest stores carbon because the amount of standing biomass is increasing. But a mature, climax forest does not store carbon and most likely contributes carbon dioxide to the atmosphere through the death and decay of old trees. If there were a massive amount of tree planting through-out the world, there would be increased uptake of carbon dioxide and storage of carbon.

When crops are grown, carbon dioxide is assimilated and stored in the plant material, but when humans or cattle eat the crops, most of the carbon is released once again to the atmosphere as carbon dioxide. Humans also eat the cattle and respire any carbon stored there as well. On an average annual basis, this results in a net exchange of atmospheric carbon dioxide of zero because there is no permanent storage of carbon.

Future concentrations

We can project future rates of CO_2 released from human activities. If we assume that the burning of fossil fuels will be the dominant source of CO_2 buildup in the atmosphere and that about 50% of the CO_2 produced will remain in the atmosphere, then estimates of future fossil fuel use will give estimates of the atmospheric CO_2 levels.

Not all fossil fuels release the same amount of CO_2 per unit of energy consumed. Table 1 shows that natural gas releases the least amount of CO_2 per unit of energy of all the conventional fossil fuels. Oil releases 40% more CO_2 than natural gas, coal 70% more, and synfuels as much as 230% more. The resource quantities listed in Table 1 are middle

TABLE 1. Conventional fossil fuel resources for the world and amount of carbon released as CO_2 if burned

Fossil fuel	Resource (quads)	Gt of carbon per 100 quads	Ratio of CO_2 produced	Gt of carbon released
Gas	9,258	1.45	1.0	134
Oil	11,600	2.0	1.4	232
Coal	140,600	2.5	1.7	3515
U.S. shale oil[a]	1,148	4.8	3.3	55
Total	162,606			3936

[a]Synfuels from coal release 3.4 Gt/100 quads for a ratio of 2.3.

values taken from Tables 2 and 3 in Chapter 4. Sundquist and Miller (1980) estimate that the full utilization of the oil shale reserves of the Green River Formation in the western United States will release about 300 Gt of carbon (see Chapter 11 for a discussion of oil shale). According to the numbers in Table 1, if all the recoverable reserves of conventional fossil fuels in the world and the U.S. oil shale reserves were burned, the total carbon released would be 3936 Gt. If 50% of this were to remain in the atmosphere, the atmospheric content of carbon would increase by 1968 Gt. Because the current atmospheric concentration is 340 ppm, equivalent to 721 Gt of carbon, the total atmospheric carbon load would then be 2689 Gt of carbon, equivalent to 1268 ppm. This would mean that the atmospheric concentration of carbon dioxide would have increased by 3.7-fold from the current level and by 4.7-fold from the estimated preindustrial level of 270 ppm.

As we shall see in the next section, the climate change consequences resulting from a fourfold increase in the atmospheric carbon dioxide concentration would be so catastrophic as to make this result completely unacceptable. Even a doubling of either the preindustrial or current levels of CO_2 would have extremely serious climatic and ecological consequences.

Figure 5 presents projected atmospheric CO_2 concentrations for the next century; the projections are based on past trends and the current level. If global fossil fuel use remains at its present level (assuming zero growth rate), the CO_2 concentration will reach twice the preindustrial level, a level of 540 ppm, around the year 2175. If fossil fuel consumption increases with a 2% annual growth rate, a doubling will occur by 2055, and at a 4% annual growth rate, twice the preindustrial level will be reached by 2030. Several factors may mitigate such growth rates in fossil fuel expenditure, including fuel costs, economic

depressions, social conflicts, energy conservation, and increased use of nuclear reactors. However, annual growth rates of fossil fuel use have exceeded 4% during the decades since 1950, and it is not certain that rates under 2% can be achieved in the future. As we shall see later in this chapter, decisions to reduce the rate of growth of fossil fuel use must be made within the next few decades if tolerable levels of atmospheric CO_2 are to be achieved by 2055.

PAST TEMPERATURES

To project future climate conditions, it is important to understand the conditions of the past. If human activities are imposing factors for climate change on naturally occurring factors, it is essential that past relationships be-

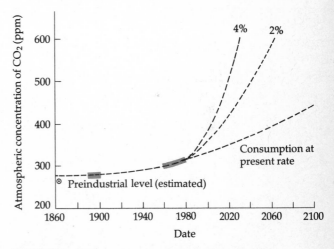

FIGURE 5. Projected atmospheric carbon dioxide concentrations for three possible scenarios: 0, 2, and 4% increase per year in the consumption of fossil fuels. These growth rates would lead to a doubling of the estimated preindustrial concentration of 270 ppm by the years 2175, 2055, and 2030, respectively. The heavy dark lines represent periods when real measurements were made of atmospheric CO_2. The black dot at the left is an estimate of the preindustrial value.

tween cause and effect be understood. If doubling of the atmospheric carbon dioxide concentration is expected to increase the mean global temperature by 3°C, it is crucial to understand whether this is a large temperature change or a small one. We can gain this perspective by considering the temperature changes of the past.

The last 140,000 years

The mean global temperature of the ocean as determined from oxygen isotope data and flora and fauna records in ocean sediments during the last 140,000 years is shown in Figure 6A. Global temperatures are higher at present than at any other time since the last interglacial period 130,000 years ago. Interglacial periods only last for about 10,000 to 15,000 years. The present interglacial period began about 10,000 years ago. Does this suggest that another ice age is about to begin?

The last 10,000 years

Records of air temperature in eastern Europe indicate the pattern shown in Figure 6B for the past 10,000 years. There have been periods somewhat warmer than the present, notably the HYPSITHERMAL PERIOD, also called the ALTITHERMAL PERIOD, around 4000 to 8000 years ago. This was followed by a very cold period about 3200 years ago and a warm period, the LITTLE OPTIMUM, from 2700 to 700 years ago, when the climates of Greenland and Iceland were ameliorated sufficiently for the Vikings to explore and settle those regions. The mean annual air temperature values in eastern Europe varied within ±1°C during the last 10,000 years except at the peak of the hypsithermal period, when it was about 1.2°C warmer than at the present time.

The last 1000 years

Midlatitude temperatures for the northern hemisphere during the last 1000 years are shown in Figure 6C. The mean annual air

temperature did not vary more than ±1°C. The LITTLE ICE AGE began about 650 years ago and lasted until the end of the nineteenth century. During the little ice age, there was great privation in Europe. One-third of the population of Finland died in the famine of 1697. Tens of thousands of people died in Europe in 1709 when crops were destroyed by cold and grain failed to ripen.

The last 100 years

Although we can question the accuracy of temperatures derived from floral and faunal records, or even from oxygen isotopes, and hence the precision of the records just described, during the last 100 years accurate temperature records have been kept at many stations in the world. Figure 7 shows five-year running means of the surface air temperature for three latitude bands and for the entire world (Hansen et al., 1981). Temperatures everywhere rose from 1883 to 1940. The global mean temperature rose more than 0.5°C during this period, the temperatures in the northern latitudes (23.6 to 90°N) nearly 0.8°C, and those in the low latitudes (23.6°N to 23.6°S) and southern latitudes (23.6 to 90°S) about 0.3°C. Then between 1940 and 1970, the air temperatures in northern latitudes (23.6 to 90°N) dropped about 0.5°C. A remarkable feature of Figure 7 is that the global temperature is almost as warm today as it was in 1940, even though it decreased by more than 0.2°C between 1940 and 1965. The often-mentioned idea that the world is cooling is based primarily on northern hemisphere observations through 1970.

CLIMATE CHANGE

Climate change factors

Interfere in any way with the stream of radiation from the sun to Earth's surface or with reradiation from the surface to outer space

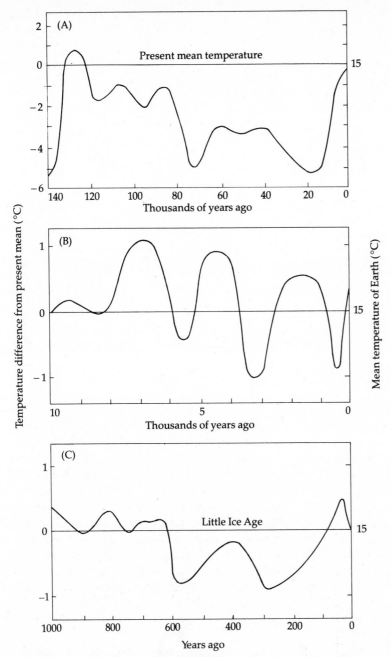

FIGURE 6. (A) The mean global temperature of the ocean during the last 140,000 years, determined from isotope data and flora and fauna records in the ocean sediments. (B) Air temperatures for eastern Europe during the past 10,000 years, determined from tree ring and other data. (C) Midlatitude air temperatures for the northern hemisphere during the past 1,000 years.

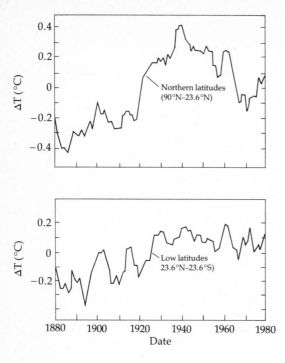

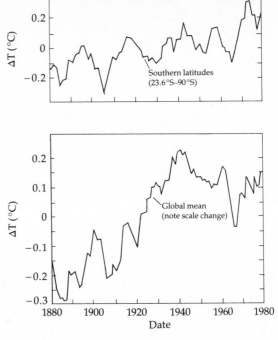

FIGURE 7. Five-year running means of the surface air temperature for three latitude bands and the entire world from 1880 to 1980. (Redrawn from Hansen et al., 1981.)

and the temperature of the atmosphere, surface, and oceans will change. Increase the dust load of the atmosphere and more incoming sunlight will be reflected to space and Earth will cool. Increase the atmospheric concentration of carbon dioxide, ozone, methane, water vapor, or other infrared-absorbing gases and less radiation will escape to space and Earth will warm. Vary the amount of radiation emitted by the sun up or down and Earth's temperature will change up or down.

Solar influence

Scientists have known that the sun exhibits considerable variability near its surface, where SOLAR STORMS and SUNSPOTS occur with some regularity. Sunspot cycles occur with periodicities of 11, 22, and 90 years. The quantity of radiation emitted by the sun is affected not only by the number of sunspots, but also by their particular size and structure. The ratio of the umbra to the penumbra areas (area of the central core to the area of the peripheral ring of the sunspot) is a measure of the radiance of the spot. When this ratio is measured for each sunspot and multiplied by the number of sunspots in each size class, the result correlates well with the surface air temperature of the Earth (Figure 8).

Volcanic influence

Major VOLCANIC ERUPTIONS eject fine dust into the stratosphere, where it may float around the world for several years. Stratospheric dust reflects sunlight and has a cooling effect on climate. The record of major

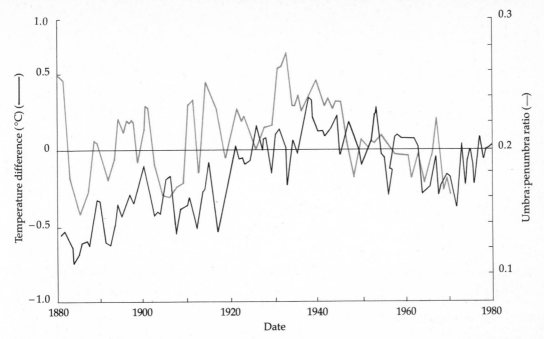

FIGURE 8. Comparison between a temperature change predicted from sunspot activity and the observed temperature change as annual means for the period from 1880 to 1980. The sunspot activity is measured by the number of sunspots of various size classes as determined by the ratio of the umbra area to the penumbra area for each sunspot. The dark line is the observed temperature change; the gray line is the umbra:penumbra ratio. (The umbra area is defined as the area of the core of the sunspot, and the penumbra area as the peripheral area.) The ratio of these areas changes with the size and intensity of the sunspot. The solar activity appears to account for some of the temperature change features.

volcanic eruptions during the last 100 years is shown in Figure 9. Many lesser eruptions are not shown. The period from 1915 through 1943 was a time of unusually low volcanic activity. Many of these eruptions are well known to most of us. Krakatoa in 1883, a major blast, produced red sunsets for the next several years. The dust from Mt. Agung in 1963 was discernible in optical measurements of atmospheric transmission above Mauna Loa, Hawaii until 1977 (Mendonca et al., 1978).

Figure 9 shows that nearly all of the eruptions were at times of cooler temperatures. However, it is also seen that the temperature cooling may have begun before the onset of most eruptions. This suggested to Rampino and Self (1979) that variations in climate lead to stress changes of Earth's crust, which may have initiated the eruptions. This hypothesis is as yet unproved, but it is a most interesting conjecture.

El Chichon, also known as El Chichonal, erupted in Mexico on 4 April 1982 and shot a massive amount of dust and sulfuric acid droplets into the stratosphere. This may have been the largest volcano to erupt since Krakatoa in 1883 and larger than either Agung in 1963 or Katmai in 1912. Modern instruments have permitted reasonably good measure-

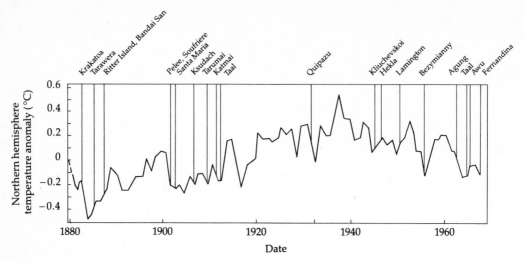

FIGURE 9. Major volcanic eruptions of the last 100 years in comparison with a change in the mean annual temperature of the northern hemisphere. It is interesting to note that cooling trends preceded as well as followed a major eruption. However, temperatures are always cooler at times of much volcanic activity and warmer at other times.

ments of the El Chichon cloud. The cloud traveled westward from southern Mexico in the flow of easterly winds that prevail at low latitudes. Scientists estimate that about 20 million metric tons of dust and 50 million metric tons of sulfur dioxide or sulfuric acid were injected into the stratosphere from El Chichon and went to a height of at least 32 km. The dust cloud has now encircled the globe, spread to higher latitudes, and crossed the equator. As with the other major volcanic eruptions, it is most probable that the dust cloud from El Chichon will reflect significant amounts of solar radiation and cool the northern hemisphere by a few tenths of a degree centigrade for a few years. This would appear as a cool transient in what otherwise seems to be a warming trend in the average annual air temperature.

Carbon dioxide influence

The mechanism is well established by which an increase of the atmospheric carbon dioxide concentration might affect Earth's climate, that is, the greenhouse effect. Knowledge concerning this mechanism indicates that an increase of the carbon dioxide concentration will produce a warming of Earth's surface and lower atmosphere.

Because the atmospheric carbon dioxide concentration has been increasing steadily from late in the last century to the present, we would expect its influence on climate to correlate with this trend. If its influence has been effective, then the mean temperature of the Earth must have been rising slowly but persistently. We would not expect the CO_2 trend to account for any of the short-term fluctuations in globally averaged climate, as we do expect from sunspots or from volcanic activities.

Climate versus weather

The climate of any region results from the weather that occurs there daily and seasonally. Climate is the average weather taken over a period of several years. The weather is the consequence of extremely complex interactions involving the sun, atmosphere, continents, and oceans, and their features (including mountain ranges, ice sheets, sea ice, snow, and vegetation) (Figure 10). The rotation of Earth affects the flow of air and is an important driving force of weather and climate.

Most of us are familiar with the difficulty of making weather predictions more than a few days in advance. Therefore, it seems like a contradiction to suggest that climate change might be predicted decades in advance. However, because the climate of any region is the average of the weather over long periods of time, all the short-term uncertainties and variations are averaged out. Thus, predic-

tions of climate are substantially more reliable than predictions of weather. Still, the uncertainties connected with the prediction of climate are considerable.

MODELING CLIMATE

The modern computer is an extremely useful tool for modeling the general circulation of the atmosphere and climate. These models are complex, costly, and very useful. Even though they contain many approximations to real world conditions, they are able to simulate quite well past climates and to suggest future trends.

A GENERALIZED CIRCULATION MODEL (GCM) is three-dimensional, has idealized

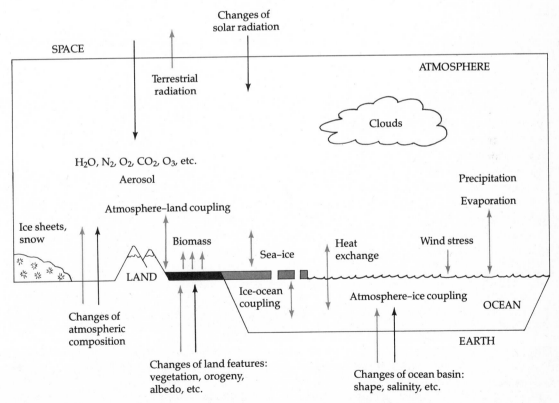

FIGURE 10. The components and processes affecting the global climate system. Black arrows represent external processes; gray arrows represent internal processes.

continents and oceans, has initial atmospheric temperatures and moisture distributions based on real observations, and contains inputs of radiation exchange, cloud cover, snow and ice cover, and atmospheric aerosols. Some GCMs include latitudinal and seasonal distributions of some of these factors.

Modeling past climates

Hansen et al. (1981) have constructed detailed numerical computerized models of global climate into which they have designed radiative and convective energy exchange involving clouds, aerosols, surface albedo, water vapor concentration, carbon dioxide concentration, and a 100-meter-mixed-layer ocean with thermal diffusion down to 1000 m. Their time-dependent model includes the inexorable increase of carbon dioxide concentration since 1880, a stratospheric dust veil resulting from highly varied volcanic activity, and irradiation changes from solar cycles. Their climate model regenerates the climate of the last 100 years and shows a good match between the predicted and the observed global air temperatures from 1880 to 1980 (Figure 11). This does not mean the model is completely correct, but it strongly suggests that CO_2 and volcanic aerosols, and, to a lesser extent, sunspots, are responsible for much of the global temperature variation during the last 100 years. Their model predicts a warming of 2.8°C for a doubling of the atmospheric carbon dioxide concentration. Hansen and his colleagues show not only that the increasing CO_2 concentration proba-

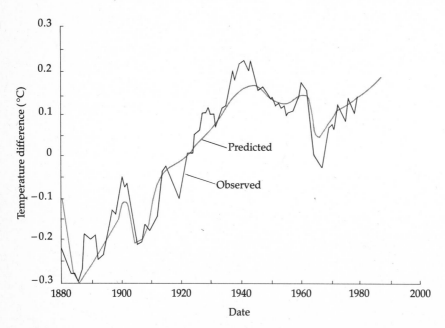

FIGURE 11. Predicted temperature change of the global climate (predicted from a combination of sunspot activity, volcanic dust, and an increase of the atmospheric carbon dioxide concentration) compared with observed temperature change as determined by five-year running means for the period 1880 to 1980.

bly has already been responsible for atmospheric warming, but also that a temperature signal produced by increasing atmospheric CO_2 levels will be clearly discernible from the climate noise during the 1980s and will be strongly discernible by the 1990s.

Modeling future climates

Manabe and Weatherald (1975, 1980) and Manabe and Stouffer (1980) have used a general circulation model and increasing refinements to compute the expected climate change resulting from increasing levels of atmospheric carbon dioxide concentration. Their model uses a realistic geographic distribution of continents and oceans for both hemispheres, and seasonal and latitudinal distributions of moisture, cloud cover, snow cover over the land, sea ice, and solar radiation intensity. Their model ocean contains a mixed layer 100 to 200 m deep.

Their model succeeds in reproducing the large-scale characteristics of the seasonal varia-

tion and of the geographic variation of observed atmospheric temperatures throughout the world. The model also gives the proper height distribution of air temperatures for most regions and seasons. Although their more recent modeling predicted a 1.9°C increase of the mean global air temperature for a doubling of the atmospheric CO_2 concentration, their earlier model estimated a 2.8°C increase. Because other climate models, including that of Hansen et al. (1981), predict an increase of about 2.8°C, we shall consider that this is the most probable temperature change to accompany a doubling of the atmospheric CO_2 concentration.

Latitudinal temperature change

The increase in the mean annual surface air temperature predicted by the general circulation model for various latitudes after a doubling or quadrupling of the atmospheric carbon dioxide concentration is shown in Figure 12. Also shown is the air temperature re-

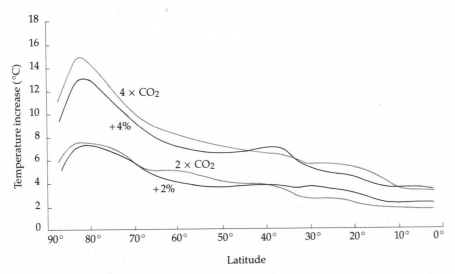

FIGURE 12. Predicted temperature increase as a function of latitude for a doubling and quadrupling of the atmospheric carbon dioxide concentration and for increases in the solar irradiance of 2% and 4%.

sponse to increasing the solar radiation intensity by 2 and 4%. The calculated air temperature changes are nearly identical as a result of increased intensity of solar radiation or of increased concentration of carbon dioxide. The low latitudes will experience an air temperature increase of less than 1°C whereas the high latitudes will have an air temperature increase of up to 7°C for a doubling of the carbon dioxide concentration.

Seasonal temperature change

Figure 13 shows the expected air temperature increases with time of year and latitude for a doubling of the CO_2 concentration. The seasonal air temperature changes are expected to be very small at low latitudes, but very large at high latitudes. The greatest air temperature increases are expected to occur in the northern hemisphere winter at high latitudes as a result of a substantial reduction in the amount of sea ice and of snow cover on the land because of the warming. The loss of ice and snow produces a big change in the surface albedo (reflection of sunlight)—ice or snow being highly reflective and water or soil being quite nonreflective. Although this produces a net increase in the absorbed sunlight during the summer, it turns out that this ad-

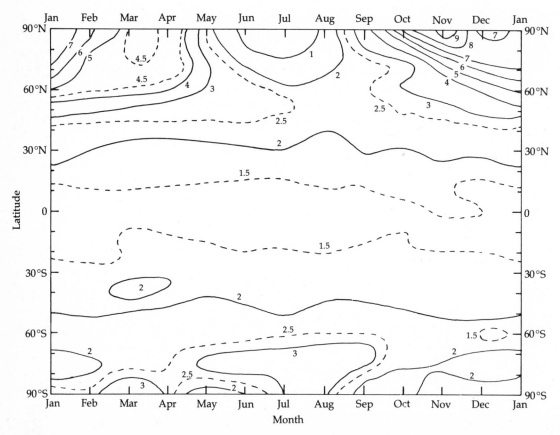

FIGURE 13. Expected air temperature increases as a function of latitude and time of year for a doubling of the atmospheric carbon dioxide concentration.

ditional energy goes into the melting of ice and snow or into the warming of the ocean. Hence, the increase of the arctic surface air temperature is relatively small in summer. However, the additional heat absorbed by the ocean during the summer delays the formation of sea ice during the autumn and that in turn reduces the thermal insulation of the sea surface during the early winter. Normal, sea ice acts as a thermal blanket between the water and the air. The lack of sea ice allows the heat stored in the ocean to warm the air above it and thereby to increase the winter air temperature.

For a doubling of the atmospheric carbon dioxide concentration, the mean arctic air temperature may increase as much as 7° to 9°C during October through January. This expected increase in air temperature will have enormous impact on the ecosystems of arctic regions. In fact, the early winter temperature increase for regions north of 60° latitude would be more than 6°C. Such regions would include all of Greenland, Scandinavia, northern Russia, northern Canada, and Alaska. There is a secondary center of relatively large warming around 65°N latitude in April. This results from a large reduction of the surface albedo in the spring when the intensity of sunlight is strong and a rapid melting of snow and ice occurs. As the highly reflective snow cover disappears from the surface, the underlying dark soil and vegetation absorbs the sunlight, the surface becomes warmer, and the air temperature increases.

The expected temperature increase in the antarctic regions is very much less than that expected in the arctic regions after a doubling of the atmospheric carbon dioxide concentration. Despite the expected smaller temperature increases, the impact of warmer temperatures on the West Antarctic ice sheet could be significant. A change of the ocean temperature in this region may produce a rapid melting of the West Antarctic ice sheet and a subsequent, serious rise of sea levels throughout the world.

Hydrologic cycle changes

The general circulation model of climate predicts that the latitudinal distribution of the water balance of Earth's surface is significantly altered with a quadrupling of the CO_2 concentration of the atmosphere. The poleward transport of moisture increases markedly in response to the warming resulting from an increase of the CO_2 concentration. The model shows that the $4 \times CO_2$ atmosphere would receive more moisture from the Earth's surface in the form of evaporated water at low latitudes and would return more moisture to the surface as precipitation in high latitudes than does the present CO_2 atmosphere. The moisture gained by the model atmosphere at low latitudes comes primarily from the tropical oceans whereas the increased, high-latitude precipitation occurs over both continents and oceans. A belt of enhanced aridity (i.e., a decrease in the difference between precipitation and evaporation) occurs around latitude 45° in both the northern and southern hemispheres.

The model calculations show the snowmelt season to end earlier in the spring for the high CO_2 concentration scenario than for the current CO_2 concentration scenario. The warm-season, soil-moisture depletion characteristic of moist midlatitude regions begins earlier because of increased evaporation rates and less snow accumulation resulting from warmer winter temperatures. The middle latitude rainbelt, where rainfall occurs during the transition period from winter to summer, shifts poleward in the high-CO_2 scenario, a result of the penetration of warm, moist air into higher latitudes. In middle latitudes, the spring maximum rainfall begins earlier and the spring to summer reduction of rainfall begins earlier and leads to a drier summer. These shifts in the precipitation patterns will be particularly serious for the great grain belts of the world. Figure 14 shows the expected precipitation changes throughout the world for a doubling of the atmospheric CO_2 concentration. Decreased precipitation is expected over central United States, eastern Europe, and central parts of the Soviet Union, whereas western Canada and Alaska will be wetter, as also will much of North Africa.

Verification of modeling projections

If the GCMs are projecting a warmer world resulting from increasing atmospheric carbon dioxide concentration and concurrent changes in precipitation patterns, it would be useful to see what the global climate has done in the past. Of particular interest are the latitudinal and regional distributions of the changes of temperature and precipitation. Do the climate events of the past resemble those projected for the future? Verification would lend credibility to the results of the model and give us confidence that the projections are reasonable.

Warmest versus coldest years

Wigley et al. (1980) analyzed the recorded temperatures and precipitation patterns for the world during the five warmest years in the period 1925 through 1974 and compared these with the five coldest years during the same period. Their results are shown in Figures 15 and 16. Indeed, high latitude regions of the northern hemisphere did experience the greatest warming. The maximum air temperatures were over northern Asia, Canada, and eastern Alaska. They also found that the maximum air temperature increase occurred in the winter and in the latitude band 50 to 70°N—definitely confirming the projections made by using the GCMs.

The warmest years were 1937, 1938, 1943, 1944, and 1953, and the coldest years were

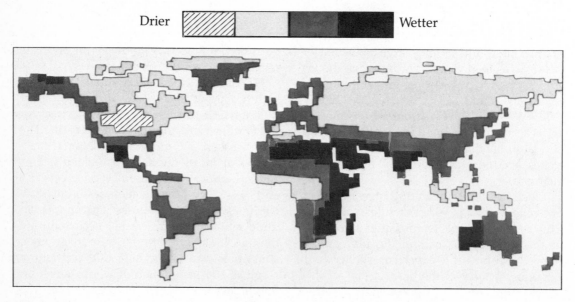

FIGURE 14. Expected precipitation changes over the world for a doubling of the atmospheric carbon dioxide concentration.

1964, 1965, 1966, 1968, and 1972. The average annual air temperature difference between the five warmest years and the five coldest years was 1.6°C for the high latitudes, but only 0.6°C for the northern hemisphere as a whole. High latitude winters were 1.8°C warmer during the warmest years when compared with the coldest years. The region that extends from Finland across the northernmost parts of Russia and Siberia had five times the hemispheric mean air temperature increase. A large part of North America, including Alaska, had positive temperature changes of two or more times the hemispheric mean. There were also regions with negative changes, including much of India, Turkey, North Africa, Spain, a small region in central Asia, the southwestern coast of the United States, Mexico, and southwestern Greenland.

Further examination of the climate between the cold and warm years showed the warm years to have intensified high-latitude westerlies, greater storm (cyclonic) activity in the arctic and subarctic regions of the eastern hemisphere, and westward displacement of the Siberian High in the winter.

Numerical modeling of the global climate showed that worldwide precipitation may increase about 5% after a doubling of atmospheric CO_2. A comparison of the five warmest years with the five coldest years discussed in the previous paragraphs showed an increase of precipitation between 1 and 2% for the northern hemisphere; a change that is sta-

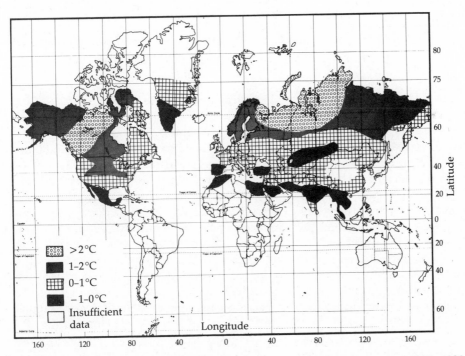

FIGURE 15. Temperature differences between mean annual surface temperature for the five coldest years of the period 1925 through 1974. The average temperature difference is 0.6°C. The warmest years were 1937, 1938, 1943, 1944, and 1953. The coldest years were 1964, 1965, 1966, 1968, and 1972. (After Wigley et al., 1980.)

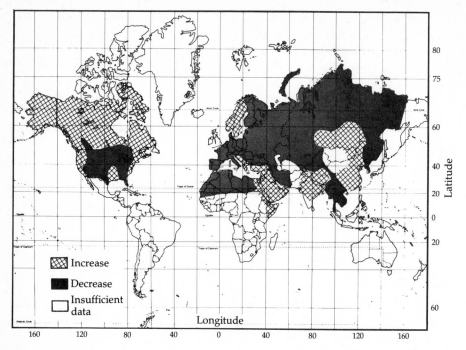

FIGURE 16. Precipitation differences between mean annual precipitation for the five warmest years and that for the five coldest years of the period 1925 through 1974. (After Wigley et al., 1980.)

tistically insignificant. The most important results are decreases in precipitation over much of the United States, most of Europe, Russia, and Japan. A decrease also occurred over western North Africa. Increases of precipitation are found for India and for the Middle East and for the eastern part of North Africa. The climate model for the future predicts the annual precipitation amounts to change for many regions of the world and significant shifts in the seasonal patterns of precipitation to occur. Reductions in precipitation are expected during the autumn and winter, and some regions, such as eastern Europe, are to be drier during the summer as well. Because air temperatures are projected to increase and precipitation to decrease in some regions of the world, there will be a double effect on soil moisture deficits, and the impact on agricultural production could be considerable.

The altithermal period

Between 4500 and 8000 years ago, the world experienced a warm climate known as the AL-TITHERMAL OF HYPSITHERMAL period. To deduce whether a region was wetter or drier than it is at the present time, various kinds of evidence were gathered by scientists. Part of the evidence comes from an identification of the species of plants growing at the time. This identification can be made by studying the types of pollen and spores found in old lake sediments and in peat bog sediments. The species profile reflects the influence of rainfall and soil moisture on vegetation during the growing season. Another source of information on average rainfall conditions is the record of lake levels and stream flows. Kellogg and Schware (1981) gathered this kind of information; and from it they were able to reconstruct the moisture characteristics of the warmer world at the time of the altithermal period. Their results are shown in Figure 17.

Wetter regions of the world during the altithermal period appear to have been Alaska, Mexico, central South America, all of Africa except West Africa, Europe, the Middle East, India, Southeast Asia, and Australia. Drier regions appeared to have been central and northeastern United States, all of Canada, northern and southern South America, most of the Soviet Union, China, and Antarctica. There is a general similarity between the inferred altithermal period precipitation pattern and that projected for a warmer world resulting from an increasing atmospheric carbon dioxide concentration.

GLACIERS, ICE, AND SEA LEVEL

The West Antarctic ice sheet

If the entire West Antarctic ice sheet melted, sea levels would rise about 5 m; and if all of the world's glacial ice melted, the ocean would rise about 60 m. Normally ice sheets melt slowly, but the West Antarctic ice sheet could melt quite rapidly. This ice sheet is far more responsive to a temperature increase than is the Greenland ice sheet because it is grounded under the ocean and extends out to sea (Figure 18). Because water is a good conductor, it would transfer heat to the glacier more rapidly than would air. Not only would the glacier melt more rapidly, but large masses of ice would break loose and drift northward as icebergs, where they would then melt quickly in the warmer waters. There is evidence that is exactly what happened during the previous interglacial period, the Eemian, about 130,000 years ago when sea levels were 6 m higher than they are today.

Kukla and Gavin (1981) report that the extent of the pack ice around Antarctica during summers has decreased in the 1970s com-

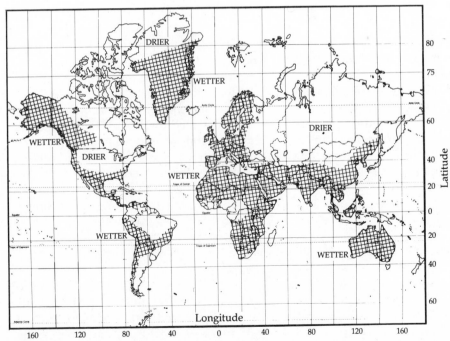

FIGURE 17. Wetness during the altithermal period (4000-8000 years ago) compared with the present. (Redrawn from Kellogg and Schware, 1981.)

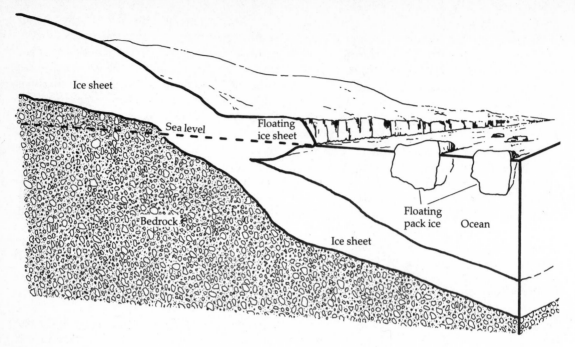

FIGURE 18. Schematic drawing of the West Antarctic ice sheet, showing how it is grounded underneath the ocean.

pared with the amount of pack ice present during the 1930s. The antarctic summer pack ice decreased by about 35% (2.5×10^6 km²) between 1973 and 1980. Satellite photographs made this analysis possible. Prior to the use of satellites, various ships made reports of the extent of the pack ice during the summer. Analysis of all available data showed that summer ice conditions were consistently heavier earlier in this century.

Arctic Sea ice and snow

Kukla and Gavin (1981) reported that although there had been no shrinking of the arctic ice pack, the mean surface air temperatures along selected northern latitudes between 55° and 80°N during spring and summer were higher by 0.9°C in the period 1974 to 1978 than in the period 1934 to 1938.

These positive temperature changes appear to be associated with a shifting belt of melting snow, although the pattern is not completely consistent. Ramanathan et al. (1979), using an energy balance numerical global climate model, predicted that the air temperature increase resulting from an increasing concentration of atmospheric carbon dioxide should be strongest along the edge of the seasonally shrinking sea ice and along the continental snow line during late spring and summer. This prediction by Ramanathan and the analysis of Kukla and Gavin appear to agree.

Sea level

Naturally, all coastal dwelling people should be much concerned with whether or not sea levels are rising or falling. If the sea level rose by 5 to 8 m, more than 11 million people in the United States alone would be affected, mainly in Florida, Louisiana, Texas, and elsewhere along the Gulf Coast. Major coastal cities such as New York, London, Leningrad, Amsterdam, Washington, D.C., Stockholm, Venice, Hamburg, Calcutta, Bangkok, Singapore, and Jakarta would be partially flooded. Higher sea levels will mean coastal flooding and increased storm erosion.

Tide gauge records of sea levels are maintained at more than 700 stations throughout the world today, but of these only about 193 stations have records of more than 20 years duration. Sea level positions prior to the times when gauges were used are determined from shoreline indictators in the geological record, such as the deposits of mollusks, corals, and brackish-water peat deposits.

Gornitz et al. (1982) reported a detailed analysis of past and recent sea level changes as inferred by these records. They find that for most of the past 5000 years sea level has been nearly constant, but that the mean sea level rose by 12 cm during the past century, or about 1.0 mm/year. They found that sea level rose in the past century in every geographic region of the world. Most of this sea level rise can be accounted for by thermal expansion of the upper ocean, resulting from the increase in global temperature during the last century. However, there is a residual sea level rise that cannot be accounted for by thermal expansion. It is entirely possible that a small net melting of ice sheets has been responsible for the more rapid rise of sea level during the period 1880 to 1980.

Sea level is now at its highest level since the previous interglacial period 130,000 years ago. There is some indication that global mean sea level has been rising more rapidly during the last 40 years than during the previous 60 years. Etkins and Epstein (1982) discuss this rate of rise and suggest that it results from a combination of thermal expansion of the ocean and the discharge of polar ice sheets. They claim that there has been a melting of more than 50,000 km^3 of ice over the last 40 years and that this has kept ocean temperatures down, slightly reduced thermal expansion, and contributed water to the ocean.

FOSSIL FUEL FUTURES

The amount of carbon dioxide released to the atmosphere, the buildup of the atmospheric concentration, and the ensuing climate change are dependent upon the rate of burning of fossil fuels. The world used in 1980 about 250 quads of fossil fuel energy.

Limiting the carbon dioxide buildup

Figure 19 shows three production curves for the global use of fossil fuels; these levels of use would lead to a buildup of atmospheric carbon dioxide of 1.5, 2.0, and 3.0 times the early-industrial level (assumed to be 290 ppm), representing increases of 50%, 100%, and 200%, respectively. If the atmospheric concentration of carbon dioxide is not to exceed 50% of the early-industrial level, global fossil fuel use will have to peak by the year 2007 at a level only 15% above the current rate of fossil fuel use. Furthermore, the annual growth rate of fossil fuel consumption must slowly decline over the next 30 years from 1% in 1980 to 0.7% in 1990, to 0% by the year 2007. Then fossil fuel use must continue to decline for the next 100 years, down to about 163 quads/yr.

If the accumulation of atmospheric carbon dioxide is to be limited to twice the early-industrial level, fossil fuel use will have to peak in about the year 2042, at a consumption level of 420 quads/yr—about two-thirds greater than the current level. In this scenario, the annual growth rate of about 1.3% in 1980 must decline to 1.1% in 2000 and to

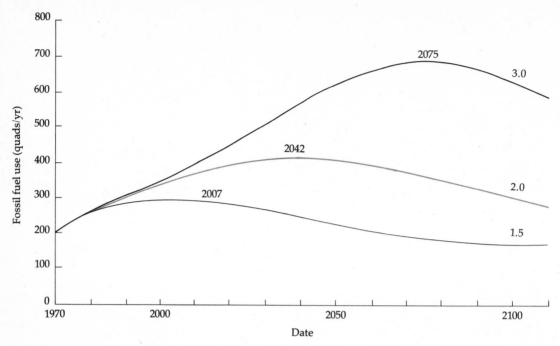

FIGURE 19. Three scenarios for the global use of fossil fuel that will yield a buildup of atmospheric CO_2 equal to 1.5, 2.0, and 3.0 times the preindus- trial CO_2 level. Year of peak use is indicated for each scenario. (Council on Environmental Quality, 1981.)

0% in 2042. Eventually, fossil fuel use must decline to a limiting value of 220 quads/yr, about 88% of current use.

If one allows the buildup of atmospheric carbon dioxide concentration to ultimately reach three times the early-industrial level, then fossil fuel use will increase to 690 quads/yr—2.75 times greater than its 1980 value— peaking in about the year 2075. In this scenario, the annual growth rate of fossil fuel use must decline from 1.4% in 1980 to 1.0% in 2040, to 0% in 2075, and ultimately approach a limiting value of about 330 quads/yr—about one-third higher than today's level.

It is important to realize that there is no single, unique production curve for fossil fuel burning leading to a fixed long-term atmospheric CO_2 concentration. Many energy curves could lead to the same ultimate amount of CO_2 in the atmosphere. For example, if we do not want the atmospheric carbon dioxide concentration to exceed the early-industrial level by more than 100%, three possible scenarios for achieving this goal are shown in Figure 20. The bottom line represents an immediate but gradual response. The middle line assumes a fossil fuel use growing at 2.5% per year from 1980 to 1990 and a reduced rate of use after that. The topmost line assumes an initial growth rate of 4% per year until 1990, followed by a reduced rate. These curves clearly show that the longer we wait to take action, the greater will be the CO_2 buildup and the more drastic must be the later response to reduce fossil fuel use to limit the total buildup of atmo-

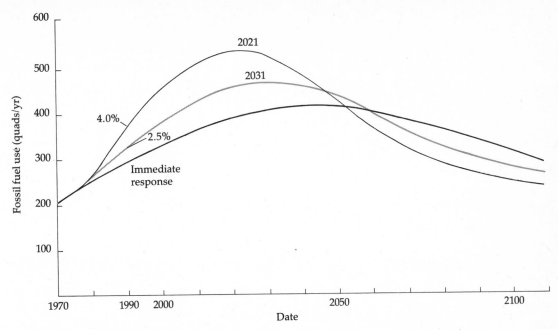

FIGURE 20. Three possible production curve scenarios leading to a 100% increase (doubling) in atmospheric CO_2 levels. The bottom curve represents an immediate response, with smaller, gradual cutback in the rate of growth of fossil fuel use; the middle curve represents a 2.5% growth in fossil fuel use from 1980 to 1990; the top curve depicts a 4% growth from 1980 to 1990. (Council on Environmental Quality, 1981.)

spheric carbon dioxide. A continuation of relatively high fossil fuel consumption growth rates for the decade 1980 to 1990 will make the transition away from fossil fuels more difficult later.

World energy demand projections

Consider the two scenarios of total world energy demand shown in Figure 21. The first (Projection A) is a high global energy demand for a world whose population levels off at 10 billion people by the year 2100 at an average per capita energy use equal to two-thirds the present U.S. consumption. A lower world energy demand (Projection B) has the world population leveling off at 8.5 billion people by 2100 at an average per capita energy use of one-third the present U.S. level. The current average per capita world energy use is about one-fifth the U.S. level.

Table 2 shows the projected total world energy demands for these two scenarios. World total energy demand today is about 276 quads. The high-demand scenario implies an energy increase of 836% by the year 2100, and about one-fourth of that is caused by the growth of population. The low-growth scenario implies an increase of 372%, and about one-half of that is due to the expected population increase. These two scenarios do not represent upper and lower limits to possible energy growth. New technologies at some time in the future might drastically change the perspec-

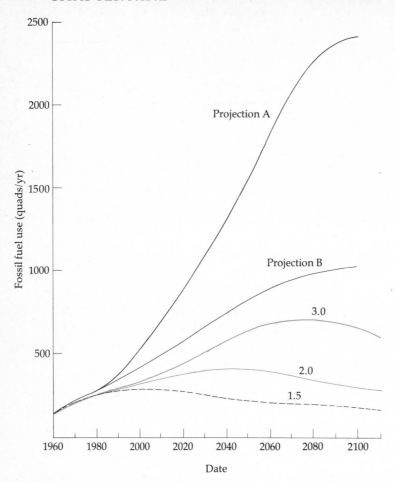

FIGURE 21. High and low projections for total world energy use and three alternative scenarios for global fossil fuel use leading to CO_2 levels that are 1.5, 2.0, and 3.0 times the preindustrial level. Projection A represents a total energy demand for an ultimate world of 10 billion people at an average per capita use of two-thirds of the present U.S. population. Projection B represents a total energy demand for an ultimate world of 8.5 billion people at an average per capita use of one-third of the present U.S. population. Projections A and B do not represent upper or lower limits possible. (Council on Environmental Quality, 1981.)

tive, particularly for low energy growth consistent with economic growth in developing countries.

Nonfossil fuel needs

We can ask how much nonfossil fuel must be used to keep the atmospheric carbon dioxide concentration from not exceeding the early-industrial level by 1.5-, 2.0-, and 3.0-fold. Estimates of the nonfossil fuel projections are given in Table 2.

To avoid exceeding a 50% increase of CO_2 and to meet the low energy demand scenario, nonfossil fuel sources would be required to increase from 26 quads at the

TABLE 2. Nonfossil fuel projections

Year	CEQ projected[a] (total energy quads)		1.5		2.0		3.0	
			Quads[b]	AAGR[c]	Quads[b]	AAGR[a]	Quads[b]	AAGR[c]
2000	High	528	235	12	200	11	184	10
	Low	417	135	9	100	7	84	6
2020	High	868	590	5	485	5	420	4
	Low	568	290	4	185	3	120	2
2050	High	1522	1282	3	1098	3	886	3
	Low	818	582	2	398	3	186	1
2100	High	2382	2214	<1	2084	<1	1758	<1
	Low	1026	844	<1	710	<1	388	1

[a]Total world energy demand. In 1980, the total world energy demand was 276 quads, of which 26 quads was supplied by nonfossil fuels.
[b]Required nonfossil fuel supply, assuming CO_2 ceilings of 1.5, 2.0, and 3.0 times the preindustrial level.
[c]AAGR, Average annual growth rate (%).

present time to 135 quads by the year 2000. To meet the high energy demand scenario, 235 quads would be needed in the year 2000. These are large numbers when we consider that currently the total world energy use is about 276 quads, of which about 250 quads is fossil fuel use. The numbers become even more staggering when carried out to the years 2020, 2050, or 2100. To limit the CO_2 increase to 50% under the low energy demand future, the nonfossil fuel growth rate would need to average 9% per year between 1980 and 2000. For the following 20 years the nonfossil fuel growth rate would have to average about 4% per year. The figures for the high demand scenario are even more sobering. The growth of nonfossil fuel must be 12% for the years 1980 to 2000 and 15% for the period 2000 to 2020. For comparison, the average annual increase in the U.S. electrical generating capacity was 7.3% per year between 1960 and 1970. Today it is much lower.

The requirement for nonfossil energy is somewhat less if we allow the CO_2 concentration to exceed the early-industrial level by 100 or 200%. However, the quantities of nonfossil fuels needed and the annual growth rates are still enormous, as seen in Table 2.

The principal nonfossil energy sources available are solar, geothermal, and nuclear. It is not at all clear how well solar energy use will develop, particularly since there are practical limitations to the use of this dispersed source of energy. Geothermal energy has limitations of geographic location, and nuclear power has not yet become socially acceptable in all parts of the world. It is likely that only about 30 to 50 quads per year of primary energy will be provided by nuclear power by the year 2000.

Hydroelectricity presently makes a significant worldwide contribution to energy supply, and by the year 2000 it could supply 35 to 45 quads. The ultimate total amount of hydroelectric energy potentially developable is estimated to be 100 quads per year. Together, hydroelectric energy and nuclear power might supply between 65 and 95 quads per year by the year 2000. However, between 135 and 235 quads per year of nonfossil fuels are needed to limit the CO_2 increase to 50% of the early-industrial level, between 100 and 200 quads per year at a 100% increase, and between 84 and 184 quads per year at a 200% increase. The use of solar energy will need to make up the differences, a requirement that becomes particularly difficult to meet if hydroelectric and nuclear energy do not develop to the maximum amount estimated.

The prospect of nonfossil fuels displacing sufficient amounts of fossil fuels to limit the increase of atmospheric carbon dioxide concentrations to tolerable levels is not good. The issue is of global proportions and therefore international in scope, but the prospect for an appropriate decision by all nations of the world is small compared with the likelihood of an appropriate decision by one or a few. It is always possible that the growth of energy demand per capita will be much less than anticipated, largely because of degrading ecological conditions throughout the world. It is even possible that the horrendous rates of population growth that we witness today will not continue as long as we think, again largely because of poor ecological conditions, such as loss of soil fertility, poor crop production, poisoning of crops and lakes by acid deposition, and depletion of ocean fish stocks, as well as the result of poor economies.

We do not wish for poor ecological, environmental, and economic conditions as a means to limit the growth of the world population and the commensurate energy demand. However, if the people of the world cannot voluntarily decide to create their own conscious limits to population growth, then the deleterious factors may dominate. If neither happens, and other means for limiting the output of carbon dioxide into the atmosphere do not occur, then we can expect a substantial warming of the Earth during the next century—a warming that will be clearly discernable within the lifetimes of the majority of people alive today.

ECOLOGICAL CONSEQUENCES TODAY

A warmer world, wetter in some regions and drier in others, will cause ecological disloca-

tions that are beyond anything modern society has ever experienced. In fact, we must go back thousands, or even tens of thousands, of years to find a change of such magnitude. Ecotones (the boundaries of major ecosystems) will shift poleward by hundreds of kilometers during the next 200 years. Steppes will turn to forest or desert, forests to desert or bog, pine forests to deciduous forests, small rivers to major rivers, and some major rivers to a trickle. Because humans have so constrained the refugia of many endangered species and put hostile habitats in their paths of movement, it is unlikely that many will survive the climate change. Our knowledge concerning expected responses by flora and fauna are very limited. Some of what we know follows.

Plant response to CO_2

All plants are carbon dioxide-limited for photosynthesis and growth. Plants will respond favorably to increased concentrations of atmospheric carbon dioxide by increasing their rates of photosynthesis and by improving their water use efficiencies whereby they will fix a greater number of carbon atoms for the number of water molecules transpired. Not all plants will respond to increased CO_2 levels in the same way, and, as a general rule, if CO_2 concentration doubles, photosynthetic rates will only increase from 20 to 50%, depending on the species and the availability of water or nutrients. Plant growth sometimes follows photosynthetic response directly and in other cases seems to have quite an indirect relationship to photosynthesis. With increasing carbon dioxide levels in the atmosphere, it is expected that the biosphere will respond positively. The biosphere will increase in total rate of carbon assimilation and of carbon storage. All agricultural crops will respond favorably to increasing atmospheric concentrations of carbon dioxide, and these responses will be affected by the benefits or difficulties caused by increased temperatures or by shifts in precipitation patterns.

C₃, C₄, and CAM plants. Plant scientists studying the ability of plants to assimilate carbon dioxide have identified three general types of plants on the basis of the biochemical pathway by which the carbon dioxide is converted to carbohydrates. These pathways are referred to as C_3, C_4, or CAM photosynthetic pathways and the plant groups are similarly named. In C_3 plants the first product of assimilation is a three-carbon acid, in C_4 plants, it is a four-carbon acid. Although this difference sounds rather simple, these two groups of plants have very different characteristics. The C_4 plants are known as high efficiency plants because their photosynthetic capacities are often two or three times greater than those of C_3 plants. C_4 plants are better adapted to dry, hot habitats than are C_3 plants. C_4 plant species use only about half as much water as C_3 species for the production of an equal amount of plant dry matter. Typical C_4 plants are pigweed and Bermuda grass and the tropical grasses, corn, sorghum, millet, and sugarcane. Most other well-known agronomic plants (such as spinach, tobacco, wheat, rice, and beans) and all trees and shrubs of temperate regions are C_3 plants.

CAM plants have a photosynthetic pathway different from those of C_3 and C_4 plants. They can open their stomates at night, take in CO_2 from the atmosphere, store it in chemical compounds, and assimilate it in light the next day. CAM plants include many desert succulents, such as cacti and agaves. CAM plants are even more efficient with respect to water use than either C_3 or C_4 plants, and therefore they are well adapted to desert conditions.

Under increased levels of atmospheric carbon dioxide concentrations, it is expected that C_3 plants will increase their photosynthetic rates the most, C_4 plants less so, and CAM plants the very least. If conditions become warmer and drier, CAM and C_4 plants will be most favored and C_3 plants least favored.

Past research, mainly with plants in greenhouses or growth chambers, suggests that with a doubling of the atmospheric carbon dioxide concentration, one or more of the following may occur: (1) an increase in average net photosynthesis; (2) a change of leaf area and leaf structure; (3) a change of shape in the plant canopy; (4) a change of the pattern of photosynthate allocation; (5) an increase in water use efficiency; (6) an increase of the plants' tolerance to toxic atmospheric gases; (7) a change of root to shoot ratios; (8) a change of flowering dates and an increase in the number of flowers per individual plant; (9) an increase in the number and size of fruits and in the number of seeds produced per plant; (10) an effect on germination of some species; and (11) an increase in the rate of nitrogen fixation.

All of the preceding changes have been observed in various plants; however, there have been contrary responses found for some plants for some of the items. Some genetic studies show that at least some of the responses listed are heritable. Plant responses differ enormously from species to species, and within a given species they depend on the growth conditions. No studies have been done on the effects on plants of long-term CO_2 increases. It is quite possible that plants may make morphological changes over extended periods of time that will reduce or mitigate the effects expected (see Lemon, 1983 for further information).

Animal response

Direct responses of animals to increasing CO_2 levels are likely to be undetectable, except if concentrations exceed 1000 ppm. Only at very high concentrations does CO_2 become slightly toxic to animals. Animals will, of course, respond to increased plant biomass by having a greater food supply available, and therefore, they are expected to increase in numbers.

With a general warming of Earth, there will be a shift of animal populations toward higher latitudes. Abundant species may accommodate to the shifting climate and dominate whereas endangered species may become extinct, being less able to accept additional stress or competition. The habitats of polar animals such as the musk ox (*Ovibos moschatus*), caribou (*Rangifer caribou* and *R. arcticus*), and reindeer (*R. tarandus*) may be so altered as to greatly disrupt their populations. Very warm summer temperatures in the arctic could be fatal to the populations of these animals. They cannot move further northward; nor would it help, even if they could because all polar regions would become warmer.

Agricultural production

The climate changes anticipated in a warmer world are quite different, region by region. Low latitudes will have the least amount of warming and high latitudes the most. Precipitation changes, as we noted earlier, will be very different from region to region. It is anticipated that the central United States will get warmer and drier. This change may reduce corn yields or shift the production from corn to wheat in some areas. Cooper (1978) suggests that U.S. corn production will decrease about 11% for each 1°C increase in average temperature during the growing season. This production loss would be further exacerbated by dryness. It is also estimated that for each 1°C temperature increase, the corn belt would shift approximately 175 km to the northeast in the United States. Soybeans grown in the Midwest would also be adversely affected. Cotton yields will increase with temperature and with increased CO_2 levels. Cotton does well under dry conditions, and increased dryness in parts of the southern United States may favor it. Wheat yields will decrease with increasing temperature, but increasing levels of CO_2 may mitigate this and stabilize yields.

An increase in the mean annual temperature will also extend the growing season for most crops, and this could be beneficial. However, it will also extend the time when insect pests can breed, and may make it possible for some species of insects to have another generation during the year. An important ecological law is that numbers of insects increase exponentially with the number of generations per season. Increased moisture in some regions may increase the number of pests and pathogens, and increased dryness may reduce the numbers.

Plant geneticists assure us that they will be able to have on hand the necessary genetic stocks of hybrid corn, wheat, soybean and other crop species that will be productive under warmer and even under drier conditions. Not everyone accepts this confident view, and some experts believe we may find ourselves seriously short of the necessary plant varieties for continued high productivity under changing climate conditions.

A warmer climate would shift agriculture further north in Canada and extend the growing season for this part of the world. Where soils are adequate, this could permit more crop production. The main problem is the fact that much of eastern Canada is covered by the Laurentian Shield of granitic rock (see Chapter 10), which is overlaid with very thin and generally poor soils. In the Soviet Union, it is probable that a northward shift of crop production would give increased crop yields, particularly in the alluvial soils of the Yakutian plain along the Lena River. Cooper (1978) points out that much of the northern Soviet Union has heavily leached soils that are more impoverished than the fertile soils of the central Soviet grain belt. A drier climate is expected over much of the Soviet Union, and this would be deleterious to crop production.

Forests

Forest trees will grow better at enhanced levels of atmospheric carbon dioxide concentration. Generally an increase of CO_2 will produce less than a proportional response in photosynthesis. The response of one species may be very different from the response of another. A doubling of the current CO_2 level (from 340 to 680 ppm) will nearly saturate the photosynthetic response of sugar maple (*Acer saccharum*), whereas bigtooth aspen (*Populus deltoides*) will not saturate until nearly five times the present CO_2 level. As atmospheric CO_2 concentration increases, the aspen may grow substantially better than the sugar maple, but because the sugar maple can grow in low light and the aspen cannot, they may continue to fill the same niche in the forest ecosystem as they do today.

Forest ecosystems will shift toward higher latitudes as climates warm. Since the last glaciation, there has been a progressive movement of tree species into the Great Lakes region from various refugia where they survived the last ice age. Davis (1980) has documented the time sequence of forest tree species succession and their dominance within various ecosystems. One can expect further movement northward into Canada of the ecotone between the deciduous and boreal coniferous forest if a climate warming occurs as the result of increasing levels of atmospheric carbon dioxide.

Permafrost

Extensive thawing of permafrost may occur if the climatic warming of 5° to 10°C occurs at high latitudes as predicted. Permafrost occu-

pies very large areas in Alaska, Canada, and the Soviet Union. During the last 10,000 years, northern ecosystems have sequestered an amount of carbon in dead organic matter equal to about 10% of all the organic carbon currently in the world terrestrial ecosystems. Another very large amount of carbon is incorporated in the permafrost as methane hydrates a few hundred feet beneath the surface. Thawing of the permafrost could result in oxidation of peat and release of methane gas, as well as a release of carbon dioxide. This would have a positive feedback on the atmospheric carbon dioxide concentration and climate change.

Freshwater ecosystems

Rivers, streams, and lakes will be affected in many ways by increasing levels of atmospheric carbon dioxide, by the ensuing climate warming, and by the accompanying precipitation changes. Strong changes will occur in the ratio of precipitation to evapotranspiration generally, and region by region these changes will be quite distinctive. Some regions will be wetter and those watersheds will have increased quantities of water flowing through them, whereas others will have diminished amounts. In some regions there could be increased flooding and in others, such as the Missouri–Mississippi river system, there may not be sufficient water for navigation. More flooding in the tropics will fill dry lakes such as Lake Chad, but will also aggravate soil erosion. Dryness in the Rocky Mountains will seriously impact the Colorado River flow and further restrict the use of this water, including its use for oil production from shale.

Increased CO_2 levels will lower the pH of unbuffered freshwater lakes and affect amphibians and fish. Increased water temperatures will affect fish spawning adversely or cause a shift from diatoms to filamentous green or blue-green algae, which are less suitable food sources for fish. As warmer temperatures occur earlier in the spring and last longer in the autumn, there will be commensurate shifts in spawning times by fish, hatching of insects, and generally in the timing of activities of one species population with another.

More flooding of tropical watersheds may lead to an increase of SCHISTOSOMIASIS (also known as BILHARZIA), a chronic parasitic disease potentially fatal to humans and other mammals. It is spread by snails that act as hosts for the larvae of the parasite. The larvae become free swimming in stagnant water. The parasites enter the skin of humans bathing in or using the water and lay eggs in the liver. Schistosomiasis is one of the most rampant and debilitating diseases in the world. More than 250 million people are afflicted. Weight loss, listlessness, and liver disorders ensue. The construction of the Aswan Dam in Egypt greatly increased the incidence of schistosomiasis in that country by forming a huge lake around the upper Nile River. Snails proliferated and millions of people using the water became ill.

The incidence of MALARIA may increase as increased rainfall in tropical and subtropical regions produces more stagnant water breeding areas for mosquitos. An increase of flowing water, on the other hand, will provide a reproductive habitat for the blackfly, and the dread RIVER BLINDNESS disease, known as ONCHOCERCIASIS, may increase in tropical countries. About 20 million people are now affected by this disease, which blinds them. The blackfly is a purveyor of this parasite.

Marine ecosystems

The present world catch of fish is approximately 65 million tons per year. Fish production is strongly dependent on primary productivity, and this in turn depends on the biological and physical processes that recycle nutrients through the food web. In some areas, such as the Sargasso Sea, this recycling may account for 90% of the nutrients used by phytoplankton in primary production. In upwelling areas (e.g., off the coast of Peru), less than 50% of the nutrients may come from recycling. In temperate regions of the continental shelf, 60 to 70% of the nutrients used by phytoplankton are recycled within the water column and 30 to 40% enter from estuaries or rivers, intrude from deeper ocean waters, or recycle from bottom sediments.

Various physical time scales affect productivity of the open ocean. On a long time scale, the general circulation of the atmosphere–ocean system is significant. On a one-year time scale, variations in the mixed layer depth from summer to winter determines the availability of nutrients for seasonal production. On a time scale of a week or of days, variations in weather will induce changes in primary productivity. Areas such as the North Sea, Georges Bank, or Gulf of Maine, have at least a twofold variation in productivity. This variability depends on tidal action and does not seem to relate as much to weather or climate.

Relatively little of the world's fish catch comes from the open ocean, but major quantities come from upwelling areas and temperate shelf regions. These areas are highly susceptible to climate variations, such as El Niño off Peru. (See Barber and Chavez, 1982 for an excellent discussion of the biological consequences of the unusually strong El Niño event of 1982.) It is in these areas of coastal upwelling, driven by the general circulation of the atmosphere–ocean system, that one can expect strong effects on fish production from climate change. If the general circulation weakened, because of the reduced temperature gradient between equator and poles on a warmer Earth resulting from increased atmospheric CO_2 concentration, we would expect a decrease in fish productivity.

It is in general rather difficult to predict changes in fish stocks. For example, in the North Sea and the Northeast Atlantic ocean, the spring bloom of plankton was delayed nearly a month between 1950 and 1970, but stocks of cod increased by threefold after 1960. During the relatively warm period between 1920 and 1960, cod flourished near West Greenland and on the Sealbard Shelf, where they previously had been absent; but since the 1960s the stock of cod off West Greenland has decreased enormously. If the westerly winds weaken or the circulation pattern shifts, there will probably be major changes in fish stocks.

As atmospheric CO_2 levels increase, there will be a corresponding increase in oceanic CO_2. Along with the CO_2 increase, a redistribution of carbonate and bicarbonate compounds in the ocean will lead to a decrease in pH. A fourfold increase in atmospheric CO_2 will result in a pH decrease of 0.6. Phytoplankton numbers may diminish with decreasing pH, and dinoflagellates, green algae, or blue-green algae may then dominate. If this happens, the entire fish food chain will be affected; fish productivity may decline because fish prefer phytoplankton to other algae. Anchovies off Peru, krill in antarctic waters, and other invertebrates may be particularly susceptible to a change in phytoplankton production.

Estuaries

Estuarine ecosystems dominate many of the world's coastlines and yield substantial productivity, including many species of fish and shellfish of importance commercially. Estuaries are quite likely to reflect changes in climate or pH resulting from an increase in the atmospheric carbon dioxide concentration. Increased CO_2 may affect the pH of these areas more strongly than other marine environments because they are poorly buffered, get heavy inputs from continental runoffs, and are strongly linked to the carbon cycle of the land. Estuarine ecosystems contain high nutrients loads and are vulnerable to inputs of toxic substances such as heavy metals, pesticides, and industrial chemicals. Increased precipitation on land would change the salinity of estuaries and cause their loading with detritus and sediments. Estuarine species generally are adapted to highly variable conditions and for this reason may be protected from the changes expected to occur with increasing levels of atmospheric CO_2.

10

Acid Deposition

Acid rain is affecting fish populations. This lake trout is shown as it appeared (in Lake 223 of the Experimental Lakes Area, northwestern Ontario) before and following a pH change from pH = 6.49 in 1976 to pH = 5.17 in 1983. A collapse of the food web in the lake was the cause of the lake trout becoming emaciated. Courtesy of Kenneth H. Mills, Freshwater Institute, Winnipeg, Canada.

INTRODUCTION

Modern industrialized societies burning coal, oil, and gas, smelting metals, and manufacturing synthetic products are sending into the atmosphere massive amounts of water vapor; carbon dioxide; hydrocarbons; ozone; oxides of sulfur, nitrogen, and other elements; metals; and a large variety of complex compounds. From the industrialized areas of the world, polluted air is flowing over vast regions of the landscape; and the global atmosphere, biosphere, and ecosystems are being systematically and inexorably modified. It has been suggested that there is not an unpolluted piece of surface or volume of air or water in the world today—not one place that does not contain the products of industrialization.

The chemical burden of the atmospheric air mass interacts with everything with which it makes contact. The ecosystems of the world are active chemical factories taking in elements and compounds, transforming them, using them, and ejecting them back to the environment. These chemical cycles have gone on since the formation of this planet. The precise chemistry of the biosphere as we know it today has evolved through synergism between living and nonliving components. Now, human activities are modifying the atmospheric chemistry. Soils and surface waters are affected, the biota of lakes and streams are modified, plant growth is retarded or stimulated, animal behavior is changed, the activities of microorganisms, pathogens, and fungi are changed, sometimes sensitive organisms are killed, and even physical structures such as monuments or bridges are eroded.

Moisture interacts with many of the compounds that enter the atmosphere, either from natural or human-generated sources. Precipitation in the form of rain or snow scavenges the air for these compounds. Normally, precipitation is slightly acidic because naturally occurring carbon dioxide gas forms carbonic acid when in contact with water. However, scientists have discovered that precipitation in some parts of the world is much more acidic than normal, an acidity that seems to be generated from human sources of pollution.

Although individual scientists have long been concerned about the deleterious effects of air pollution on ecosystems, a general awareness by the public about acid precipitation and its consequences has come only in recent years. This lag resulted from a need for a decade or more of data collection of precipitation chemistry before scientists could describe what changes were occurring over time. Gorham (1958) made measurements during the 1950s on the chemistry of precipitation and the quality of lakes in northern England. A network of monitoring stations in Sweden was started in 1948, extended to the rest of Scandinavia in 1952 to 1954, and to other parts of Europe in subsequent years. The USSR established an atmospheric chemistry network all across Asia to the Pacific Ocean in 1958, a network still in operation. In the United States, the situation has been less satisfactory. Chemical data was collected from precipitation at 24 sites in the eastern United States in 1955 to 1956 by Chris E. Junge. From 1959 through 1966 the U.S. Public Health Service, and then the National Center for Atmospheric Research, operated a precipitation chemistry network throughout the contiguous states. In 1976 the Canadian Network for Sampling Precipitation (CANSAP) was begun; and the U.S. Deposition Program (NADP) was started in 1978. Unfortunately, there was over a decade (1966 to 1978) when relatively little data were being gathered except at a very few specific sites. Nevertheless, from the European data and, with more uncertainty, from the U.S. data, trends concerning the chemical composition of precipitation in several parts of the world could be described. In addition, some rather startling results were obtained for single rainfall events at certain locations.

A rain falling on Kane, Pennsylvania on 19 September 1978 had a pH value of 2.32, and a storm in Scotland in 1974 dropped rain of pH value 2.4, values more acidic than vinegar. The pH of precipitation over Scandinavia decreased from values of around 5.4 in 1955 to 4.6 or less by 1975. Some of the best data for the United States come from the Hubbard Brook Experimental Forest in New Hampshire. It would appear that there was a small decrease in pH between 1964 and 1970 followed by a gradual increase since 1970; overall, there seems to be no significant trend. Despite this fact, there are many claims asserting increasing acidity of precipitation throughout the northeastern part of the United States, as well as for areas in Michigan, Florida, Colorado, and California. There are also numerous data to support claims of changing chemical composition of precipitation with time (information to be described later in this chapter).

Early measurements

Lest we believe that concern about acid precipitation is a recent idea, we should remind ourselves that Lavoisier in the early 1820s exposed baskets of chalk (calcium carbonate) to air in Paris for a year and found that significant quantities had converted to calcium nitrate. He concluded that this was caused by nitric acid from the atmosphere. Scientists in Britain noted the presence of nitric acid in rainwater and found that there were greater amounts during thunderstorms. During the 1850s researchers in France and England found that the nitric acid content of precipitation reached a maximum during the summer and a minimum in the winter. In 1863 the British parliament passed the famous Alkali

Act that gave the government authority to inspect industrial facilities suspected of emitting pollutants and to negotiate with the industry to find the best way to control the emissions. Angus Smith in Britain in 1872 found that the concentration of nitric acid and sulfuric acid in precipitation around cities was greater than in the countryside and concluded that the activity responsible was coal burning. The concentration of sulfuric acid was 35 times greater in Glasgow than in 12 places in the Scottish countryside. He also found that the concentration of sulfuric acid in precipitation was always greater than that of nitric acid and that the chemistry of nitric acid formation seemed more complex.

Much of the interest in England concerning the chemistry of precipitation centered on the amount of nutrients reaching soils from precipitation and the resulting effects on agriculture, and much of the research took place during the nineteenth century. Unfortunately, the results of this work and the earlier work of Smith cannot be used to calculate the acidity because the chemical methods employed were inadequate. In 1911, Crowther and Ruston (1911) published measurements of the precipitation on the industrial city of Leeds, England and on a farm six miles away; they were concerned with the effects of pollutants on farm crops. They noted that free acid in precipitation inhibited plant growth by direct action of acid on leaves and that it reduced the nitrogen-fixing activity of soil organisms.

The first attempt to correlate large-scale weather patterns and wind trajectories with acid precipitation was made by Anderson (1915). He noted that nitric acid was formed from the action of lightning. Atkins (1922) found that trout were not found in lakes or streams with a pH less than 5.5. In 1927 Dahl reported that fish hatcheries in southwestern Norway were having difficulty rearing trout fingerlings. He did a series of tests on juvenile trout and found some difficulty with rearing them below a pH of 6, some mortality at pH 5.1, and total mortality at pH below 5.0. When the pH of one hatchery was raised from 5.2 to 6.0, the mortality among trout fingerlings ceased. Heavy kills of salmon occurred in the Kuina River in 1911, the Mandal River in 1914, and the Frafjord in 1921. Dahl speculated that acid precipitation was the cause of the low pH because each event occurred following prolonged drought and subsequent heavy precipitation. In 1925 and 1926 similar kills occurred, and Dahl found the stream water to have a pH of 5.2 each time.

There was little interest in precipitation chemistry during the 1930s and 1940s. However, Erickson (1952, 1959, and 1960) in Sweden reported on the geochemical cycles of sulfur, chlorine, and iodine; and Barrett and Brodin (1955) reported increasing acidity of precipitation in Scandinavia, measurements coming from the Scandinavian network of atmospheric chemistry established a few years previously. (See Gorham, 1982 for a more recent review of the acid deposition issue. Also see Likens et al., 1979; Oden, 1976; and Toribara et al., 1980 for earlier reviews.)

ATMOSPHERIC CHEMISTRY

Compounds emitted into the atmosphere exist in either an aqueous phase or a gaseous phase. The chemical reactions that occur are usually very similar in either phase. Water vapor will react with compounds of carbon, sulfur, or nitrogen, whether or not actual droplets are formed. Acids will be present as gases, particulates, or as very small liquid droplets, or aerosols. The circulation of the atmosphere wafts these gaseous substances, particulates, and fine droplets through the air, from which they slowly fall out as DRY DEPOSITION on vegetation, soils, or human habitation. In addition, the compounds are dissolved in water droplets and form precipi-

tation as rain or snow, known as WET DEPOSITION. Dry deposition occurs all the time, whereas wet deposition is sporadic (Figure 1). The proportions of dry or wet deposition depend on the distance from the sources of pollution and the wetness of the climate. Dry deposition contributes 10 to 30%, and at times more than 50%, of the total acid deposition.

Acid formation

Natural precipitation may be slightly acid because of the presence of both strong and weak acids in it. Sulfuric, nitric, and hydrochloric acids are known as strong acids, whereas carbonic, carboxylic, clays, aluminum and ferrous hydroxides, and ammonium ions constitute weak acids. Strong acids undergo complete dissociation and ionization when in aqueous solution; weak acids undergo partial dissociation. Biological activity can generate H_2S and NO_x, compounds that may be converted into the strong acids H_2SO_4 and HNO_3 in the atmosphere. Volcanoes emit H_2S and SO_2, gases that convert to H_2SO_4 upon contact with water vapor. A large number of chemical reactions occurs among various compounds in the atmosphere, and many of the reactions are poorly understood. The following represent only some of the simpler atmospheric chemical processes.

Carbon dioxide gas dissolves in water to form carbonic acid and the acid ionizes according to the following reactions:

$$CO_2 + H_2O \rightarrow H_2CO_3 \qquad (1)$$
$$H_2CO_3 \rightarrow HCO_3^- + H^+ \qquad (2)$$

Carbonic acid only partly dissociates in water, so that only a fraction of the molecules undergo separation into bicarbonate ions (HCO_3^-) and hydrogen ions (H^+). Carbonic acid is classified as a weak acid for this reason. If equilibrium is reached, at the current level of atmospheric carbon dioxide concentration, the pH of rainwater will be about 5.6.

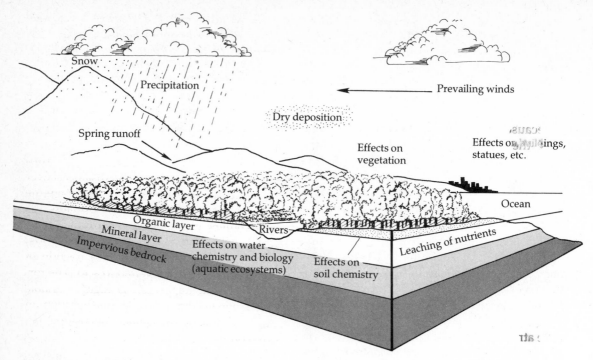

FIGURE 1. Acid deposition on the landscape affects vegetation, soils, aquatic ecosystems, and man-made objects such as buildings, bridges, highways, and monuments.

Natural sources provide most of the hydrogen sulfide found in the atmosphere, but anthropogenic sources are the origin of most of the sulfur dioxide emissions. By a series of complex chemical reactions involving oxygen and sunlight, the hydrogen sulfide converts to sulfur dioxide and this then converts to sulfuric acid by reacting with water molecules. This, of course, is the same reaction that transforms sulfur dioxide emitted from power plants, industrial furnaces, and automobiles into sulfuric acid. It is

$$2SO_2 + O_2 \rightarrow 2SO_3 \qquad (3)$$
$$SO_3 + H_2O \rightarrow H_2SO_4 \qquad (4)$$
$$H_2SO_4 \rightarrow SO_4^{2-} + 2H^+ \qquad (5)$$

Because sulfuric acid is a strong acid, nearly all of the molecules ionize into sulfate ions and hydrogen ions as shown. The actual reactions that occur in the atmosphere are substantially more complex than shown here, and some involve the presence of metal catalysts.

Furnaces and automobile engines burning fossil fuels at high temperatures emit substantial quantities of oxides of nitrogen. These are designated as NO$_x$. If ozone is present in the atmosphere at a concentration greater than 0.1 ppm, the following reactions may occur to form nitric acid:

$$2NO_2 + O_3 \rightarrow N_2O_5 + O_2 \qquad (6)$$
$$N_2O_5 + H_2O \rightarrow 2HNO_3 \qquad (7)$$

If a catalyst is present (such as a trace metal in the atmosphere or SO$_2$, either of which could come from fossil fuel-burning power plants), nitric acid will form:

$$4NO_2 + 2H_2O + O_2 + M \rightarrow 4HNO_3 + M \qquad (8)$$

A catalyst, M, is an element or compound that facilitates a reaction but is not itself changed while generating the final product.

Nitric acid in dilute aqueous solution ionizes as

$$HNO_3 \rightarrow NO_3^- + H^+ \tag{9}$$

Be___ e nitric acid is a strong acid, nearly all o___ molecules dissociate into nitrate ions and hydrogen ions. The rate of the above reactions depends on the exact composition and quantity of pollutants in the air and the temperature. They may take a few minutes to hours to react in highly polluted air, or even days in clean air. There are other reactions involving the oxides of nitrogen that are significant, particularly those involving NO. The nitrogen cycle described in Chapter 1 showed that various oxides of nitrogen are generated by microbial activity in the soil and released to the ___nosphere.

Many different chemical reactions may occur in the atmosphere converting NO_x and SO_2 to strong acids. These heterogeneous processes involve the interaction of strong oxidants and small droplets. The concentration of strong oxidants is dependent on the presence of NO_x. It therefore appears that NO_x is playing a dominant role in the formation of ___ deposition. NO_x emissions are as much ___sociated with transportation as they are with stationary sources. The oxidants that play a role in the tropospheric chemistry are atomic oxygen (O), hydroxyl radical (OH), hydrogen peroxide (H_2O_2), and ozone (O_3). NO_2 may be directly oxidized to nitric acid (HNO_3), or it may be a precursor to the formation of hydrogen peroxide and ozone. The chemical reactions that form strong oxidants depend on the energy from sunlight, $h\nu$, where ν is the frequency of the light and h is Planck's constant. The following reactions occur in the lower atmosphere:

$$NO_2 + h\nu \rightarrow O + NO \tag{10}$$
$$O_3 + h\nu \rightarrow O_2 + O \tag{11}$$

The hydroxyl radical, OH, plays a very significant role in the photochemistry of the troposphere. It is produced by the following reactions

$$O + H_2O \rightarrow 2OH \tag{12}$$
$$HONO + h\nu \rightarrow OH + NO \tag{13}$$

This, at least, shows that the pollutants we put into the atmosphere undergo chemical change by the action of sunlight. This in turn suggests that the greater the distances over which they are transported, the greater are the photochemical changes. These oxidants, as well as others, in combination with water react on both NO_x and SO_x to form strong acids. Under most circumstances only a fraction of these pollutants are converted to acid. The formation of acid deposition is highly variable and essentially episodic. At one location in England, 80% of the sulfate deposition occurred on 20% of the wet days.

The pH scale

The pH scale is used to define the degree of acidity or amount of alkalinity (base) of water; the logarithm (base ten) of the reciprocal of the hydrogen ion concentration is the pH value. Pure, distilled water has a concentration of H^+ of 10^{-7} and of OH^- of 10^{-7} moles per liter or a pH value of 7. The pH scale runs from 0 to 14, with acidic values less than 7 and base values greater than 7. When the pH value drops one unit, the acidity of the liquid rises tenfold. Therefore, a solution pH 4 is one hundred times more acidic than a solution of pH 6. When we express acidity directly in concentration of hydrogen ions, we can see that it takes 10 times more hydrogen ions to cause a shift in acidity from pH 6 to 4 than from pH 6 to 5. It is useful to give the hydrogen ion concentration in a dilute aqueous solution in terms of micrograms per liter. Table 1 gives the relationship between pH and the hydrogen ion concentration. Table 2 gives the pH values for many substances of the everyday world.

Typical pH values are baking soda, 8.2; drinking water, 6.5 to 8.0; beer, 4.0 to 5.0; soft drinks, 2.0 to 4.0; and vinegar, 2.4 to 3.4. Carbon dioxide dissolves in water and forms carbonic acid (H_2CO_3); at equilibrium, it has a pH of 5.6. When water condenses on small dust particles in the atmosphere, a variety of base cations (positive ions) of calcium, magnesium, potassium, and sodium go into solution and raise the pH. In regions where there is abundant windblown dust, precipitation may have pH values in the range 7 to 8. Ammonia gas coming out of the soil goes into solution in water droplets and raises the pH. Normally, snow or rain has a pH of 6.0 or greater. Ice cores from Greenland deposited before the industrial revolution have a pH of 6.0 to 7.6. Analysis of old glacial ice from the Cascade Mountains in Washington State showed minimum pH values of about 5.6. Analysis of rainwater from Geneva, New York during the years 1919 to 1929 showed that it contained large quantities of bicarbonate, a situation suggesting a pH of 7. Rainfall in areas downwind of industrial sources may have pH values of 6.0, 5.0, or even less.

TABLE 1. Relationship between pH and hydrogen ion concentration

pH	Hydrogen ion concentration moles/liter	μg/liter
7.0	10^{-7}	0.1
6.0	10^{-6}	1.0
5.6	$10^{-5.6}$	2.5
5.0	10^{-5}	10
4.6	$10^{-4.6}$	25
4.3	$10^{-4.3}$	50
4.0	10^{-4}	100
3.0	10^{-3}	1000

ACID PRECIPITATION

The European experience

From the industrialized Ruhr valley, the English midlands, and the coal areas of East Germany, southern Poland, and Czechoslovakia, sulfur emissions are wafted across the North Sea over Scandinavia. Emissions of sulfur dioxide in Europe were estimated at about 25 million tons of sulfur in 1973. More than 77% of the sulfur being deposited in Sweden is estimated to come from outside the country. In the month of April 1974 rainfall of pH 2.4 was reported at Pitlochy, Scotland, 2.7 on the west coast of Norway, and 3.5 in Iceland. The lowest annual pH recorded was at DeBilt, Nether-

TABLE 2. The pH scale

		pH value	Substance
Basic		14.0	
		12.4	Lime (calcium hydroxide)
		11.0	Ammonia (NH_3)
		10.5	Milk of magnesia
		8.0–8.5	The Great Lakes annual mean
		8.3	Seawater
		8.2	Baking soda
		7.4	Human blood
Neutral		7.0	Distilled water
		6.6	Milk
		6.0	Normal rain
		4.0–5.0	Beer
		4.2	Tomatoes
		3.0	Apples
		2.4	Vinegar
		2.0	Lemon juice
Acid		0.0	

lands in 1967, a value of 3.78. Maps of volume-weighted mean annual pH of precipitation over Europe showed that the region of highly acidic (pH 4 to 4.5) precipitation had spread from Belgium, the Netherlands, and Luxemburg in the late 1950s to include most of Germany, northern France, the eastern British Isles, and southern Scandinavia by the late 1960s. Nitrogen oxide emissions had approximately doubled between 1959 and 1973, reaching about 2 million metric tons per year for western Europe only. Norway and Sweden receive particularly heavy acid precipitation because of their mountainous terrain against which moist air masses condense as they travel from southwest to northeast with the prevailing winds. In order to reduce emissions near the ground in industrialized areas of high population densities, power plants in Europe started building tall stacks to eject the emissions high above the ground. This action contributed to the reduction of serious smog episodes in London but allowed the pollutants to travel on the prevailing winds to Scandinavia and elsewhere. Dry deposition of sulfur dioxide gas and particulate sulfate drops a lot of acidic material close to the industrial sources throughout central Europe where the deposition rate is four times the rate of acid precipitation in rain and snow.

Scandinavia. In Sweden, where acid rain has been most intensively studied, the wind trajectories were analyzed. The chemistry of the precipitation clearly reflects the degree of pollution coming from different parts of Europe. The least contaminated air trajectory has the highest pH values (about 5.2) and the lowest amounts of total sulfur and total nitrogen. This trajectory comes out of the North Atlantic, across the northern British Isles and the North Sea, bringing relatively clean air to Scandinavia. However, this air mass has the highest chloride concentrations. Air masses sweeping further south over central and east-

ern Europe bring the highest contamination to Scandinavia—air that is highly acidic (pH value 4.7) and highest in total amounts of sulfur and nitrogen, but least in chloride.

Yearly average pH values at stations in Norway and Sweden dropped from about 5.4 to 5.6 in 1955 to around 4.4 in 1975. There is considerable year to year variation. It appears that sulfur plays a major, but complex, role in the acidity of precipitation. At every station the sulfur content is steadily increasing and the pH is decreasing. However, the correlation pattern is often very irregular between sulfur and acidity. Even though sulfur emissions from central Europe and the Soviet Union have increased year by year, the fallout of sulfur over Scandinavia has not always corresponded to the increased emissions. There are many reasons for this because the behavior of sulfur in the atmosphere depends on many chemical and physical factors, such as the rate of photochemical oxidation of SO_2 to H_2SO_4, the oxidation by O_3, the effect of catalytic dust particles, the presence of NH_3 or other bases in the air, the lifetime of the chemical products, the mixing of different air masses with a variety of chemical compositions, and finally the gas exchange at the surface. The point is that what appears at first to be a simple cause and effect relationship turns out to be a very complex situation that is difficult to understand.

The Netherlands. In the Netherlands pH values fell below 4 on an average yearly basis about 1966. The acidity of drizzles is usually greater than for heavy rains. Over 80% of the drizzles in North Holland had pH values between 3.0 and 3.5 and sometimes values as low as 2.5 to 3.0. When the wind had passed over large land areas of Europe it was decidedly more acidic than when it came off the sea. At all Dutch stations there was a good correlation between the sulfate and nitrate concentrations in precipitation, with the sulfate being about double the nitrate concentration. There was a significant correlation between the hydrogen ion content of precipitation and the sum of the ions of NH_4^+ + NO_3^- + SO_4^{2-}, with NO_3^- being the dominant variable 45% of the time and SO_4^{2-} 20%; 35% of the time there was no dominant variable. Again this example indicates the complexity of the chemistry involved and the difficulty of drawing simple conclusions.

The history of fuel use in the Netherlands is an interesting case history as it relates to the amount of SO_4^{2-} emission. After World War II, about 85% of Dutch fuel was coal. Then oil use came in between 1950 and 1960. By 1970 coal was only 10% of the total fuel supply, and 4% by 1977. In 1959 the largest natural gas field in the world was discovered in the Netherlands. When the price of oil went up, the use of natural gas from this field was accelerated. Natural gas provided 18% of the total Dutch energy consumption in 1967 and 60% in 1976. Dutch power stations were using 85% natural gas by 1975.

In 1946, when the Netherlands burned mainly coal, the SO_2 emissions were 200,000 tons per year; by 1956 these were 400,000 to 500,000 tons per year. Cheap oil came in, but it had an even higher sulfur content than coal, and SO_2 emissions went up to nearly 1 million tons per year during the early 1960s. Frequently during stagnating weather conditions, ambient SO_2 concentrations were 500 to 2000 g/m^3 in many places. In 1968 the Dutch Clean Air Act was passed and became effective at the end of 1970. Desulfurization units were installed on power plants, lower sulfur oil was imported, and sulfur emissions began to decrease. In fact they had started to decrease in 1967 to 1968. Despite these necessary controls as mandated by the Clean Air Act, the absolute decrease was caused mainly by the increasing use of natural gas.

The Dutch are concerned about the ultimate depletion of their natural gas supply.

They must eventually quit exporting natural gas, as they are now doing, and reserve it for domestic heating. Electric power stations will then need to go to coal or oil, and sulfur emissions will go up. Nuclear power could contribute substantially to a reduction of SO_2 emissions, but there has been strong opposition in the Netherlands to building such plants. Only two nuclear power plants are operating today and there are no prospects for more before the year 2000.

While SO_2 emissions have been going down in the Netherlands, the emission of NO_x has continued an upward trend, although the rate of growth has slowed. The reason NO_x emissions still continue to increase is that NO_x is a product of combustion irrespective of the type of fuel used, whereas SO_2 is characteristic of the fuel. As energy consumption grows, so also will NO_x emissions.

The United States experience

The record of changes in the precipitation chemistry over North America is much less complete than for Europe. The sampling network of stations is much newer. Nevertheless, there is sufficient data to draw some conclusions, but clearly with the idea that as more substantial information becomes available the conclusions may need modification.

The most detailed chemical analysis of precipitation in the United States has been done at the Hubbard Brook Experimental Forest in New Hampshire by Gene E. Likens, F. Herbert Bormann, and their colleagues. Their measurements (see Likens et al., 1980) of pH values show no discernible trend between 1964 and the present when reduced to weighted annual averages. The weighting is by volume of precipitation. Sulfate ion concentration shows a discernible decrease from about 1964 to 1974 and since then there has been no significant change. The annual nitrate concentration in precipitation increased markedly after 1964, but since 1970 there has been no discernible trend. The ammonium ion concentration has not changed significantly since 1964. However, data from the Cornell University Agricultural Experiment Stations at Ithaca and Geneva, New York show that nitrogen concentrations increased very sharply after 1945. Prior to that date (back to 1915), there was no discernible change. Apparently, before 1930 relatively large amounts of bicarbonate were found in precipitation samples taken in Virginia, Tennessee, and New York. The presence of bicarbonate in these samples shows that they could not have had a pH of less than 5.6.

Likens and Butler (1981) and Likens et al. (1976) estimated the geographical distribution of the hydrogen ion concentration of precipitation over the eastern United States for the periods 1955 to 1956, 1965 to 1966, and 1975 to 1976. These average annual values are shown in Figure 2A–C, which shows the acidity of precipitation over a vast area of the northeast-

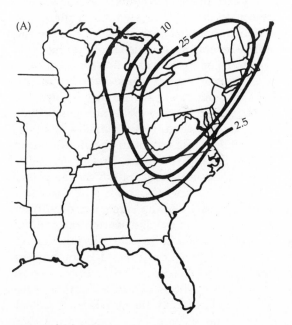

FIGURE 2. Yearly average concentration of hydrogen ions in $\mu g\ H^+$/liter in precipitation in the eastern United States during three periods: (A) 1955–1956; (B) 1956–1966; and (C) 1975–1976. (From Likens and Butler, 1981.)

ern region of the United States. These figures show a considerable spreading into the Southeast and westward and an intensification of acidity in the northeastern states. Although it is not discernible from the maps, the highest H^+ concentrations increased from about 38 μg/liter (pH 4.44) in 1955 to 1956, to 105 μg/liter (pH 3.98) in 1975 to 1976. The southeastern United States has undergone the most rapid percentage increase in acidity of any region. The emissions of both SO_2 and NO_x have approximately doubled there between 1960 and 1978. The data from Hubbard Brook show no statistically significant trend for the volume weighted average H^+ concentration from 1964 to present. It has remained constant at about 100 μg/liter (pH 4.0). Some industrialists have argued that this shows there has been no increase in the acidity of precipitation during this period. However, some atmospheric chemists suggest that there is a natural limit to the average acidity of precipitation because of the nature of the precipitation processes and the chemistry of converting gaseous emissions to strong acids. They also suggest that the average annual pH of precipitation will never go lower than 3.5 to 4.0 (about 100 μg/liter or a little greater). It is interesting to note that the lowest average annual value reported in Europe was pH 3.78 at DeBilt, Netherlands (mentioned in the European experience section above).

Stensland and Semonin (1982) have reassessed the U.S. precipitation data for the mid 1950s and found excessively high values of calcium and magnesium compared with current values. They consider the most likely explanation to be the severe drought and dust storms that prevailed in the United States during the 1950s. When the excess soil loadings of the atmosphere and precipitation are adjusted to nondrought conditions, the corrected pH trend, suggested to have been produced by acid-forming emissions since the mid 1950s is much smaller than previously estimated.

Sulfur oxide emissions in the midwestern industrial states have decreased by about 17% over the past 15 years (prior to 1981). During that time the acidity of precipitation continued to increase in the Northeast. During the same period the average height of new stacks installed in electric power-generating plants increased about 60%. According to Patrick et al. (1981), this action has facilitated the dispersion of pollutants over a wide geographic area. Emissions of NO_x in the eastern United States increased by 46% between 1960 and 1978, but at a lower rate in the period 1970 to 1978. SO_2 emissions increased by 22% from 1960 to 1970 in the eastern United States, but decreased by 10% from 1970 to 1978 even though acidity continued to increase. This finding suggests that NO_x plays a significant role in determining the acidity of precipitation. In the New England states and New York, the absolute quantities of SO_2 and NO_x have decreased during the last 10 to 15 years. Because the acidity of precipitation has continued to increase, long-range transport of pollutants is strongly suggested.

In Florida the weighted average annual pH values are now below 4.7 over the northern three-quarters of the state. Junge's atmospheric chemistry samples taken in the 1950s at five locations indicate precipitation with pH values above 5.6. Large increases in sulfates and nitrates have occurred since the 1950s (Brezonik et al., 1980). Summer rainfall is more acidic than winter rainfall in Florida; this pattern also occurs in the northeastern United States. There are 56 electric utility power plants in Florida, of which 52 burn oil or oil and natural gas and 4 burn coal as a principal fuel. Many more coal-burning plants are to come online during the coming decade. The origin of acid precipitation in Florida is not well understood at the present time.

There are reports of acid precipitation occurring in California. In the northern part of the state, the mean pH of rainfall during 1978 to 1979 was 4.4 to 5.3. Nitrates dominate sulfates at most sites. In southern California, the mean pH during 1978 to 1979 ranged from 4.4 to 5.4. The ratios of NO_3^- to SO_4^{2-} are less than unity at inland locations. On the average, the nitrate levels in southern California precipitation are 62% higher than in the north, and sulfate levels are 275% higher. There are few, if any, coal-burning power plants in California, but primarily there are oil-burning plants. The high ratios of NO_x^- to SO_4^{2-} suggests that automobiles make a contribution. However, for the northeastern United States, transportation is estimated to contribute less than 14% of the atmospheric NO_x.

ECOLOGICAL IMPACTS

Sensitive regions

Acid deposition, whether wet or dry, will affect all components of any ecosystem in multiple, far-reaching ways—changing them physically, chemically, and biologically. Parts of the landscape differ enormously in their manner of response to acid deposition. Some watersheds are highly sensitive to acidic input, whereas others are highly insensitive. The degree of sensitivity depends very much on the nature of the rocks and soils in the watershed. We know that regions with granitic, highly siliceous bedrock exposed at the surface are very sensitive to acid deposition. Most mountainous regions of the world can be characterized in this manner. Much of Scandinavia, most of Ontario and Quebec provinces in Canada covered by the Precambrian shield, the Adirondack and Appalachian mountains, and much of New England are sensitive regions. The regions of sensitivity in the United States and Canada are shown in Figure 3. These regions have no calcium carbonate ($CaCO_3$) in their sedimentary rocks. Calcium carbonate provides a buffering action or an ability to neutralize acids

in solution. The chemical reaction between sulfuric acid (H_2SO_4) and calcium carbonate is

$$H_2SO_4 + CaCO_3 \rightarrow Ca^{2+} + SO_4^{2-} + H_2O + CO_2$$

The ions released, Ca^{2+} and SO_4^{2-}, neutralize one another and therefore buffer the system against acidification.

Aluminum concentrations

Waters affected by acid deposition have elevated aluminum concentrations. Aluminum concentrations found in lakes of southern Norway and the northeastern United States are 5 to 10 times higher than in nearly neutral (circumneutral) waters in these same areas. Aluminum hydroxide compounds are common in podzolic soils. These acidic soils are characteristic of humid temperate climates having high precipitation. The soils are subject to severe leaching and are usually covered by coniferous or hardwood forests. The high leaching leads to a low nutrient content in the soils and this in turn makes for poor conditions for soil microorganisms that are so vital to nitrogen fixation. The decomposition of plant residues is retarded by the low activity of decomposers and by a thick organic layer that forms on the forest floor. The solubility of aluminum hydroxide compounds is a sensitive function of pH: as hydrogen ion concentration increases, the solubility of aluminum increases. There is a complex interaction between organic compounds in podzolic soils and the aluminum concentration. The aluminum cations neutralize the negative charge carried by the organic compounds. The result is that both the aluminum and the organic material precipitate out in the soil. However, as more organic matter forms at the surface, it must go deeper into the soil to find additional aluminum with which to interact. When a soil becomes more acidic as a result of acid deposition, aluminum cations (and iron cations) are released from the soil and

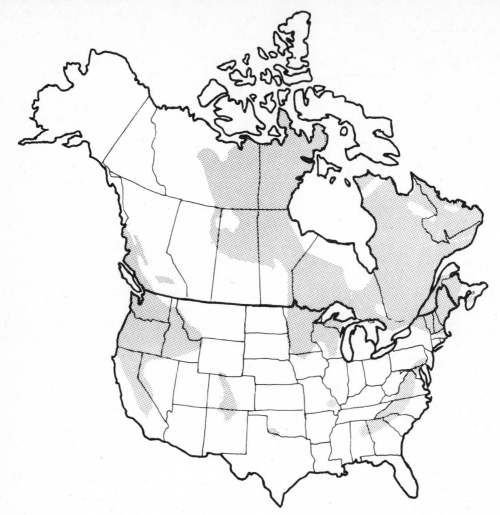

FIGURE 3. Areas (shaded) in the United States and Canada most sensitive to acid deposition.

enter the water percolating into streams and lakes. These aluminum ions are highly toxic to fish. When aluminum concentrations reach 0.1 to 0.3 mg/liter, reductions in the growth of brook trout occur and mortality ensues from severe necrosis of gill epithelium.

Mercury poisoning

Acid deposition releases mercury from the watershed or lake sediments. Although inorganic mercury is somewhat toxic to higher organisms, methylmercury (CH_3Hg, an organic form) is far more toxic. Inorganic mercury is converted to methylmercury by anaerobic microbial activity and, once produced, accumulates rapidly in the aquatic food web and in the tissues of higher organisms. It is also toxic to humans. Many fatalities occurred in Japan at Minamata and Niigata between 1955 and 1962 when residents ate fish and shellfish contaminated by mercury released by industries.

General effects of acid deposition

The altered chemistry of ecosystems, altered by acid deposition, has affected organisms at every trophic level. Streams and lakes have been particularly vulnerable to alteration, but terrestrial ecosystems are affected as well. Objects constructed by people are also damaged. The potential effects of acid deposition are many:

1. Acidification of a lake or stream results in disruption of the food web. The species diversity of the lake or stream decreases, organism behavior is affected, and organisms may become weakened or die. Acidophilic algae, mosses, or fungi grow abundantly, and litter accumulates in the bottom of the lake or stream.
2. Acidification and demineralization of soils occur.
3. Nutrient cycles are disrupted as a result of the die-off of soil bacteria and fungi. Litter accumulates because decomposers are absent.
4. Damage to crops and forests occur.
5. Deterioration of limestone monuments, bridges, highways and buildings takes place as sulfate ions replace the calcium ion in calcium carbonate, thereby forming water-soluble compounds.

Although these are all negative effects, there can be positive effects from acid deposition. Cowling and Linthurst (1981) state: "Even a given molecular species, such as SO_4 or NO_3, may be absorbed and utilized as a beneficial nutrient at one concentration and time of development, whereas it may be toxic to the very same plant, animal, or microorganism at another time and higher concentration."

Acid impacts on streams

As a stream becomes more acidic, a shift to acid-tolerant plants, such as the filamentous green algae, occurs. The density of algae on the stream bottom will increase because of decreased decomposition rates and decreased grazing by invertebrates. Diatom densities decrease as the stream becomes more acidic. Invertebrates vary enormously in their responses to acidification. Sensitive species of snails, clams, and amphipods are eliminated at pH values between 5 and 6 and some mayfly larvae and midge larvae die.

As streams become acidic, increased levels of Al, Fe, Ca, Mn, Mg, and K occur in the water. Episodic events, such as snowmelt or storms, can create a sudden decrease of pH by one or two units, at which time there will be a serious loss of organisms, even though the stream may be tolerable to them the rest of the time. High aluminum concentrations are toxic to many invertebrates and fish. Aluminum damages the gills of fishes. Increased aluminum in the water leads to increased sedimentation of organic matter. It is then unavailable as a primary food source for many stream organisms.

Nine rivers in Norway are now very acidic (pH values 4.5 to 5.5), and the acidity is still increasing. In these rivers, salmon reproduction no longer occurs. Research on Ramse Brook in Norway showed that the filamentous green algae *Mougeotia* sp. and the diatom *Tabellaria flocculosa* were acid tolerant. However, at low pH *Mougeotia* was more abundant, and at pH 6.1 or above *Tabellaria* was more abundant. Acidification of the stream to a pH value of 4.0 lead to accumulation of algae and increased chlorophyll levels. The chlorophyll *a* concentration and the amount of biomass produced increased by twofold and nearly sixfold, respectively, as the pH in Norris Brook went from about 6.0 down to about 4.2.

In the river Duddon in England, all mayflies (Ephemeroptera), some caddisflies (Trichoptera) in the genera *Wormaldia* and *Hydropsyche*, the river limpet *Ancylus fluiratilis*, and the amphipod *Gammarus pulex* only existed at pH values greater than 5.7. Stoneflies (Plecoptera), caddisflies of the genera *Plectrocnemia* and *Rhyacophila* and the family Limnephilidae, midges (Chironomidae), blackflies (*Simulium*), and craneflies (Tipulidae) were all present in acid waters at pH 5.7 and appeared insensitive to pH. Other water quality factors may have been more important to these organisms. It is not only the pH value that affects these organisms but the entire chemical and physical environment of the stream. Their responses to pH under one set of conditions may differ from their responses under another set of conditions.

Acid impacts on lakes

In clear, pristine, freshwater lakes, Ca^{2+} and Mg^{2+} are the dominant cations and bicarbonate anions (HCO_3^-) exceed sulfate anions (SO_4^{2-}). The acid-neutralizing capacities of natural waters depend upon the carbonate–bicarbonate buffer system. When an aquatic ecosystem is depleted of bicarbonate, the sulfate anion becomes dominant and the concentrations of aluminum and lead become elevated. The ratio of HCO_3^- to SO_4^{2-} is a very sensitive index of the acidification of natural waters. When this ratio reaches 0, bicarbonate disappears. This occurs when the pH value is about 5.5. Lakes with pH values in the range from 6.5 down to 4.5 are sensitive to acid input. The alkalinity of water is derived from the weathering of carbonate materials. The Ca^{2+} and Mg^{2+} concentration should balance the HCO_3^- concentration in a well-buffered aquatic ecosystem. Further buffering is provided by the presence of organic acids, amino acids, and colloidal organic matter, such as humus. Once the proton-accepting anions are lost from the soil or watershed, free hydrogen (H^+) and its surrogate (Al^{3+}) rise quickly to toxic levels, and biota are seriously affected.

Small lakes in high-altitude watersheds are generally the most vulnerable to acid deposition. A seasonal decrease in the pH values of lakes in Scandinavia and the Adirondacks takes place every spring as snowmelt water rushes into the lakes. Sometimes this runoff has a pH value of 3.0. Unfortunately, runoff coincides with the spawning time for many species of fish, a time in their life cycle when they are most vulnerable. Small lakes at high altitudes may actually have pH values as low as 4.0.

In the lakes of Scandinavia, decomposition of organic matter is retarded at low pH levels and fungal growth increases. Organic debris and fungal mats cover the lake sediments and inhibit nutrient exchange. Reduced recycling of nutrients leads to lower productivity and to a more oligotrophic lake. Many invertebrates are affected by falling pH, as also are species of phytoplankton and zooplankton. Reproduction by amphibians also is impaired by low pH.

Many species of fish feed on plankton. As the food supply for fish is altered, fish populations decline. When the pH value drops below 5.5, the decline of fish populations is severe. In southern Norway 2083 lakes were surveyed in 1975 and 741 were reported to be without fish. Thousands of lakes in Sweden are reported devoid of fish. Hundreds of lakes in the eastern United States are purportedly without fish. In the Adirondacks alone there were more than 190 lakes without fish in 1975 (Schofield, 1976). The Ontario government estimates that 48,000 lakes in Ontario are threatened with acidification.

It is possible to use the presence or absence and relative abundance of the skeletal remains of various plankton in lake sediments as a means to estimate the pH of the lake at the time these organisms were alive. Davis and Berge (1980) have done this using certain diatom species from the sediments of Lakes Hovvatan and Blavatn in Norway. For 200 years prior to 1945, pH values in Lake

Hovvatan were inferred to be in the range 4.7 to 5.3, and then slowly decreased until 1964 when the rate of decrease accelerated down to a pH value of 3.9.

An approximate ranking of the ecological consequences of lake acidification is given in Table 3. To be even more specific concerning the effects of acidification on organisms, the following paragraphs describe the responses of specific categories of organisms. More details may be found in the report by Flamm and Bangay (1981).

Aquatic macrophytes. Increasing acidity in lakes causes plant succession with reduced abundance of normally occurring macrophytes and replacement of initial species with dense growths of *Sphagnum* mosses. This changeover begins at a pH of about 6.0 and increases markedly when the pH is below 5.0. *Sphagnum* moss has excellent ion-exchange capacity and is able to remove cations such as Ca^{2+}, Mg^{2+}, Fe^{2+}, Na^+, and K^+ from the water in exchange for H^+. This ion-exchange process further increases the acidity and reduces the biological productivity by making these cations unavailable to other plants.

In five lakes on the western coast of Sweden, the bottom areas covered by *Sphagnum* increased from 8 to 63% between 1967 and 1974, while the pH decreased 0.8 units to a value of 4.8. According to Conway and Hendrey (1981), all of the available inorganic carbon is in the form of CO_2 or H_2CO_3 in these lakes. *Sphagnum* can utilize these molecules of carbon, but not bicarbonate (HCO_3^-), which is more readily used by most aquatic plants. During the years of *Sphagnum* increase, the macrophyte communities dominated by *Lobelia dortmanna* and *Isoetes* decreased. *Lobelia* is able to absorb CO_2 from the sediment through its roots and very often becomes the dominant macrophyte in many acid lakes.

Acidification of some lakes in New York State has caused the growth of acid-tolerant species of algae that completely cover the macrophytes and screen them from sunlight; the macrophytes then die. In Scandinavian lakes, acidification has caused the accumulation of organic material in the bottom. This

TABLE 3. Ecological consequences of lake acidifications

pH	Condition and consequences
8.0	Lakes are fairly alkaline
7.0	Alkalinity drops; calcium content declines; and eggs of salamanders fail to hatch (may be a lack of calcium)
6.6	Snails die
6.0	Tadpoles and shrimp die; more salamanders disappear; anaerobic methane-producing bacteria show little activity below pH 6
5.9	Mercury converts to monomethyl mercury and is absorbed by fish and poisons the top predators
5.5	Diversity of species drops; leaf litter accumulates on lake bottom as decomposers disappear; plankton start to drop out as food chain begins to fall apart; serious effects on fish; calcium concentration across gill membranes disrupted in bass and trout; egg production of walleye pike prevented; toxic metals—aluminum, mercury, lead, cadmium, tin, beryllium, nickel—released from watershed soils and lake sediments
5.5	Acidophilic mosses, fungi, and filamentous algae grow abundantly and choke out other plants; northern pike, suckers, and perch die; fish eggs do not hatch
5.0	Sphagnum grows abundantly and withdraws calcium from the lake water; gill damage and mortality in brook trout from aluminum ions in the pH range 4.4 to 5.2
4.5	All fish die; most frogs and many insects die; lake becomes clear

accumulation may be the result of macrophytes dying and of fungi dominating the bacteria, thus retarding the rate of decomposition and delaying the recycling of nutrients.

Microorganisms. Microbial processes in the water column and in sediments are responsible for nutrient cycling and in particular for the decomposition of carbon compounds. Acidification of the water decreases the rate of decomposition. In laboratory studies, Bick and Drews (1973) showed that as pH decreased, the number of bacteria and protozoans decreased, the populations of fungi increased, and the rates of decomposition and nitrification decreased. Traaen (1980) reported measurements of decomposition rates of birch litter and glucose–glutamate mixtures. When the pH was decreased from 7.0 to 3.5, litter decomposition dropped to 30% of control levels, and fungal dominance replaced that of bacteria. As a general rule, when rates of decomposition are reduced, organic matter accumulates, and nutrient cycling is inhibited. This outcome interferes with nutrient supplies for plants. In addition, many invertebrates prefer to feed on detritus that has been acted on by bacteria. Therefore, a reduction in bacterial activity reduces invertebrate populations.

Phytoplankton. Acidification of waters usually results in diminished populations of some algal species and increased populations of others. Usually, one or two highly dominant acidophilic species will increase enormously, while other species will decrease in numbers. Decreasing species diversity goes with increasing acidity, but sometimes total biomass production will remain unchanged, or even increase, because of less competition for available nutrients. Hydrogen ion concentration does not control phytoplankton biomass directly, but indirectly through its affect on phosphorus cycling. Phosphorus is a growth-limiting nutrient. Phosphorus concentrations may be controlled by altered biogeochemical processes associated with acidification, including decomposition of organic matter, nutrient recycling, and aluminum release from soils. According to Conway and Hendrey (1981), "aluminum removes dissolved phosphorus from the water column by flocculation and precipitation." (Almer et al., 1978, also discuss this process.) This precipitation process is greatest in the pH interval between 5 and 6 and could result in higher phosphorus concentrations at the lowest pH levels. On the other hand, the aluminum concentration in lakes increases when the pH is below 5 and chemical combination of aluminum with phosphorus may actually increase.

Almer et al. (1978) reported that for 58 Swedish lakes studied, biomass levels were minimum at intermediate pH values (5.1 to 5.6) and much higher in lakes with pH values less than 4.5 and greater than 6.2. Although they did not measure the phosphorus concentrations in these lakes, it is possible that the highest concentrations existed in the most acidic lakes.

In Sweden and in Ontario, acidified lakes containing dense and extensive beds of periphyton (*Mougeotia* and *Zygogonium* sp.) in littoral zones have been found. Whereas nearby nonacidic lakes had as many as 50 species of algae in a single sample, the highly acidic (pH 3.8) Lumsden Lake in the LaCloche Mountains of Ontario had only 9 species. Diversity indices decreased sharply as pH fell below 5.6. Dinoflagellates dominated in strongly acidic lakes, whereas nonacidic oligotrophic lakes in eastern Canada were dominated by chrysophytes, or diatoms. (Because zooplankton cannot ingest dinoflagellates, this shift in populations with acidification affects the zooplankton population and in turn affects the fish populations.) In Sweden, dinoflagellates formed 85% of the biomass in lakes of pH 4.6–5.5. Dinoflagellates formed between

30 and 70% of the phytoplankton biomass in four acidic lakes (pH 4.2–4.8), but only 2 to 30% of the biomass in 10 lakes with pH levels of 5.8–6.8 (Yan, 1979).

Although chrysophytes and dinoflagellates dominate the plankton of acidic lakes, there are many exceptions to this. In Florence Lake near Sudbury, Canada, a lake with a pH range of 4.4 to 4.9, blue-green algae constituted 70% of the phytoplankton community. In six other lakes in the Sudbury area, blue-green algae were found in large numbers. This makes it difficult to be categorical as to what may happen with increasing acidity. So many synergistic factors are operating that numerous responses are possible as pH values decline.

Zooplankton. Zooplankton are the smallest metazoans in lakes, ranging in size from about 0.05 to 10 mm. Protozoans, rotifers, crustaceans, and small insects constitute most zooplankton communities. They provide food for many species of fish and therefore are vital parts of the food web of lakes. Of these four groups making up zooplankton communities, crustaceans usually form more than 90% of the biomass, whereas rotifers, which have short generation times, make up 50% or more of the zooplankton productivity. Studies from acid mine-drainage lakes indicate that some species of rotifers survive when all crustacean zooplankton have been killed. From this observation, we would expect rotifers to become more dominant among zooplankton populations as lakes undergo acidification. However, observations in the field appear contradictory because both decreases and increases of rotifer populations are found with increasing acidity. However, acidification brings on other changes in lakes besides increases in the H$^+$ concentration. Acidification increases the transparency of a lake by reducing algal densities, thereby increasing light intensities; and changes in nutrient cycling occur. These changes affect zooplankton populations. In addition, microbial decomposition is inhibited with increasing acidity, thereby reducing the recyling of essential nutrients.

Aquatic macroinvertebrates. In certain instances an entire phylum of macroinvertebrates may be affected by acidification, whereas in other situations susceptibility is highly species specific. Mollusks in general are highly susceptible to low pH, often being restricted to habitats with pH greater than 5.8 to 6.0. No crustaceans were found in waters with a pH less than 4.6. *Gammarus lacustris* was absent from waters with pH below 6.0, and the crayfish *Astacus astacus* are rare in lakes where the summer pH value was less than 6.0.

Insects exhibited a wide range of tolerances to low pH. Larvae of Chironimidae, Hemiptera, and Megaloptera were often abundant in acid lakes (Almer, 1978), whereas species of Ephemeroptera and Plecoptera seem to drop out with decreasing pH.

The cause and effect relationships for most of these species have not been determined in relation to varying pH. Other factors vary with pH, such as the concentration and availability of nutrients, of bicarbonate and sulfate ions, and of various metals and organic compounds. However, it appears that shell-bearing organisms and molting crustaceans, because of their requirement for calcium, are most sensitive to low pH levels.

Amphibians. Frogs, salamanders, and toads breed in small ponds or in temporary pools. These are particularly vulnerable to acid runoff from snowmelt in the spring. Pough and Wilson (1977) found pH values of about 4.5 in temporary pools even though permanent pools nearby had pH values of 6.1. Amphibians are particularly sensitive to acidity during embryonic stages. This fact may account for the enormous decreases in amphibian populations that have occurred throughout much of eastern North America during the last 30 years. Likens et al. (1980) have shown that the initial runoff from melting snowpack may be appreciably more acid for a few days in the spring than the average monthly concentration of hydrogen ions. This shot of high acidity could be particularly lethal to embryonic amphibians.

Amphibians are important predators of aquatic insects, and in turn they serve as a high-protein food source for birds, snakes, and mammals. Pough (1976) reported embryonic mortality and deformities in yellow-spotted salamanders at pH values less than 6.0. This observation was confirmed by Strijbosch (1979) in the Netherlands, where he observed increasing percentages of dead and molded egg masses of frogs and toads with decreasing pH values below 6.0. No adult newts were caught from ponds in England when the pH values were less than 4.2, and the smooth newt rarely occurred at pH values less than 6.0. Gosner and Black (1957) studied the response of 11 species of frogs and toads to increasing acidity. Embryos were more sensitive than larvae or adults. When embryos of the cricket frog and the northern spring peeper were exposed to water of pH less than 4.0 for a few hours, more than 85% mortality occurred. Although many of these tests were in the laboratory rather than in the field, they do indicate a high probability for lack of reproductive success among amphibians in ponds receiving acid deposition. As with any group of organisms, there is a range of sensitivity to acidity— some species are more sensitive than others.

Fishes. Many reports exist concerning the loss of fish species from lakes and streams undergoing increasing acidity. Some specific fish kills are attributed to direct acidification. For example, Jensen and Snekvik (1972) re-

port the mass mortality of Atlantic salmon (*Salmo salar*), and Leivestad and Muniz (1976) report a brown trout (*Salmo trutta*) kill, both attributed to low pH levels. Harvey reported a kill of several species of fish in Plastic Lake, Ontario during the spring snowmelt.

Laboratory studies indicate that fishes are particularly sensitive to acid water during their reproductive stages. Low pH can inhibit gonadal development, reduce egg production and viability, reduce sperm viability, and cause spawning failure. Juvenile fishes are often more sensitive to acid stress than are eggs, particularly near pH 4. Fish species differ considerably in their sensitivity to acid water at different pH levels. Smallmouth bass (*Micropterus dolomieui*), walleye pike (*Stizostedion vitreum*), and burbot (*Lota lota*) have reproductive failure at pH values between 5.5 and 6.0. Lake trout (*Salvelinus namaycush*) and trout-perch (*Percopsis omiscomaycus*) have reproductive failure at pH values between 5.2 and 5.5, whereas lake herring (*Coregonus artedii*), yellow perch (*Perca flavescens*), and lake chub (*Cousius plumbeus*) only fail at pH values of 4.5 to 4.7. There is even considerable variability in sensitivity to acidification among strains of the same species. The question arises: Can genetic selection occur for more tolerant varieties? It very well might except for the fact that acidification of many lakes and rivers in North America, as in Europe, is occurring too rapidly for genetic selection to take place.

The loss of genetic diversity within fish species is one of the great ecological disasters of modern times. As lake after lake becomes acidic, more genetic diversity will be lost. This loss in the gene pool is irreversible.

In Nova Scotia, nine rivers can no longer support salmon runs because of low pH levels. The angling catch as a percentage of the average for the years 1936 to 1940, for 12 rivers with pH greater than 5 in 1980 and for 10 rivers with pH less than 5 in 1980, is shown in Figure 4 for the period 1936 to 1980. The dramatic decline since 1950 in the angling success in Nova Scotia for rivers with high acidity is clearly evident.

Increased acidity of lakes or streams produces an increase in the mobility of heavy metals and aluminum. The metal ions are released from both the sediments and the watersheds. Certain metals are more toxic to fishes than others. Metals can affect osmoregulation and ionoregulation, calcium metabolism, oxygen transport, and possibly the susceptibility of fish to disease. Aluminum toxicity affects the epithelium of gills, as well as osmoregulation. Aluminum seems to be most toxic at around pH 5, although it begins to be mobilized from the watershed at about pH 5.5. Aluminum is most toxic to fishes when in hydroxyl form; when complexed by organic matter it is not toxic at all. Mass mortalities of fishes, observed during episodic, acidic spring runoff, are most likely the result of elevated concentrations of inorganic aluminum, mobilized from soils by the strong acids contained in snowmelt. Increases in aluminum to 0.2 mg/liter or greater are sufficient to induce stress and mortality in fishes.

The Adirondack region is one of the most sensitive areas of the eastern United States to ecological impact by acid deposition. This is a region of granitic rocks and shallow soils, so there is little buffering capacity against acidification. It is also a region receiving some of the most acidic precipitation in the United

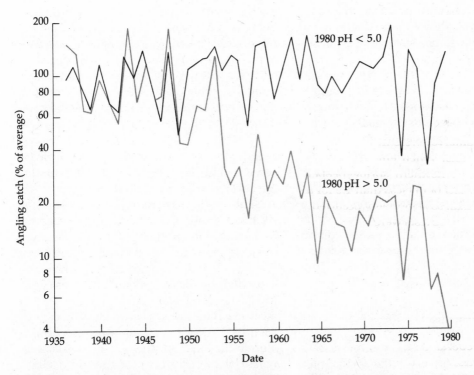

FIGURE 4. Angling catch as a percentage of the average for 1936–1980 for streams in Nova Scotia. The solid line shows the catch in streams whose mean pH in 1980 was less than 5.0; the dotted line charts those streams whose mean pH in 1980 was greater than 5.0.

States. Forty high-elevation lakes in the Adirondack Mountains were surveyed for fish during the period 1929 to 1937. According to Schofield (1976), only three of the lakes had a pH below 5 and were devoid of fishes at that time. Entire fish communities, consisting of brook trout (*Salvelinus fontinalis*), lake trout (*Salvelinus namaycush*), white sucker (*Catostomus commersoni*), brown bullhead (*Ictalurus nebulosus*), and several chubs (Cyprinidae) were eliminated over the 40-year period from 1935 to 1975, and the pH level of all of the lakes declined. In 1975 Schofield (1976) surveyed 214 Adirondack lakes, including the 40 mentioned above. He found that 52% had surface water pH levels below 5 and that 90% of these acidified lakes were devoid of fishes.

Birds. The effects of lake acidification on birds is indirect because it causes changes in the food supply. Almer et al. (1978) wrote that "fish-eating birds, such as mergansers and loons, have been forced to migrate from several acid lakes, with decreasing fish stocks, to new lakes with ample food supply. In this way, many territories will become vacant, and this will lead to decreasing stocks." A reduction in the populations of young fish in a lake will affect aquatic birds very sharply. In the Adirondack Mountains of New York, the common loon (*Gavia immer*) has declined over the last 15 years, a time when fish populations in this area also declined. This decline has been attributed more to human disturbance in the area, and the probable relationship of food depletion has not been adequately investigated.

Terrestrial ecosystems

In contrast to the situation for the aquatic ecosystems, there is very little documented field evidence of visible or detectable damage to terrestrial ecosystems from acid deposition. An exception is the heavy fallout of pollutants from smelters; for example, in the vicinity of Sudbury, Ontario, all vegetation for miles around has been killed. Here the effects of acid deposition and heavy-metal poisoning are truly horrifying. By acid deposition, as used in this chapter, we do not mean the heavy localized effects, such as at Sudbury, but rather what is happening to terrestrial ecosystems well removed from locations with extreme local effects. Acid mine drainage and the consequent acidification of streams are discussed in Chapter 8 (see Hutchinson and Havas, 1980 for further information).

Potential impacts of acid deposition on plants include (1) damage to the leaf cuticle; (2) interference with normal stomate opening and closing; (3) poisoning of plant cells after diffusion of acid compounds through the stomates or cuticle; (4) interference with normal metabolism or growth processes; (5) interference with flowering, germination, and other reproductive processes; (6) interference with nitrogen fixation by inhibiting microbial or fungal activities; and (7) possible synergistic interactions with other environmental stress factors.

Soils. Most of the regions of the world receiving acid deposition at the present time are covered with north temperate forests, the soils of which are naturally somewhat acidic. Of these forest soils, those most susceptible to acidification are those that are noncalcareous, are often sandy and well drained, and are not strongly acid. Such soils are nearly saturated by basic cations, such as Ca^{2+}, Mg^{2+}, K^+, and Na^+, and these are leached away as they are replaced by acid cations, such as hydrogen, aluminum, or hydroxyl-aluminum ions. As the soil becomes dominated by the acid cations, the pH declines. On the other hand, finely textured soils having a high clay content have a much greater cation exchange capacity and are much better buffered against acid deposition than are the coarse-textured, more sandy soils described above. Strongly acid, podzolic soils are only slightly susceptible to further acidification because their exchange sites are already dominated by the acid cations, and the further addition of hydrogen ions really makes no difference.

Agricultural soils are usually limed to maintain pH levels in a range suitable for crops. The addition of fertilizers and lime to agricultural fields will far outweigh any inputs of hydrogen ions by acid deposition. Therefore, acid deposition should not affect crops through the soil linkage but may do so by direct physiological damage to the crop canopy. Soils that are not limed may be affected, and if crop production is dependent on biological nitrogen fixation, rather than on nitrogen fertilizers, then productivity would be seriously impaired.

Soils have a complex chemistry, and it is not always obvious what the effects of acid deposition will be. Further information concerning the properties of soils may be found in Clapham (1973), Ricklefs (1979), or Etherington (1982).

Forests. Our perception of the impact of acid deposition on forests is changing rapidly. A report by Aber et al. (1982) states that

> No changes in forest production have been reported even under severe, although short-term, acidifications in both laboratory and field trials. Actual increases are occasionally reported. Attempts to measure changes in tree growth through time under natural conditions have also shown little or no effect.

Vogelmann (1982) reports widespread damage to conifers throughout parts of North America and Europe and implicates acid deposition as the primary causative factor. Between 2.5 and 5 million acres of forest in Central Europe are reported damaged and die-

off is frequent. According to Vogelmann, the red spruces (*Picea rubens*) of the virgin forests on the slopes of Camel's Hump Mountain in the Green Mountains of Vermont are dying from acid deposition (Figure 5). Some of these trees are over 300 years old, although many younger trees are also dying. Dead and crown-damaged evergreens are common in the Adirondack Mountains of New York, the White Mountains of New Hampshire, the Laurentian Mountains of Quebec, and the Appalachian Mountains of West Virginia. Attempts to link the mortality of trees in both

FIGURE 5. Dead spruce trees on Camel's Hump in the Green Mountains of Vermont. Conifers in high-elevation forests of the northeastern United States are dying in increasing numbers, a phenomenon thought to be caused by acid deposition. (Courtesy of David Like, *Natural History Magazine*.)

Europe and America with disease or insects are unsuccessful and the inference that acid deposition is the cause is very strong.

Acid deposition falling on plant canopies is modified during its transport from canopy to soil. Contact of precipitation with vegetation results in repartitioning of some of the chemical compounds in the precipitation. Some substances can be absorbed by the foliage, and substances on or in the foliage can be leached and added to the throughfall. Studies indicate an enrichment of inorganic ions in the throughfall after contact with vegetation, with increasing losses of cations as the hydrogen ion concentration increases. Laboratory experiments show that organic compounds, such as hormones, amino acids, carbohydrates, and organic acids, are leached from leaves by throughfall water. The nature of leachate or throughfall depends on plant characteristics such as tree species; leaf structure and morphology; stand density, age, and composition; and rate of precipitation.

Spruce canopies have been found to filter dry deposition better than deciduous canopies. This is partly because the needles of conifers are present during the winter. During rain or snow, the SO_2 is dissolved in water adhering to the needle surfaces. Throughfall from coniferous canopies results in decreasing the pH relative to the pH values of precipitation in open areas. Generally, soils beneath coniferous stands are acid, and it is believed that the conifers themselves add to the acidity. An increase in soil acidity is considered detrimental to the chemical availability of several essential nutrient elements, such as nitrogen, phosphorus, potassium, calcium, and magnesium.

In Germany, the evidence is largely circumstantial that acid deposition is responsible for the die-off of silver fir, spruce, pine, and some deciduous trees. The problem is compounded by other environmental factors such as drought and cold, or by fungi, bacteria, and insects. It is possible that when trees are weakened by the stress of acid deposition, they then succumb to extremes of moisture or temperature. The symptoms are quite distinctive. In fir, the needles at the bottom of the crown turn yellow or dull and begin to fall off. Needle loss begins at the base of the crown and progresses upward. In spruce trees, crown die-off begins at the top and moves downward as well as from the outside to the inside. In beech, there is disruption of natural regeneration. Seedlings sprout but soon die. This seems to relate to poor root development. In mature beech trees, crown die-back occurs as yellowing and necrosis moves from the edge of the crown inward.

Eight percent of the forests in West Germany are afflicted by the syndrome, according to a government survey. Sixty percent of Germany's fir trees are affected and much of the damage is severe. Nine percent of the spruce are affected, 5% of the pine, and about 4% of beech, oak, and other deciduous species. The spruce are dying within their normal habitats on both well-buffered alkaline soils and acidic soils. Trees are dying in areas of relatively low air pollution as well as in highly polluted areas. For this reason, it is possible the combination of several factors acting synergistically is the cause of tree die-off. The combined effect of sulfur compounds and oxidants together may be more pernicious than when in isolation. Trees suffering from nutrient deficiency, brought on by nutrient leaching in acid soil, will lose their buffering capacity and become more susceptible to direct damage by sulfur compounds. More research on this issue is necessary before all the answers are clear.

Crops. Crops can be affected in many ways by wet and dry acid deposition. The canopy of the crop and the soil can be affected in ways similar to those discussed in preceding sections.

Crop production is always very dependent on the available nitrogen supply. Nitrifying bacteria are very sensitive to acidity. Their ability to fix nitrogen diminishes rapidly with decreasing pH below 6.0. Experiments show that the average number of nodules containing nitrogen-fixing bacteria on beans and soybeans decreases from 65 at pH 6 to 25 at pH 3.2. Soil acidification causes a reduction in the populations of bacteria and actinomycetes and an increase in the abundance of fungi. The increase of fungi may be the result of decreased competition from other heterotrophic microorganisms at low pH. Increased acidity can affect plant diseases by either inhibiting certain pathogens or by inhibiting the antagonist or predator to the pathogen. Microorganisms that break down plant litter are reduced by increased acidity, and as a result there is increased accumulation of soil organic acids.

Soybeans are a major economic crop throughout the world and are known to be sensitive to SO_2 pollution. Muller et al. (1979) and Irving and Miller (1981) report the results of both greenhouse and field experiments with soybeans subjected to SO_2 fumigation and to acid precipitation. They showed that the acid precipitation simulant produced no statistically significant effect on seed yield and a slight increase in seed size. The simulated acid rain may have contributed to the nutritional requirements of soybeans by providing sulfur and nitrogen to the plants during the pod-filling period. Fumigation of soybean plants with SO_2 always reduced gross photosynthesis at all concentrations of SO_2 (Figure 6), except at concentrations below 100 ppb, when the photosynthetic response of the fumigated plants exceeded the response by the control plants. Sprugel et al. (1980) reported further studies of SO_2 fumigation on soybean yields in the field. They found that the yield of the fumigated plants was 5 to 48% lower than that of the control

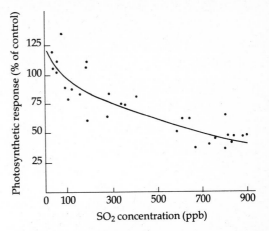

FIGURE 6. Relation between gross photosynthesis of soybean plants fumigated with SO_2 (expressed as percentage of control) and the mean values of SO_2 concentration during the fumigation period. Fumigations usually lasted from 4 to 6 hours and were done 24 times between 13 July and 17 August. (Redrawn from Muller et al., 1979.)

plants. Seed quality was affected less than seed yield. Apparently seed yield was more strongly reduced by SO_2 fumigation than was the photosynthetic response. These studies also showed that seed yield could be reduced by SO_2 concentrations at levels below that causing visible leaf damage. This finding is significant because secondary air standards currently in effect are designed to prevent visible injury to vegetation. These standards may not be adequate to protect against a reduction of crop yield.

SOURCES OF ACID DEPOSITION

The sources of air pollution contributing to acid deposition are many, including electric utilities, industrial and commercial power plants and boilers, residential furnaces, automobiles, trucks, and other vehicles (such as locomotives and heavy duty construction machines) (Figure 7). In addition, there are natu-

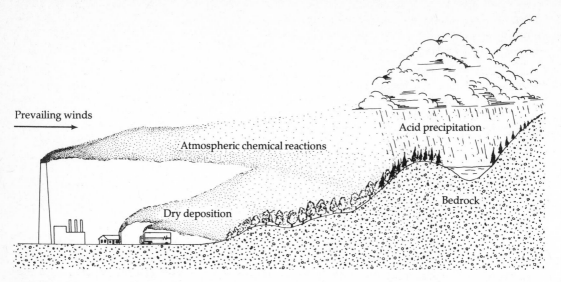

FIGURE 7. Sources of emission that contribute to acid deposition are electric power generating plants; smelters; automobiles, trucks, and other heavy duty vehicles; and residences.

ral sources of sulfur and nitrogen compounds entering the atmosphere, as described under the topic of biogeochemical cycles in Chapter 1. Emissions of SO_2 in the United States rose from about 14 million tons per year in 1950 to 29 million tons per year in 1965 and have been stabilizing near that amount since then. NO_x emissions were about 8 to 10 million tons per year in 1950 and about 26 million tons per year in 1980. Canadian emissions contribute significantly to North American air pollution. Canadian anthropogenic sources in 1979 contributed about 5.3 million tons of SO_2 per year and 2.2 million tons of NO_x per year.

Emissions by source

Table 4 shows the emissions of SO_2 and NO_x from human-made sources in 1978 for the United States. This table shows that electric utilities contribute 65% of all U.S. SO_2 emissions, but only 31% of all NO_x emissions. Transportation contributes about 40% of the nonbiological NO_x emissions in the United States. Coal burning accounts for 90% of the utility industry's sulfur emissions and 71% of its NO_x emissions. Environmental Protection Agency data show that a disproportionate share of these electric utility emissions is concentrated in the Ohio River valley, which includes all of Kentucky, most of West Virginia, and major portions of Illinois, Indiana, Ohio, and Pennsylvania. About 18% of the total U.S. electricity is produced in this six-state region, but it apparently contributes nearly 47% of the total SO_2 emission by utilities and 32% of those from all sources. Many of the electric power plants in this region are older, burn eastern coal, and are less stringently regulated than newly installed plants. Older power plants on the average emit more than 80 pounds of sulfur dioxide for each ton of coal consumed, whereas new power plants with scrubbers emit as little as 12 pounds per ton of coal.

Whereas in the United States about two-thirds of the total domestic SO_2 emissions

TABLE 4. 1978 Emissions (thousands of tons per year) from energy sources in the United States[a]

Source and fuel	SO_2	NO_x
Electric utilities		
Coal	17,490	5,632
Oil	1,892	1,353
Natural gas	0	946
Total	19,382	7,931
Industrial		
Coal	2,079	671
Oil	1,265	363
Natural gas	0	3,949
Other fuels	165	429
Total	3,509	5,412
Commercial		
Coal	44	11
Oil	990	297
Natural gas	0	154
Total	1,034	462
Residential		
Coal	66	0
Oil	286	154
Natural gas	0	231
Total	352	385
All stationary sources		
Total	24,310	14,190
Transportation		
Highway vehicles	440	7,370
Other	440	2,970
Total	880	10,340
Nonenergy sources		
Total	4,510	1,100
TOTAL	29,700	25,630

[a]From a report by the Comptroller General of the United States, "The Debate over Acid Precipitation: —Opposing Views — Status of Research." EMD-81-131, Sept. 11, 1981. Available from the U.S. General Accounting Office, Document Handling and Information Services Facility, P.O. Box 6015, Gaithersburg, MD 20760.

come from power plants, in Canada about 40% comes from nonferrous smelters and about 15% from power plants.

Future projections

Emissions of SO_2 from electric power plants in the United States are expected to remain roughly constant, and may even decline, during the next two decades; whereas NO_x emissions from U.S. power-generating stations are expected to increase by about 50% by the year 2000. Emissions from transportation in the United States during the next 20 years are expected to remain near their present levels, unless the automobile air quality regulations are relaxed. Emissions of SO_2 from industrial, commercial, and residential sources are expected to increase about 50% in the United States over the next two decades.

In Canada during the next two decades, SO_2 emissions from power-generating plants are expected to increase from 0.8 million tons per year in 1980 to 1.4 million tons per year by the year 2000, unless emission controls are instigated. With controls, Canadian SO_2 emissions from electric power plants could decrease. Canadian copper and nickel smelting is expected to contribute no more SO_2 per year in 20 years than now, that is, about 2 million tons per year. Uncontrolled NO_x power plant emission in Canada is expected to rise from 0.33 million tons per year today to around 0.70 million tons per year in 20 years. Transportation in Canada may contribute as much as 50% more NO_x in 20 years unless strict vehicle emission standards are adopted.

11

Petroleum

Not all oil comes from wells. Oil-bearing rocks of the western United States contain enormous quantities of extractable oil. Photograph by Donald C. Duncan, U.S. Geological Survey; courtesy of John Donnell, U.S.G.S.

INTRODUCTION

Petroleum is a fossil fuel that, like coal, is a substance transformed from dead organisms into combustible hydrocarbons during millions of years under heat and pressure. The source of most petroleum is aquatic organisms (algae and plankton) that settled to the bottom of the shallow, nutrient-rich seas around the edges of continents. Many of these organic deposits are 500 million years old. Since their deposition, they have been overlaid by layer after layer of accumulated sediments. These sediments subjected the organic deposits to enormous pressure and heat. As these organic substances decomposed and formed various hydrocarbons, the lighter gaseous and lighter liquid materials migrated toward the surface where natural traps, such as shale, kept the oil and gas within the more porous underlying rocks. The most common porous rock is limestone, which was formed by coral reefs on the ocean floor. Not only did impervious sediments form over the reefs, but warping and folding further entrapped the oil and gas.

Oil

The first U.S. oil well was drilled by Colonel E.L. Drake near Titusville, Pennsylvania in 1859. However, as early as 6000 B.C. asphalt was used as fuel in Mesopotamia. Records show that about 3000 B.C. natural gas was used to light temples in Mesopotamia. The Chinese drilled wells to depths of 3000 feet or more to produce natural gas that was transmitted in bamboo pipelines for lighting and space heating. The Indians in North America dug oil wells by hand prior to 500 A.D., and the Indians of Mexico and Peru used oil and asphalt as fuel. By 1300 there was some production of oil at Baku in what is now the Soviet Union. Since then, oil production has spread to many parts of the world, including all of America, the Middle East, and many offshore coastal waters of continents. World oil consumption in the past has grown largely in the major industrialized countries. In 1978 almost 75% of the world's total oil consumption of 52 million barrels per day was in the United States, Europe, and Japan. However, projections show that by 1990 this percentage will be 63% of a consumption rate of 60 million barrels per day and by the year 2000 57% of 65 million barrels per day.

The United States produced 79% of the oil it consumed in 1978, but by 1990 it will produce only 40%. Domestic oil production includes conventional production of crude oil and natural gas liquids but not synthetic oil and gas. It is anticipated that more oil will continue to be produced than discovered, with the result that reserves will continue to decline. For example, the amount of crude oil discovered in the onshore lower 48 states has declined at an accelerating rate since the peak was reached around 1930 (taken as a 10-year average). From this decade to the decade around 1940, discovery dropped 15%, another 34% in the decade around 1950, and 59% in the decade around 1960. Once the big fields had been discovered, it was all downhill. Conventional oil production is expected to decline in the United States and Canada through the 1980s and then to remain fairly constant for the next decade. Discovery may moderate the decline, but most of the new fields are likely to be in deep water or in frigid climates, where production costs are very high. Unfortunately, many of the large trapping geologic structures in the eastern Gulf of Mexico, the Gulf of Alaska, and the midAtlantic outer continental shelf have turned out to be dry. A reevaluation of 12 of the top outer continental shelf prospects of 1975 led to decreases for 10 of them. Estimates for the Beaufort Sea went up from 4.7 to 13.6 billion barrels of oil. But the rest of the outer continental shelf accounted for a net decrease of 13 billion barrels of oil equivalent. Estimates for onshore Alaska dropped by 6 billion to 18.1 billion barrels of oil equivalent. Estimates are difficult to make; and there are wide discrepancies because various methodologies are used to make the estimates.

Oil is expected to remain the largest single source of supply for meeting the world energy demand over the period extending from now to the year 2000. Availability will depend on new discovery. Prior to 1970 discovery rates of oil worldwide exceeded demand, so that the inventory of reserves was increasing. Since the early 1970s, a decline in oil discoveries and a continuing rise in consumption have reversed the situation, so that reserves are declining. World reserves of conventional oil are estimated to be between 1.45 and 2.12 trillion barrels.

Oil is easily mined, transported, stored, and controlled. Once burned, it is gone. There is generally a short supply in the world; and there are many environmental dangers associated with its production and use. The burning of oil produces carbon dioxide, carbon monoxide, nitrogen oxides, sulfur dioxide, particulates, and aldehydes. Oil spills, although serious on land, are particularly troublesome on seas, lakes, and rivers.

Gas

The gaseous phase of petroleum is natural gas, which is a mixture of hydrocarbons and is mainly methane. Natural gas is of biogenic origin, having formed as a result of chemical and bacterial decompositions of larger organic molecules. It is most often found in association with pools of oil.

Gas is the cleanest burning of all fossil fuels. It is reasonably free of sulfur, but does release carbon dioxide, carbon monoxide, and nitrogen oxides, particularly when burned in large power plants. Gas is a convenient energy form because it can be transported through pipelines and compressed or liquified in tanks. It provides heat for half of all U.S. homes and industries and is a vital raw material for the production of fertilizers. Thirty percent of U.S. energy consumed comes from gas.

Many dire projections have been made about the expected lifetime of natural gas reserves in the United States. Linden (1981) suggests that at current rates of consumption there is a 50-year supply remaining if the total U.S. natural gas recoverable reserve is 1000 trillion cubic feet, equivalent to 1000 quads of energy. This is a conservative figure as the combined total of conventional and unconventional quantities of natural gas from Tables 1 and 2 in Chapter 4 would indicate that 917 plus 671, or 1588, quads are potentially available. Gas experts optimistically assume that conventional natural gas will be substituted for the large quantities of oil now used in stationary heat applications, such as industrial boilers and power plants.

Composition of petroleum

CRUDE OIL is one of the most complex of natural substances on Earth. Basically, oil is made up of hydrocarbon molecules (constituted of hydrogen and carbon), but small amounts of sulfur-, oxygen-, and nitrogen-containing compounds also are present. Sulfur, hydrogen sulfide gas, and organic sulfur compounds are usually present and represent between 0.1 and 5% of the total by weight. Excess sulfur must be removed from crude oil. The heavier the crude oil, the higher the sulfur content. Sodium chloride and heavy metals such as vanadium and nickel also occur in crude oil.

Crude oil contains four principal types of hydrocarbons. ALIPHATIC COMPOUNDS are straight- and branched-chain molecules in which each carbon atom is linked to four other atoms—the so-called saturated molecules. These compounds often make up the bulk of crude oil and are common in gasoline and other fuels. They are also referred to as PARAFFIN hydrocarbons.

ALICYCLIC COMPOUNDS are also saturated molecules, but some of the molecules join together in closed rings. These compounds include the naphthenes.

AROMATIC COMPOUNDS are unsaturated, closed-ring molecules. They contain the so-called benzene ring and include many one-, two-, or multi-ring compounds. These compounds are often carcinogenic.

OLEFINIC COMPOUNDS are unsaturated hydrocarbon molecules in which double or triple chemical bonds are formed between carbon atoms; however, they are not of the benzene-ring type. Olefins do not occur in crude oil but are formed during the refining process and occur in many oil products.

The physical properties of crude oil are often described in terms of the boiling points of the hydrocarbon compounds within it. The boiling points may range from below room temperature to over 500°C. The compounds with the lowest boiling points are the simpler ring- and chain-type, saturated hydrocarbons—the aliphatic and alicyclic compounds. Intermediate fractions are represented by the toxic aromatic compounds, and the higher boiling point compounds include the complex polycyclic aromatics that are carcinogens.

Various hydrocarbon products are obtained from crude oil by fractional distillation. The products distilling in different temperature ranges are collected separately. For example, the product that distills below 200°C is known as NAPHTHA; that distilling between 200 and 300°C is a refined-oil distillate containing KEROSENE; next is an oil used in some engines and referred to as GAS OIL; and at higher temperatures one gets LUBRICATING OIL, RESIDUAL FUEL OIL, ASPHALT, and PARAFFIN. The naphthas are used for making gasoline, a product derived by further refining. The naphthas are broken down into lighter and simpler molecules by a process of distillation (boiling at high pressure and temperature) known as CRACKING. Increased yields of gasoline are accomplished by the use of a catalyst, usually a synthetic clay, and by distillation at low pressure.

PETROLEUM IN THE OCEANS

The annual contamination of the oceans by oil is horrendous. The quantity of oil from human activities is variously estimated at 1 to 10 million metric tons per year, with the most probable rate at about 5 million metric tons per year. This is greater than or equal to the amount of oil seepage from underground sources into the ocean, estimated to be in the range of 0.2 to 6 million metric tons per year. Table 1 is a summary of the sources for this ocean oil contamination.

The main sources of oil contamination in the oceans are estimated to be motor vehicles at 29.4%, or 1,440,000 metric tons; tanker cleaning operations at 19.75%, or 967,000 tons; industrial machinery at 15.3%, or 750,000 tons; bilge pumping from all other vessels at 12.25%, or 600,000 tons; and refineries and petrochemical plants at 6.12%, or 300,000 tons. Tanker and tank barge accidents together account for only 5.76%, or 282,000 tons, although vessel casualties get a great deal of notoriety. Other vessel casualties account for 5.11%, or about 250,000 tons. It is surprising that the estimated contribution to ocean pollution from motor vehicles is so large, but onshore waste oil disposal practices are poor. The American Petroleum Institute estimates that 1.25 billion gallons (about 4.2 million metric tons) of crankcase oil are released to the environment each year. Table 1 shows that about one-third of crankcase oil released to the environment gets into the oceans. (See Travers and Luney, 1976.)

TABLE 1. Amounts of oil released by human activities into the marine environment

Operations	Metric tons per year	Percentage of yearly total
Tankers		
Cleaning operations	967,000	19.75
Bilge pumping, leaks, and spills	100,000	2.04
Vessel accidents	250,000	5.11
Terminal activities	70,000	1.42
Tank barges		
Leaks	20,000	0.41
Vessel accidents	32,000	0.65
Terminal activities	18,000	0.38
All other vessels		
Bilge pumping, leaks, and spills	600,000	12.25
Vessel accidents	250,000	5.11
Offshore operations	100,000	2.04
Nonmarine operations		
Refineries and petrochemical plants	300,000	6.12
Industrial machinery	750,000	15.31
Motor vehicles	1,440,000	29.41
TOTAL	4,897,000	100.00

Although most of the concern about oil pollution has been with the marine environment, the impact on freshwater ecosystems may be more serious because of their small size, which limits their ability to assimilate the oil. The smaller dilution volume of a lake or stream limits the dispersion of oil, and the relatively calm surface allows the oil to spread over the entire surface. Freshwater sources of oil contamination include refineries, petrochemical plants, industrial and farm machinery, boats, and automobiles. The oil gets into the water directly or indirectly through sewers.

Oil transport at sea

Approximately 4000 tankers are plying the seas of the world, moving petroleum from sources to users. Tankers are the source of the highest volume of oil spilled at sea. Tanker sizes have been increasing during recent years. Where once most of the world's oil transported at sea was by tankers of deadweight tonnage (dwt) under 50,000, today a large number of SUPERTANKERS of at least 100,000 dwt are in operation. Deadweight tonnage is the total weight of the ship's cargo, fuel, and stores, excluding the weight of the ship itself. For tankers, deadweight tons are nearly equal to the amount of petroleum carried, because fuel and stores are minor by comparison. Tankers over 200,000 dwt are called VERY LARGE CRUDE CARRIERS (VLCCs); those over 500,000 dwt are called ULTRA LARGE CRUDE CARRIERS (ULCCs). The oil companies prefer the larger tankers because their use considerably reduces the cost of shipping oil. For each 1000 tons of petroleum delivered in a 250,000-dwt tanker over 20,000 km, only 20 tons of fuel are consumed to propel the ship. A 50,000-dwt vessel requires three times the fuel to move the same tonnage. The length, breadth, and depth of a 200,000-dwt tanker is about twice that of a 20,000-dwt tanker, but it will carry 10 times as much oil. However, supertankers, although cheaper to build and operate than the smaller tankers, have limited maneuverability and deep drafts. This in turn decreases their ability to avoid collisions and running aground. The crash stop distance for a 200,000-dwt tanker is about 3.2 km, and the time it takes to stop is about 11 minutes. (A crash stop distance is determined with the engines in full reverse and the ship without steering.)

It is probable that the estimates of oil spilled from tankers are too low. Ninety percent of the oil spilled at sea from tankers is spilled deliberately. The unloading procedure for tankers is so inefficient that substantial petroleum residues remain in the tanks. The tankers then use seawater as ballast on the return trip to the oil fields; and the contaminated seawater is pumped out prior to reloading. Many tankers haul grain on the return trip from the United States, so they go offshore to flush out their tanks with seawater prior to taking on the grain.

A system called LOAD-ON-TOP (LOT) is used for most modern tankers and greatly reduces the amount of oil discharged at sea. In this procedure the oil–water mixture in cargo tankers is allowed to settle so that the oil rises and lies on top of the water. The water is pumped out from under the oil and the remaining oil is retained on board to combine with the next shipment. The use of LOTs has reduced the amount of pollution put into the oceans by tankers.

The principal crude oil routes at sea are shown in Figure 1; nearly 2 billion tons of oil are transported annually along these routes. It is evident that the major amount of oil transport is from the Middle East to Europe via the Cape of Good Hope and to Japan via the Straits of Malacca between Sumatra and Malaysia. The greatest hazards occur as the tankers get near shore or into narrow channels, where a failure of the steering mechanism or poor navigation can result in an accident and a spill. Steps are being taken to improve tanker operations and to reduce the risks of collisions through better navigational equipment and improved traffic control systems. Although the world may never be free of tanker accidents, we can minimize the opportunities for them to occur. (See Figures 2, 3, and 4, which illustrate tanker accidents and their effects.)

Morphology of a spill

Several processes interact to alter the distribution and composition of oil released into the marine environment. Physical processes dominate first over biological processes with respect to the movement and degradation of an oil spill. Spreading is the earliest and most critical process, followed by intermediate-time processes such as evaporation, emulsification, and dispersion. Sedimentation and biodegradation largely determine the ultimate fate of the oil and may be active for many years following a spill. Oxidative processes also are important with respect to the final decomposition of the oil.

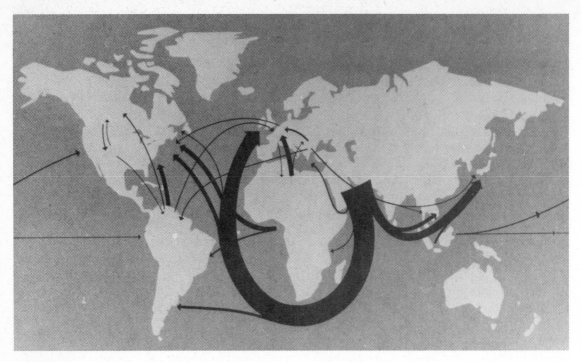

FIGURE 1. Principal crude oil routes in 1978. Arrow thicknesses indicate relative amounts of crude oil traveling the route. (From Exxon, 1980.)

Spreading. SPREADING, enhanced by wind, is the most dominant process for about 10 hours after an oil spill. Spreading is retarded by the inertia of the oil and by the oil–water surface friction. Inertia is a function of thickness, density, and temperature. Cold water temperatures cause oil to spread more slowly than warm water temperatures. Spreading increases the surface area of an oil slick and accelerates the weathering and degradation processes.

Evaporation. EVAPORATION is the most substantial initial degradative process of an oil spill. Fifty percent of an oil volume may be evaporated within the first 24 hours. During evaporation oil moves from the liquid to the vapor phase. The evaporation rate for a hydrocarbon is a function of its vapor pressure, which tends to increase with decreasing molecular weight. Aliphatic compounds in the oil have the highest vapor pressures and aromatics the lowest. As a result, an oil slick is depleted of the lighter aliphatics and enriched in aromatics as evaporation proceeds. Changes in the physical properties of a slick resulting from the loss of volatile hydrocarbons inhibit spreading and diffusion. Spills of refined products, such as kerosene and gasoline, may evaporate completely.

Dissolution. DISSOLUTION is the mass transfer of hydrocarbons from the floating or suspended state into the water column. The rate and extent of dissolution depend on oil composition and other factors such as tempera-

FIGURE 2. The Liberian registered tankship S.S. *Sansinena* buckled 19 December 1976 while alongside the Union Oil Terminal in San Pedro, California. (Courtesy of U.S. Coast Guard.)

ture, turbulence, and dispersion. Solubility in water determines the rate of dissolution. The solubilities of organic compounds decrease from the more polar molecules containing nitrogen, sulfur, and oxygen to the aromatics and aliphatics. Branched, saturated hydrocarbons have a higher solubility than straight-chain paraffin molecules because of their higher vapor pressures. Low-molecular-weight compounds are more soluble than those with higher molecular weights. The most volatile and toxic hydrocarbons, such as benzene and toluene, are the most soluble in water. However, these volatile compounds also evaporate rapidly.

Dissolution is a chemical degradative process of significance only within the first hour following a spill. Although continuing for several hours, the process becomes insignificant compared to evaporation. The envi-

FIGURE 3. The barge *Ethel H.* is seen surrounded by a barrier boom that contains the No. 6 oil spilled when she collided with the tanker *South-* *west Cape* on 5 March 1980. (Courtesy of U.S. Coast Guard.)

ronmental threat from dissolution of various toxic substances is limited to a relatively short time span.

Dispersion. DISPERSION is the incorporation of small globules of oil into the water column. Dispersion begins soon after the oil spill occurs and is most active around 10 hours following the spill. Once formed, the globules continue to break down and disperse throughout the lifetime of the spill. Dispersion has overtaken spreading as the primary process for distributing the spilled oil at about 100 hours after the spill. Dispersion progresses as oil drifts from the source and is greatest where there is much wave action, particularly in surf.

Emulsification. EMULSIFICATION is a process whereby oil in water forms a viscous cream that is referred to as CHOCOLATE MOUSSE. In this case the oil is the continuous phase into which tiny droplets of water are incorporated. During the first 10 hours following an oil discharge, the more volatile components evaporate, and subsequent increases in density and viscosity occur. At about this time emulsification also occurs.

Sedimentation. SEDIMENTATION takes place when the specific gravity of the petroleum residue exceeds that of seawater. This happens when adhesion occurs between the petroleum and detrital mineral particles, such as silts and clays, or the exoskeletons of

FIGURE 4. Cleaning up tar and oil from the beaches of Sandy Hook, New Jersey on 18 March 1980. The tarballs were the result of the major oil spill after the collision on 5 March of the barge *Ethel B.* and the tanker *Southwest Cape* in New York Harbor. (Courtesy of U.S. Coast Guard.)

plankton and algae. Common detrital minerals have specific gravities of 2.5 to 3.0; seawater is 1.025 and petroleum 0.82 to 0.92. Once adhesion occurs, the sediment sinks slowly to the bottom, or it may be transported to shore. The sedimentation process is most active during a period of a week to several months or longer. Petroleum sediments may persist for years in a state of slow degradation. Oil stranded on low-energy beaches is usually buried with little chance of reexposure, whereas on a high-energy beach the oil is frequently buried, reexposed, and transported about the shallow water environment. Oil stranded along shorelines is subject to severe weathering, autooxidation, or biological oxidation. Petroleum sediments that sink to

the bottom may be buried and effectively isolated. Some, however, are laterally transported by seasonal or permanent currents.

Autooxidation. AUTOOXIDATION is a degradative process that occurs when hydrocarbons in floating or dispersed oils react with oxygen molecules in the water column. The products of oxidation vary, depending on the petroleum properties and composition, water temperature, solar radiation intensity, abundance of inorganic components in the oil or water, and extent of diffusion and spreading of the oil. Weathering processes, such as evaporation and dissolution, act more rapidly than oxidation reactions to remove the more volatile compounds. Through the oxidative process petroleum hydrocarbons form alcohols, aldehydes, or ketones, and these are quickly oxidized to form carboxylic acids. These acids dissolve rapidly in the water. Another oxidative pathway is the formation of high-molecular-weight compounds by polymerization of aldehydes and ketones by phenols or the formation of esters by reactions between alcohol and carboxylic acids. High viscosity induces the formation of persistent tarry residues rather than allowing degradation to occur. The opaque nature of the tar inhibits photooxidation and degradation. The types of hydrocarbon groups in petroleum determine the extent and rate of autooxidation. Alkylated hydrocarbons oxidize more readily than other hydrocarbons because of weaker tertiary $C-H$ bonds. Oxidation of petroleum enhances dispersion and emulsification. Chain-breaking reactions of sulfoxide formation produce surfactants that in turn encourage the formation of emulsions. This process contributes more to the breakdown of an oil slick than the simple formation of water-soluble oxidation products.

Biodegradation. The preceding processes are primarily physical ones. Physical weathering may reduce the destructive potential of an oil spill. However, even after all those events occur, many petroleum hydrocarbons persist unchanged and some remain highly toxic. Petroleum residues, asphalt, and nonvolatile hydrocarbons may persist indefinitely. About 100 hours after an oil spill, biological and chemical degradative processes set in. Marine microorganisms, and some macroorganisms, ingest and metabolize the petroleum as a carbon source. The rate and extent of BIODEGRADATION depend on the nature and abundance of microbial species, predators, available nutrients, oxygen, temperature, and oil distribution and composition. Hydrocarbon-utilizing microbes are everywhere in the oceans of the world, as well as in fresh waters. These petroleum-loving microbes are often more abundant in polluted waters than in more pristine waters. In the North Sea it was noted that there was a dominance of oil-degrading bacteria in the water column and sediments near the Ekofisk production field and the Elbe River mouth in West Germany. This was also the situation for natural oil seeps in the Santa Barbara Channel (described later in this chapter). Organisms that are able to metabolize or degrade hydrocarbons include 28 genera of bacteria, 12 genera of yeast, 30 genera of filamentous fungi, and some genera of algae. The particular assemblage of organisms available varies greatly from place to place in the oceans, and so will the rate of biodegradation. Aromatics and n-paraffin hydrocarbons are the most readily biodegraded, but other components can support microbial activity. Different microbial assemblages dominate different crude and fuel oils, depending on their composition. Biodegradation rates are consistently slower in cold water than in warm water and oil trapped in ice is isolated and preserved indefinitely.

Cleanup processes

Effective efforts to clean up an oil spill depend on a good understanding of the characteristics of the oil and the dominant physical, chemical, and biological processes acting on the spill as a function of time. Physical methods of containment, such as placing floating booms around the spill (Figure 3) allow the oil to be treated chemically, or even burned. However, it is often the case that most of the cleanup methods do not work under open ocean conditions. For example, for the spills from the *Arrow* on 4 February 1970, the *Argo Merchant* on 15 December 1976, and the *Kurdistan* on 15 March 1979, none of the oil was effectively removed by these methods.

Various detergents and emulsifiers have

often been used to clean up oil spills, but their biological impacts can be more serious than those produced by the oil itself. Surfactants may be used to enhance the formation of emulsions, but as a general rule, emulsification inhibits cleanup. New methods are being developed for retarding emulsification or for removing emulsions and pelagic tar from the water. Such developments will reduce the risk of nearshore damage.

Hydrocarbons in the marine environment

Sediments and suspended particulates act as temporary and also as long-term sinks for many organic compounds. Transport of sediments and particulates to more pristine areas from heavily impacted areas by petroleum results in the dispersal of pollutants over a much wider region than that of the initial insult.

The highest concentrations of hydrocarbons, excluding sites of large petroleum spills, are in areas most affected by human activities. These include lake sediments in places like Lake Constance, Lake Zug, and Lake Washington; coastal sediments along tanker routes; regions of sewage and dredge spoil disposal; and sites in or near major harbors. According to Wakeham and Farrington (1980), hydrocarbon compounds found at these sites resemble weathered petroleum.

In sites far removed from direct anthropogenic sources, hydrocarbon concentrations are two to three orders of magnitude lower than in sites near human sources. Aliphatic and polycyclic aromatic hydrocarbons decrease in concentration with increasing distance from cities. Surface sediments of some freshwater and marine environments are enriched in aliphatic and aromatic hydrocarbons—concentrations that reflect industrialization.

Biogeochemical processes very much affect the composition of hydrocarbons in the marine environment. Microbial degradation removes *n*-alkanes but leaves behind the more resistant isoprenoids, cycloalkanes, and aromatics. It is important to distinguish hydrocarbons in the marine environment that are of petroleum origin from those of biological or biogenic origin. Petroleum contains a much more complex mixture of hydrocarbons with a much greater range of molecular weight than biogenic hydrocarbons have. (See Farrington, 1980 for a more detailed discussion of this topic. An excellent bibliography on the effects of petroleum in the marine environment was compiled by Samson et al., 1980; Walton, 1981.)

ECOLOGICAL IMPACTS

The most serious ecological and environmental impacts of oil exploration, drilling, and production occur in aquatic habitats. However, even terrestrial ecosystems are disturbed and must be restored, or the impacts must be mitigated, particularly in more fragile environments such as arctic and arid habitats. Exploration, drilling, and production on land each require roads and often pipeline corridors. Noise and dust are associated with the movement of vehicles. Other human activities, such as recreation or hunting, can ensue after access to wilderness areas is provided by roads. Animal movement, such as that of caribou in Alaska, may be much interrupted; and small animal activity, such as that of hares and rodents, may be seriously disturbed. Oil spills can contaminate soils and wash into streams and rivers.

Impacts on biota

Oil released to the aquatic environment will produce a variety of impacts depending on quantity, chemical composition, time, and duration of release, location, weather conditions, and cleanup operations. Oil spilled will either float on the surface or settle to the bottom if adsorbed on sediment. The impact of an oil discharge will be greatest on the organisms at or near the surface and on the benthic organisms at the bottom. Oil can cause death to aquatic biota by some or all of the following mechanisms: (1) coating and asphyxiation; (2) poisoning through ingestion of toxic petroleum compounds; (3) poisoning from exposure to toxic water-soluble petroleum compounds; (4) disruption of body insulation and loss of buoyancy. The impact of these mechanisms on biota is usually more acute in juvenile forms than in adults.

Birds. Thousands of birds have been the victims of oil spills. The Santa Barbara oil rig blowout in 1969 is estimated to have killed more than half the population of loons and grebes in the Santa Barbara Channel. Oil reduces the insulating quality of a bird's feathers sufficiently to cause a rapid loss of body heat and the onset of pneumonia or of hypothermia. Buoyancy of a bird's feathers can be sufficiently reduced to cause drowning. Preening by the bird of oil-contaminated feathers can cause death by ingestion of the oil into the digestive tract, where inflammation or poisoning occurs. Short of immediate mortality through the loss of insulation, a bird's metabolic rate increases, preening interferes with normal food gathering, and starvation results. Reproduction can be affected when oil hampers the laying and hatching of eggs. One of the important properties of eggs is their porosity to oxygen and carbon dioxide but not to water. A coat of oil on an egg interferes with this delicate gas exchange and causes death to the unborn chick.

The most susceptible species of birds are those that are attracted to an oil slick and attempt to land on it. Diving birds are especially vulnerable. Casualties from oil spills in western Europe have been auks, puffins, razorbills, guillemots (murres), and some sea ducks. Penguins are highly susceptible because they swim by porpoising (leaping out of the water) and thereby pass through the slick. The jackass penguin, an inhabitant of South Africa, faces extinction because of oil slicks left by tankers rounding the Cape of Good Hope. Gulls and shearwaters seem to be the least vulnerable to oil spills, partly because they seem to avoid them. In general, the widespread destruction of bird life during an oil spill seems inevitable.

Mammals. The effects of oil pollution on sea mammals are not well documented. However, generally they seem to be less vulnerable to damage by oil than are birds. For example, there is no evidence that mammals were seriously injured at the time of the Santa Barbara oil spill. If, however, the spill had occurred at the time when the elephant seals had their pups, the consequences could have been more serious.

Fishes. Fishes may be less vulnerable to oil spills than other marine animals because of the protection given by the slimy mucous on their surfaces and because of their ability to move away from contaminated areas, except during spawning. Laboratory experiments show that oil can be toxic to fishes. Many fish species feed on benthic invertebrates that can be contaminated by oil on sediments. It is also known that fish larvae concentrate near the surface, where they are susceptible to damage from toxicity or by entrapment. Oil tainting is a frequent complaint of humans using fishes, clams, oysters, and mussels. There is some evidence of carcinogenic effects of oil (for example, lesions and cancerous growths) on marine organisms.

The *Argo Merchant* tanker piled up on Nantucket Shoals in the winter of 1976 and dumped a full cargo of crude oil. Some of the oil coated fish eggs and larvae, killing about 20% of the cod eggs and 46% of the pollock eggs in the immediate area. Fish abundance decreased after oil was released from the wreck of the *Torrey Canyon* off the British coast in 1967, and 50 to 90% of the eggs of the pilchard, a herring, were killed. Fish kills have been reported for oil spills off Puerto Rico, Wake Island, and West Falmouth, Massachusetts. The Santa Barbara oil spill did not affect any significant number of fishes because it occurred at a time when most fishes were outside of the channel.

Plankton and mollusks. The spill of oil from the *Torrey Canyon* caused some destruction of phytoplankton, but zooplankton were not affected.

Extensive barnacle kills occurred at Santa Barbara in 1969 and in San Francisco Bay in 1971 following a tanker collision. The *Torrey Canyon* spill killed limpets along the Cornwall coast. These kills are the result of suffocation by crude oil rather than by toxicity. Nelson-Smith (1977) describes the toxicity of various hydrocarbons on the limpet *Littorina littorea*. Gasoline is highly toxic, kerosene and Kuwait crude oil are moderately toxic, and diesel fuel is nontoxic. So, depending on the nature of the spill material, the impact on mollusks may vary from toxicity to smothering. Barnacles and limpets are intertidal mollusks clinging to rocks and other objects. They are covered by a hard calcereous shell that affords them some protection.

CASE HISTORIES

Santa Barbara oil spill

At 10:45 a.m. on 28 January 1969 an oil blowout occurred from well No. 21 on Union Oil's platform A in the Santa Barbara Channel. The flow was unchecked until the well was finally plugged on February 8. After that, the well leaked about 500 barrels per day or less until May 31. The amount of oil released into the Santa Barbara Channel between January 29 and May 31 was estimated at 70,000 barrels. This oil eventually polluted the entire channel and over 230 km of mainland and island shoreline.

According to Foster and Holmes (1977), the greatest known damage occurred in surf-grass communities and among barnacle and bird populations. An estimated 14.6 tons of leaves and attached algae and invertebrates were killed; in addition, 9 million barnacles, 52,000 limpets, 30,000 mussels, and 9000 birds (60% loons and grebes) were killed. Cleanup procedures resulted in more damage on rocky shores and sandy beaches. Many uncertainties remain concerning the full ecological impact of this oil spill, partly for lack of adequate data on the populations of organisms prior to the spill. Unusually heavy rains in December 1968 and January and February 1969 caused flooding and introduced large quantities of sediment and fresh water into and onto the nearshore environment, thus confusing and compounding the situation. The result is that we do not have, nor shall we ever have, a full assessment of the biological consequences of the Santa Barbara oil spill.

The *Amoco Cadiz* accident

"The World's Biggest Environmental Hangover," so wrote John Clark (1979), director of the Conservation Foundation's coastal resources program. On 16 March 1978, the *Amoco Cadiz*, a new American supertanker carrying 68,000,000 gallons of crude oil and flying under Liberian registration, went on the rocks off the coast of Brittany near Portsall, France. A black tide, Maree Noire, floated forth, coated the beaches with a chocolate-mousse-like material, and drifted around the coastal bays, estuaries, and English channel for weeks afterward. A unique characteristic of the type of crude oil being carried by the *Amoco Cadiz* was the ready manner in which it combined with water. The compound, its consistency that of a gooey chocolate mousse, had an oil-to-water ratio of 2 parts of oil and 3 parts of water (for 5 parts of "mousse"). This substance was buoyant, penetrated sand, and adhered to particles. It penetrated the beaches of Brittany to depths of 1 meter.

On 15 April 1978 it was estimated that 16 million razor clams, sea urchins, and cockles were dead along a limited stretch of Brittany beaches. John Clark writes, "Farther east, at the great tidal marshes and flats behind Isle Grande, near Perros Guirec, we found 3,000 acres of scarce and vital wetlands habitat obliterated." At Les Sept Isles, a group of islands off the French coast, there is a bird preserve. According to Colonel Philippe Milon, then the curator of the preserve and president of the French League for the Protection of Birds, the kill of cormorants, guillemots, puffins, and auks was extreme. Over 1000 bodies were recovered, many more were dead. Probable total deaths were 5000. The number of birds rescued, cleaned, and released alive was 150.

The Atlantic puffin is an endangered species in France. In 1966 there were about 2300 nesting pairs on the remote offshore islands of the Les Sept Isles preserve. The *Torrey Canyon* spill on 18 March 1967, which had occurred just as puffins were landing and rafting up in the water before going ashore for breeding, knocked the population down to about 400 pairs. In the ensuing 11 years, the population had recovered to about 700 pairs. Then came the *Amoco Cadiz* accident at a critical rafting time for the puffins, and their population crashed again. Cormorants, guillemots, auks, and gannets survived better than puffins because they were not rafting but were on land or in the air. The total bird kill caused by the *Amoco Cadiz* disaster for both east and west sectors of the Breton coast was estimated at 15,000 to 20,000 birds.

After the spill French scientists studied two major bays along the coast and found that the oil had penetrated as much as two to three feet into the bottom sediment and that the concentration of oil in the sediment was extremely high, about one part of oil per thousand parts of sediment. They estimated that 40,000 tons of oil, one-fourth of the entire spill, had settled into the bottoms of two bays, Baie de Morlaix and Baie de Lannion. Of the 270,000-ton cargo, it is estimated that 70,000 tons of oil evaporated and about 150,000 tons remained in the sea and on the beaches.

Studies made 2½ years after the oil spill showed that ecologically things were recovering pretty rapidly. Amphipods had started to return to the Baie de Morlaix, but were not in their pre-spill density. Flat fish, inside the estuaries, of the year class 1978 were killed off, but in 1979 a normal stock reappeared. The Ile Grande salt marsh seemed to be permanently affected, as much as anything by bulldozing to remove the oiled sediments. Gundlach et al. (1983) report on the fate of *Amoco Cadiz* oil.

The story concerning the cause for the *Amoco Cadiz* accident is hair-raising. The English Channel is the world's busiest shipping gateway. It is a great funnel, 350 miles long, through which squeezes most of the ship transport to and from Europe. The Straits of Dover, the narrowest part of the English Channel, is 22 miles wide. Approximately 350 ships pass through the Straits daily and another 200 ferry crossings go at right angles. To add to the congestion, there is the Varne Sandbank in the middle of the Straits. Because of all the traffic, there are very definite rules-of-the-road for all ships entering the Channel.

The *Amoco Cadiz* entered the Channel on a heavy swell with west-southwesterly Force 7 winds and stronger yet to come. At 9:30 a.m. the captain was forced to turn the ship to starboard to avoid an unidentified rogue tanker steaming the wrong way down the lane. The *Amoco Cadiz* ended up pointing directly toward the French coast approximately 8.5 miles off the island of Ushant. At this moment the rudder jammed as the hydraulic control failed. The ship was drifting helplessly and a distress signal was sent. A German oceangoing tug, the *Pacific*, was in the vicinity and answered the distress call. The captain of the *Pacific* immediately made contact with the captain of the *Amoco Cadiz* and offered a Lloyd's of London open contract—a standard agreement for one ship to help another in an emergency, details to be settled later by arbitration in London. However, this is a salvage contract. The captain of the *Amoco Cadiz* did not feel he needed to agree to a salvage contract and the two captains, one Italian and one German, began to argue, mainly through radio operators ashore acting as intermediaries.

The *Amoco Cadiz* steamed steadily on toward the coast of France, as argument followed argument. The two captains not only disagreed as to terms, which were finally settled, but they disagreed as to the measures to be taken to keep the *Amoco Cadiz* off the rocky shore. Tow lines were secured and broke. An anchor was dropped, the flukes were torn off, and the anchor dragged. At 8:28 p.m. the port winch blew up and the anchor chain ran free. At 9:04 p.m. the *Amoco Cadiz* ran aground on the Rocks of Portsall on the northwestern coast of Brittany. The farce described here in only the very sketchiest of details is much more elaborate and most fascinating. (For more details, see Bennet, 1982.) This comedy of errors and human frailties dumped sticky black oil along the coast of France. Standard Oil Company of Indiana ultimately faced damage claims totaling $1.6 billion—an expensive environmental hangover indeed.

The *Torrey Canyon* accident

On 18 March 1967 the *Torrey Canyon* tanker carrying 117,000 tons of Kuwait crude oil piled at full speed into the Seven Stones Reef that lies between Land's End and the Scilly Isles. The master of this giant tanker had set his course 1400 nautical miles further back while in the vicinity of the Canary Islands. His ship was on automatic pilot and was supposed to pass well west of the Scilly Isles. A west wind, however, blew the ship eastward. When the captain finally discovered that they were seriously off course he called to the helmsman to change course. The ship would not change direction; finally it was discovered that the selector switch was in the control position and not in the hand position. When the switch was thrown and hard rudder to port was applied, it was too late, and the ship struck Pollard Rock, thereby rupturing all six starboard tanks. Human error once again resulted in a major ecological disaster.

Scientists from Liverpool University had been doing experiments concerning rates of recovery from disturbance by mollusks and algae on the rocky shores of the Cornwall coast in England. Rocky shores around the North Atlantic are dominated by acorn barnacles, limpets, mussels, and marine snails. Limpets feed by rasping algae or the immature offspring of larger plants off the surfaces of rocks in the intertidal zone. The effectiveness of this feeding action on the plants of the shoreline was amply demonstrated by the scientists' experiments in Cornwall. They had manually cleared rock surfaces of limpets and other mollusks. Within weeks of clearance all the rocks became covered by green algae (*Euteromorpha*) and by sea lettuce (*Ulva lactuca*). Sporophytes of a large brown seaweed became established and grew into moderate-sized plants. These in turn provided shelter for a greater diversity of small brown and red algae. But then limpets rapidly moved into the strip from either side and the small sporophytes were increasingly grazed by them. Winter storms ripped the larger seaweeds from the rocky shore and the whole area reverted to its predisturbed state, and the state of the surrounding ledges, in just over three years.

Following the *Torrey Canyon* oil spill a large number of limpets and many other organisms were eliminated along the Cornwall coast partly by the oil, but mostly by the cleanup efforts in which very toxic detergents were used (Nelson-Smith, 1977). The less seriously disturbed coastal areas regained a normal appearance after about three years. However, those areas where the limpets had been killed became densely colonized with algae. By 1972 algal colonization at the most damaged sites had been halted by the feeding of limpets. Enormous numbers of limpets were advancing, almost shell-to-shell, from bare rocks into areas covered by green algae. After six or seven years these rocky shores were indistinguishable from normal except for a continuing high density of limpets. Some experts estimated that 15 years would be necessary for the complete original species diversity to be reestablished. An interesting side effect of the *Torrey Canyon* disaster was the spilling of toxic emulsifiers, which were used during cleansing operations and caused serious damage to flowering vegetation along cliff-top areas. Since then, less toxic emulsifiers have been developed.

Milford Haven, Wales

Milford Haven in southwestern Wales was a new oil port in 1960. Most of it lies within the boundary of a national park. There are five marine steamship terminals and several adjacent refineries located there. Milford Haven is a deep inlet that undergoes vigorous natural flushing. The result is that there has been absolutely minimal impact of the oil spills on the biota of the rocky shoreline. No significant changes have been found in the abundance (density) and composition (species present) of the plankton in the waters of the inlet during the first dozen years of major oil transport.

Hazelbeach on the north side and Llaurcath Beach on the south side of Milford Haven had been studied by scientists for several years to establish the pattern of normal seasonal biological changes prior to the discharge of heated cooling water from the nearby Pembroke power station then under construction. Then both beaches received heavy doses of oil, first from the hull of a damaged tanker in January 1967 and two years later from an overflow of a refinery tank. All species of limpets on the rocky shores declined in density following each oil spill. *Littorina neritoides* was greatly reduced after the first spill and disappeared completely after the second spill. *L. obtusata* on the lower shore disappeared after each spill. *Monodonta lineata* is near its northern limit of distribution in Milford Haven. During the coldest winters its population is reduced naturally. *Monodonta* showed severe reductions following each spill. These reductions may have been enhanced by the winter weather on each occasion. This limpet recovered rapidly the following year or two. On the Cornish coast, about 120 miles farther south, this limpet was the only common mollusk surviving the *Torrey Canyon* disaster. It fed well on the bloom of green algae there when its competitors were eliminated. At Hazelbeach the large limpet *Littorina littorea* played the role of flourishing survivor.

In August 1973 the large tanker *Dona Marika* hit rocks in a storm near Milford Haven and spilled about 300 tons of high-octane gasoline, which washed ashore and evaporated almost immediately. Gasoline is much more toxic to winkles and limpets than is crude fuel oil. A sudden reduction in density of limpets and other grazers occurred after the spill. A bloom of green algae followed the destruction of herbivores, and brown algae invaded the rocks after that. Limpets gradually returned and eventually restored a balance between algae and mollusks.

Buzzards Bay, Massachusetts

A barge spill of No. 2 fuel oil in September 1969 contaminated Wild Harbor and another spill on 12 October 1975 polluted Winsor Cove of Buzzards Bay, Massachusetts. Teal et al. (1978) report on the biological consequences of these two spills to organisms living in the shoreline sediments.

The marsh grass, *Spartina alterniflora*, was killed back initially by the spill but partially recovered later. Aromatic compounds behaved about the same for the two spills. Light-molecular-weight compounds, such as naphthalene, decreased in concentration more rapidly than did the heavier compounds, such as phenanthrenes. The heaviest compounds seemed to be in larger total amounts in Wild Harbor after several years, although less concentrated than they were immediately following the spill. Oil slicks floating on the surface of the ocean in the bay were observed for several years after the spill, thereby demonstrating that there was a continuous washing of oil out of the sediments.

Krebs and Burns (1977), in their studies of fiddler crabs in the Wild Harbor marsh, found an inverse relationship between the amount of aromatics in the sediments and the size of the crab population. The lighter aromatic compounds seem to be the cause of most of the toxic effects; and because the lighter compounds diminish with time, the toxic impacts reduced even though the total amount of aromatics increased as more oil continuously came ashore. However, a crab overwintering in the sediments was in potential contact with oil and lighter aromatics for several months. Observations show that considerable mortality of crabs occurred during this period. All the crabs were in long-term contact with the heavier naphthalenes and phenanthracenes, but the extent to which these compounds may be taken up by crabs is not yet well documented.

A West Falmouth oil spill on 16 September 1969 in Buzzards Bay did extensive damage to local shellfish resources. Massive, immediate destruction of fish, shellfish, worms, crabs, and other crustaceans and invertebrates occurred within a few days of the spill. Bottom-living fish and lobsters washed up on shore in considerable numbers. Ninety-five percent of the marine animals recovered from trawls in 10 feet of water soon after the spill were dead or dying. The bottom sediments contained dead snails, clams, and crustaceans. Oil received by tidal rivers and marshes seriously damaged fish, crabs, and shellfish living there.

Eight months after the spill, oil that was unaltered in its chemical composition could still be recovered from the sediments of the most heavily polluted areas. Bacterial degradation of the oil was slow; the bacteria attacked the less toxic hydrocarbons first, leaving behind the more toxic compounds in the sediments. Hydrocarbons taken into the body fat and flesh of fish and shellfish are not removed by metabolic processes nor expunged through body wastes. The West Falmouth spill caused organism mortality and chronic pollution, oil in the sediments, damage to a fishery resource, and continuing disruption of the ecosystem for a long time afterward. (A fine discussion of the West Falmouth spill and its biological impacts is given by Blumer et al., 1971.)

Kurdistan tanker accident

On 15 March 1979 the oil tanker *Kurdistan* broke in two in rough seas just off Cape

Breton Island and released 7900 tons of Bunker C fuel oil into Cabot Strait. Much of the oil mixed with the pack ice and eventually began to drift ashore along Cape Breton Island, into Chedabucto Bay and on the shore of Isle Madame. Bird mortality related to this spill was estimated at more than 26,000 birds during the spring of 1979. Most susceptible to the oil were diving species (e.g., oldsquaw, eiders, and scooters), the alcids (e.g., murres and dovekies), and the grebes. Harbor seals (*Phoca vitulina*) and the gray seals (*Halichocrus grypus*) were also exposed to the oil. Most of the animals did not seem to be physically impaired. Eye lesions in seals reported after other spills did not seem to occur here because this oil did not contain a large proportion of volatile compounds. Preliminary reports indicated minimal effect on the fish populations.

Chedabucto Bay, Nova Scotia

In February 1970, the tanker *Arrow*, loaded with Bunker C oil, grounded on Cerberus Rock in Chedabucto Bay, Nova Scotia, and large quantities of oil were spilled. Extensive areas of shoreline in the bay were contaminated. Exposed rocky shores were generally free of surface oil after three years, but considerable amounts of oil persisted in sheltered lagoons at high tide level seven years after the spill.

On rocky shores the seaweed *Fucus vesiculosus* was killed at upper levels at first and then returned to its normal distribution by 1975. *Fucus spiralis*, confined to a narrow zone at mean high tide level, disappeared and did not return by 1976. Populations of barnacles and periwinkles did not show changes in abundance or distribution, except where affected by changes in algal densities. In sheltered locations, along sand and mud shores, the salt marsh cord grass *Spartina alterniflora* suffered heavy mortality about one

year after the spill. Recovery was evident two years later, and populations were normal by 1975. At lower levels of these sand and mud shores, the softshell clam *Mya arenaria* was abundant prior to the spill, but suffered heavy mortality from the oil. There was a slow recovery during subsequent years as oil concentrations declined.

Broughton Strait, British Columbia

On 24 January 1973 the freighter *Irish Stardust* ran aground on a reef in Broughton Strait between the northern end of Vancouver Island and the mainland of British Columbia. The accident released about 200 tons of fuel oil (classed as No. 5), which drifted ashore. Cleanup operations were begun immediately, but a decision was made by the Canadian government to leave one of the bays in a contaminated state for the purpose of study. This bay was then dubbed Reserve Bay. The hydrocarbons in this fuel oil were such that evaporation and dissolution did not cause much change in the composition of the oil. The cold water temperatures, persistent cloud cover, and lack of sunlight also tended to leave the oil composition relatively unchanged. However, various degradative processes did proceed more rapidly during the summer than during the winter.

One year following the oil spill most of the intertidal biota were well on the road to recovery, by which time the ecosystem had cleansed itself of most of the oil. Some oil was found in Reserve Bay four years after the spill, and it had a composition quite similar to that of the original oil.

Straits of Magellan

The oil tanker V.L.C.C. *Metula* became grounded in the Straits of Magellan on 9 August 1974 and spilled 51,500 tons of light Arabian crude oil, an oil similar to Kuwait oil.

What is particularly interesting about this site is that the geomorphology of the Straits of Magellan area is very much like the southeastern coast of Alaska, especially Cook Inlet. The western portion of the straits is similar to the northeastern coast of the United States. The impact of the spill on the beaches of the Straits of Magellan and of Tierra del Fuego was significant for more than two years after the spill. Although microbial degradation, weathering, and mechanical action of the waves and surf slowly removed the residual oil, some oil remained for many years afterward.

There appeared to be increased populations of heterotrophic bacteria at oil-impacted beaches. A cold-tolerant population of petroleum-degrading bacteria was found. Temperature did not appear to be a limiting factor for petroleum degradation in the antarctic marine environment. The oil was degraded very slowly by microorganisms, most probably as a result of low concentrations of nitrogen and phosphorus in the seawater. (Further information concerning this oil spill may be found in Colwell et al., 1978.)

Arctic ecosystems

Considerable development of offshore oil reserves in the Beaufort Sea threatens arctic marine ecosystems with oil contamination. Offshore well and pipeline structures must withstand enormous forces from ice, and ship transport of oil is highly hazardous. Failures of buried pipelines would first contaminate benthic ecosystems. Oil well blowouts will pollute open water as well as over-ice and under-ice ecosystems.

Prudhoe crude oil was found to degrade very slowly in most of the arctic ecosystems (Atlas et al., 1978). During the first month following a spill on water, the oil lost all light hydrocarbons; but during the following month almost no further changes in composition occurred. Essentially no biodegradation of over-ice oil occurred during the winter and only slow changes occurred during the summer. Under-ice oil degraded at even slower rates. More light hydrocarbons remained in the under-ice spills after three weeks than in the over-ice spills. The ice cover clearly limits evaporative losses, and dissolution probably accounts for the slow loss of light hydrocarbons. Prudhoe Bay oil had considerable impact on a variety of invertebrates in the benthos. The oil was toxic to some amphipods, but the isopods were as abundant in oil-contaminated sediment as in uncontaminated sediment. Some polychaete species seemed to be attracted to oil whereas others were killed by the oil. Some invertebrates recolonized contaminated sediments within a couple of months; however, the community composition was markedly different from what it was prior to contamination.

OFFSHORE OIL

Potential oil deposits lie off the coastlines of all continents. Drilling is going on in many of these continental shelf areas at the present time. Several thousand oil and gas wells have operated in the Gulf of Mexico nearshore waters for many years.

Drilling at sea may be done from rigs or ships. Rigs may be mounted on the sea bottom (Figures 5 and 6) or may be floating, as is the semisubmersible rig (Figure 7). Rigs mounted on the seabed are limited to depths of less than 100 m. The semisubmersible rig is used in deep water. Drill ships, which have a hole through the hull for the drilling apparatus, have mobility but must be moored by cables and anchors when drilling. They have drilled wells in water depths of 800 m and, in at least one instance, of 2000 m.

In 1968, scientists aboard the *Glomar Challenger*, a specially designed ship for scientific exploration in the Deep Sea Drilling Project

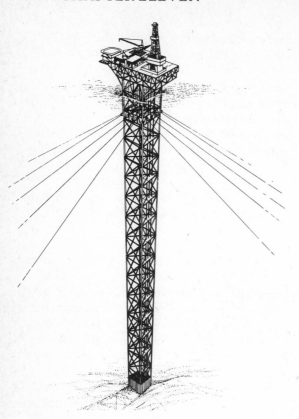

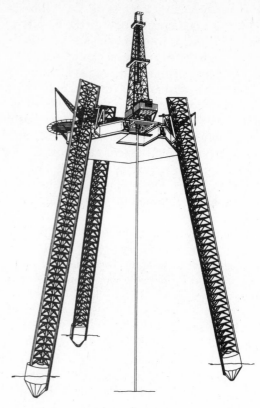

FIGURE 5. The guyed tower is a new platform for drilling or production designed to move with the waves. It can be used in depths of 200 to 600 m. (From Exxon, 1978.)

FIGURE 6. This jackup rig can be floated to location and raised (or "jacked up") to an appropriate height above the water. This rig's utility is limited to depths of about 100 m. (From Exxon, 1978.)

sponsored by the National Science Foundation, found oil and gas beneath 12,000 feet of water in the middle of the Gulf of Mexico. It was one of the first pieces of evidence that there is oil at great depths in the ocean. Now the oil companies are collaborating with the National Science Foundation in a 10-year, $700-million ocean margin drilling program at depths up to 13,000 feet. The U.S. Geological Survey estimates that the United States has between 12.5 and 38 million barrels of undiscovered, but recoverable oil offshore

along the edge of the continental shelf.

When exploratory drilling operations discover petroleum, permanent platforms are installed. These platforms are huge structures from which 30 or more development wells may be drilled, many at an angle. The platforms contain within their superstructures living quarters, storage areas, and operating space. Most oil platforms must be guyed and moored in position. Each anchor must weigh 20 tons. Offshore structures in the Gulf of Mexico are built to endure 140-mph winds

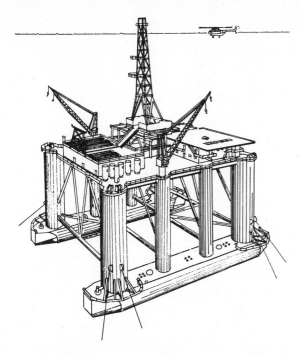

FIGURE 7. A semisubmersible rig is anchored in place and used to drill exploratory wells in water depths of 300 m or more. (From Exxon, 1978.)

and 70-foot waves; in the arctic they must withstand the enormous pressure of sea ice as well as high winds and waves. In Labrador, drilling contractors have had to tow icebergs away from rigs. Wells in the ocean have subsurface safety valves installed in the well hole beneath the ocean floor. These valves are designed to close automatically and shut off the flow of oil or gas in the event of a blowout. Despite precautions, some blowouts do occur. When this happens, an auxiliary well is drilled at an angle until it comes close to the bottom of the blowout well. Mud is then pumped under high pressure down the auxiliary well until it gets caught up in the uncontrolled well, where it stops the flow. Often blowout wells will stop flowing of their own accord. (See Ellers, 1982 for further information concerning offshore oil platforms.)

During the years 1971 through 1978, there were 7553 new wells drilled on the U.S. outer continental shelf, and 30 blowouts occurred. The total oil lost from these blowouts was less than 300 tons. In June 1979 the Ixtoc-I blowout occurred in the Bay of Campeche, in offshore Mexico, during the drilling of an exploratory well; 450,000 tons of oil were released into the sea before the blowout was capped in March 1980.

The *Ocean Ranger* disaster

The world's largest offshore oil rig, the *Ocean Ranger*, was moored 170 miles east of Newfoundland in the North Atlantic. It was struck by 80-mph winds and 50-foot waves, and the gale whipped up a blinding snowstorm at night in late February 1982. The oil rig collapsed, and its 84 crewmen were thrown into the ocean, with a loss of 53 lives. A Russian container ship went to its rescue but foundered in the pounding seas with a loss of 37 of its 42 crew members. The *Seaforth Highlander*, a standby supply ship, came to the rescue of the few survivors. The oil rig, built in Japan, was designed to withstand 130-mph winds, 110-foot waves, and a 3-knot current simultaneously. The 10-story submersible rig was operated by the Mobil Oil Company of New Orleans. It has been suggested that poor maintenance of the ballast systems and possible metal fatigue caused the collapse of the rig under conditions substantially less than its design specifications.

Offshore pipelines

Oil from many offshore fields can be pumped directly into pipelines that are laid on the bottom of the ocean. This is often the most efficient and safest way to get the oil ashore. A large pipeline may bring oil or gas ashore from several platforms and smaller gathering lines collect it to one point. Pipelines are made up of long sections of high-quality steel pipe welded together in a continuous string. Protective coating is applied and often a concrete cover is added to protect against damage and to provide more weight under water. Pressure pumps and compressors are placed along the line at intervals to maintain a constant flow of gas or oil. In shallow water near shore, at depths less than 200 feet, trenches are dug to contain the pipeline. A special trenching barge is used. A jetting sled is lowered to straddle the pipe. As it moves along the pipe, powerful jets of air and water flush out the bottom mud and sand, thereby forming a trench into which the pipeline settles. This operation disrupts plant and animal habitats. Spoil accumulates on either side of the pipeline ditch, extending to 50 m or more to either side. Most organisms in the disturbed sediments are smothered or uprooted. Studies in Chesapeake Bay show that dredging destroys about 70% of the bottom organisms in the disturbed area. Primary productivity in the water column is temporarily reduced by the increased turbidity caused by suspended materials. Recovery of the bottom organisms takes at least two years. Pipelines are thoroughly tested before oil or gas is put through them. Periodically a pipeline may be cleaned by using a plug or scraper that has blades, disks, or bristles and is pulled through the line. Deep-sea divers may inspect the line, or automatic equipment can be used to test for any problems. Pipelines are being laid at depths up to 2000 feet. Accidents will occur, but, in general, pipelines for offshore oil have proved to be extremely reliable and environmentally safe.

GULF COAST OF LOUISIANA AND TEXAS

The petroleum industry began onshore activities along the Louisiana and Texas coast in the early 1900s, and offshore operations began in the 1940s. The coastal zone of this region is known as the Chenier Plain ecosystem. It is a complex region of beaches, saltwater and freshwater marshes, tidal passes, cypress swamps, and deep offshore waters; and it is the major wintering habitat for migratory waterfowl in the United States. This region is rich in renewable marine resources and simultaneously rich in nonrenewable petroleum resources. Oil and gas production is currently valued at more than $500 million annually, but this figure is slowly declining as this resource is depleted. Sulfur and salt are also mined in the coastal zone.

Both onshore and offshore the petroleum and mining activities have caused a variety of environmental disturbances with ecological consequences. Onshore disturbances are from brine water and spoil disposal, road building, canal digging, drilling, exploring, pipelaying, process plant and refinery siting, oil spillage, and product storage. Offshore, the impacts are those of oil spills, submarine pipelaying, dredging, transport, and competition for space with commercial fishermen. Many important sport fishes concentrate around the offshore oil platforms.

Exploration activities involve the use of explosives for seismic surveys and the use of marsh buggies, boats, and barges within the marshes. All of these activities destroy vegetation and animals and produce erosion and sediment instability. Dredging (for pipelines) and digging (for canals for boat traffic) destroy the natural wetlands. In the coastal marshes it takes two to five years for a ditch to recover its original state, but in the forest swamp areas it takes 10 to 50 years for recovery. When flying over this coastal zone, one sees extensive evidence everywhere of channels, canals, pipelines, wellheads, marsh buggy tracks, spoil heaps, and other disturbances.

Coastal ecosystems

The offshore continental shelf in the Gulf of Mexico lies at a depth of less than 200 m. It is narrow near the mouth of the Mississippi River and is over 170 km wide off southwestern Louisiana and southeastern Texas. Food webs here are based on phytoplankton. Many burrowing organisms (filter-feeder types) are found in the bottom muds. The region is an important winter spawning ground for sport and commercial fishes (such as menhaden, Atlantic croaker, and mullet) and for invertebrates (such as brown and white shrimp). This zone is also the year-round habitat for ocean sunfish, oarfish, swordfish, king mackerel, seals, and whales.

The beach spawning area is an intertidal region continually buffeted by waves, tides, and currents. It possesses rich floral and faunal communities. Many mollusks, annelids, and crustaceans are found there, as well as juvenile pompano, Gulf kingfish, banded drum, longnose killifish, and rough silversides. It is used as a nesting area by boobies, brown pelicans, terns, gulls, and other shore and water birds.

The tidal passes are critical zones for the movement of organisms, nutrients, minerals, and detritus between inshore and offshore regions. The water is about 3 m deep in these passes, the bottom is of hard clay or sand and tidal currents are very strong. Young fish and shrimp move through the passes in early spring into the estuarine nursery grounds. Movement offshore through the passes occurs in the autumn and winter for spawning and for overwintering.

Open water bays along the coast are fairly turbid, have a clayey-silt bottom and contain excellent oyster beds and populations of anchovy and croaker.

The saline intertidal marsh zone is of special importance ecologically. Through it the detritus produced by the decomposition of shoreline emergent vegetation is transferred by tides to consumer organisms in the open bays offshore. It is the primary habitat for shallow shoreline fishes such as tidewater silversides, killifishes, sleepers, gobies, and oysterfishes. Marsh periwinkles (*Littorina irrorata*) and mussels (*Modiolus* sp.) are abundant, as are blue crabs. Oystergrass (*Spartina alterniflora*) is the dominant vegetation.

An intermediate or transition intertidal marsh zone between the saline and brackish intertidal marsh zones is dominated by wiregrass (*Spartina patens*). Ridges along the bayous in these areas contain marsh elder (*Iva* sp.) and buckbrush (*Baccharis* sp.). Fiddler crabs (*Uca* sp.) forage along the shoreline.

The brackish intertidal marsh zone of low salinity is an important nursery area for Gulf menhaden, red drum, Atlantic croaker, and black drum. The sediments of this zone are rich in peat and organic matter. Plants here are wiregrass (*Spartina patens*), black rush (*Juncus roemerianus*), and sedges (*Scirpus* sp.). Muskrat are abundant in this zone.

The freshwater marsh zone, about 1 foot above mean sea level, is the prime habitat for the alligator, bullfrog, nutria, and numerous waterfowl. Vegetation here includes maiden cane (*Panicum hemitomon*), water hyacinth (*Eichorina crassipes*), bulltongue (*Sagittaria* sp.), and alligatorweed (*Alternanthera philoeroides*). Deep peat deposits and floating mats of vegetation occur here.

The bald cypress (*Taxodium distichum*) and tupelo gum (*Nyssa aquatica*) swamps occur between the freshwater marshes and levees and ridges in the upper ends of the basin systems. These areas are seasonally flooded and dry, depending on the amount of rainfall. Animals here include mink, otter, nutria, white ibis, great blue heron, and bald eagles.

Finally there is the marsh–water interface zone where primary production of marsh grasses is extremely high, probably among the highest ever measured in coastal areas of the world. Fish here are extremely abundant. All of the marsh zones are very important for nutrient cycling. The sediments in them form ion-exchange beds that hold the nutrient element in equilibrium with the water. The marsh grass (*Spartina alterniflora*) acts as a pump in nutrient cycling by moving phosphorus, nitrogen, and carbon out of the deeper sediments and into the water by means of bacterial degradation of the plant material. The bacteria are in the marsh soils and on the roots of the grasses. The detrital material is the initial food source on which all the estuarine animals depend. Shrimp, oysters, and menhaden are detritus feeders. Wading birds, such as egrets and herons, feed largely on snails and minnows, which also feed on detritus. If the bird populations are damaged, the snail and minnow populations will grow unrestrained, thereby leaving less detritus for the shrimp, oysters, and menhaden. If the *Spartina* of the marshes are damaged, the entire food chain will be affected.

Clearly these ecosystems of the Gulf Coast are diverse, complex, and vulnerable to disturbance by petroleum operations. The mechanical disturbances created by blasting, filling, road building, pipelaying, dredging, drilling, or vehicle movement are obvious. Changed tidal action or interrupted flow of fresh water will seriously affect these ecosystems. Spilled oil is inevitable and some effect on organisms can be expected.

Ecosystem effects of oil

DeLaune et al. (1979) reported a study in which they added Louisiana crude oil to a uniform stand of *S. alterniflora* in a salt marsh in the Barataria Basin of Louisiana. Crude oil was added at rates of 1, 2, 4, and 8 liters m^{-2} to circular 0.25 m^2 plots enclosed by metal cylinders pushed 15 cm into the marsh environment. There were five replications of each treatment along with four control plots to which no oil was added. The crude oil was not toxic to *S. alterniflora* roots or stems and did not inhibit the flow of oxygen to the roots. Stem density was not significantly different in oil-treated and control plots, nor did the oil have any influence on the numbers of new shoots generated in the spring. The added oil did not remain on the water surface, but adhered to dead plant material on the marsh surface and to organic matter within the marsh soil. Neither the biological reduction of nitrate, manganese, iron, and sulfate nor the production of methane and ammonium in the sediments was affected by oil added to the soil. Crude oil was apparently not toxic to an assortment of anaerobes involved in these reduction processes.

A Shell Oil Company platform had a spill from 1 December 1970 to 16 April 1971, and between 90,000 and 119,000 barrels were spilled at a site seven miles south of Timbalier Bay. Observations along several transects from the platform outward showed that both benthic populations and fishes near the platform were stressed by the hydrocarbons. Numbers of each decreased as one approached the platform. Within a mile of the platform, benthic organisms were 13,000 m^{-2}, but outside a 2-mile radius the density was between 32,000 and 48,000 m^{-2}. Fishes had swollen gills, apparently as a result of the oil. Stomatopods, the adults of which are burrowing benthic organisms and the larvae of which are planktonic, were absent as adults from within a few miles of the platform. No dead organisms were found near the spill area, but some dead ducks were found near the coast about 48 km due west, where oil had drifted.

Yeast populations were monitored during the Shell spill. In oil-free waters the yeast populations were relatively sparse, but in water receiving oil from the spill concentrations increased 20-fold or more. After 13 weeks, the concentrations of yeast populations returned to the oil-free water baseline level, but species of the *Rhodotorula/Rhodosporidium* complex were absent. These species are known as oil-decomposing types. It is presumed they disappeared despite the oil enrichment because nutrients (nitrogen and phosphorus) became limiting. In other places where nutrients were not limiting, the presence of oil increased the concentrations of the *Rhodotorula/Rhodosporidium* complex. Such studies also were done in Barataria Bay, where controlled spills were generated.

Toxicity studies made on four species of marine shrimp showed that dispersants used to clean up the oil spill were more toxic than the crude oil. Concentrations of crude oil of 0.1 to 7.5% caused 100% mortalities of the four shrimp species. Mixtures of crude oil and emulsifiers were considerably more toxic than either the crude oil or the dispersant alone. Tests made on guppies showed that Louisiana crude oil at 4% concentration was lethal within 24 hours, but a high-asphalt Mississippi crude oil had no adverse effect at 4% concentration by volume over a 30-day period. An oil dispersant, the same as that used to clean up several Gulf Coast spills, killed guppies within eight hours.

OIL SHALE

Vast quantities of oil-containing shales exist throughout the world. They are present on all continents. The crude oil in shales is in the form of a waxy organic polymer known as KEROGEN. When the kerogen content of the rock material is sufficiently high, it is referred to as OIL SHALE even though in many cases other sedimentary rocks are involved rather than shales.

Commercial extraction of oil from kerogen-containing rocks began in France in 1838, in Scotland in 1850, and later in Estonia, Australia, Sweden, Spain, Manchuria, South Africa, Germany, and Brazil. However, as of a few years ago, only the Manchurian and Estonian plants were operating commercially. In the United States the American Indian burned the rock and pioneers distilled shale to make axle grease for covered wagons.

United States deposits

One of the largest deposits of oil shales in the world is in the Eocene Green River formation of western Colorado, eastern Utah, and southwestern Wyoming (Figure 8). The thickest and richest deposits are in the Piceance Creek basin of Colorado. Most of the high quality shale is in a formation called the MAHOGANY ZONE, a strata about 23 m thick that can be seen on the exposed faces of canyon walls. This zone is a sedimentary deposit that occurred in Lake Uinta, a freshwater lake covering the area during the Eocene Epoch of the Tertiary Period about 45 million years ago. In the eastern United States there are also vast oil shale deposits equivalent to as much as 2 trillion barrels of oil. These deposits are in older shales formed during Devonian and Mississippian periods about 345 million years ago. The eastern U.S. shales have generally been considered to be inferior to the western shales because they yield less oil per ton. However, new extraction methods make the eastern shales commercially useable.

According to Schurr et al. (1979), an estimate of the U.S. oil shale resources containing 10 to 65% organic matter (kerogen) is over 300,000 quads of energy; for shale with 5 to 10% kerogen, the estimate is 1.64 million quads. Table 2 in Chapter 4 lists the quantities of unconventional oil and gas resources. Notice that the quantity of oil recoverable from shale is estimated at 198 billion barrels of oil at a 60% recovery factor and would yield 5.8 million Btu/barrel, or a total of 1148 quads of energy. Heavy crude oils available would add another 174 quads of unconventional oil, or a total of 1322 quads. These

FIGURE 8. Cross-section of oil shale strata in the Piceance Basin of Colorado, showing the Mahogany Zone that contains the kerogen from which oil may be extracted.

numbers are based on shale containing 30 gallons of oil per ton. If the economics allow oil recovery at 15 gallon per ton, the potential recoverable oil would increase from 198 to 1000 billion barrels. The U.S. Department of Energy claims that the United States has a total of 4 trillion barrels of oil equivalent in shales. However, at least half or more of this will not be economically recoverable.

Extraction processes

The extraction of oil from shale is simple in principle. The oil-bearing rock is mined and brought to the surface, crushed, and heated to above 480°C to drive the oil out of the kerogen. An alternative method is to burn the rock IN SITU underground and drive off the oil and gas from the kerogen. In situ methods have not proved to be workable as yet. However, recent progress on in situ conversion techniques is promising. Usually the Mahogany Zone has an overburden of rock of 400 m or greater. If in situ methods are not used, then the ROOM AND PILLAR method of mining should be used instead of strip mining. This method creates large underground rooms carved out of the Mahogany Zone; pillars of shale are left behind to hold up the roof. Large equipment, shovels, and trucks can be used in the mine because the ceiling is 20 m high or more. Because it does not require special preparation, high pressures, or difficult catalytic procedures, processing oil shale is easier than either coal gasification or coal liquefaction. The cost of extraction has been a major limiting factor; because of continuing fluctuation in the cost of oil only limited production of shale oil is getting underway. The oil shale industry has been on a rollercoaster for the past decade or two. Oil shale has been described as a researcher's dream and an economist's nightmare.

Several different retorting methods are used. The simplest retort systems heat the crushed shale and separate the gas and oil by means of gravity and temperature differentials. A more complex system involves heated ceramic balls that are mixed into the crushed shale and tumbled together in a pyrolysis drum. The oil is literally beaten out of the crushed rock at high temperature. Because two solids are in contact, heat transfer is very efficient and the yield of shale oil is nearly 100%. By contrast, in situ techniques for oil recovery are likely to be less efficient, thereby leaving more oil in the burned-out shale. On the other hand, the enormous costs and energy expenditure associated with mining, hauling, and crushing are avoided. In situ processes require only about one-third the total number of people required for mining and surface retorting of shale, and the water use necessary is only about 25 to 33%. In situ processing reduces the necessity to dispose of spent shale by at least 80% or more and thereby reduces the environmental impacts.

In situ extraction of oil from shale requires fracturing the rock so that a flame front can pass through the bed and release the kerogen. The process is similar to that used for the underground gasification of coal. The shale must be fractured hydraulically or with explosives, but this is proving difficult. A modified in situ process avoids some of these difficulties by first mining out about 20% of the volume of shale and then fracturing the remaining shale with explosives. The mined shale is processed at the surface by the usual retorting methods. The explosion fractures the underground shale and the rubble expands to fill the void left by the mined-out shale. The shale is ignited and air and steam are injected at the ceiling to promote burning; oil flows downward ahead of the flame front and collects in a sump at the bottom of the chamber, from which it is pumped. The yield of oil is much lower than in aboveground processing.

Oil shale has less energy per ton than any substance used for commercial fuel. At least 1.5 or more metric tons of shale are required to produce a barrel of oil. With coal, only about 0.5 ton is needed to produce a barrel of synthetic crude oil. In order to be commercially worthwhile, an oil shale plant must process at least 50,000 to 100,000 tons of shale per day to produce between 33,000 and 65,000 barrels of oil per day. Studies by the U.S government indicate that the upper limit on oil shale production in the West would be 1 million barrels per day or 10 major processing plants. Daily U.S. oil consumption is about 18 million barrels per day. Therefore, oil shale will supply only a fraction of our total demand, about 5%. To produce 1 million barrels per day, 1.5 million tons of shale would need to be mined and retorted and about 1.3 million tons of spent shale would need to be disposed of each day. This is equivalent to 455 million tons per year, a truly staggering figure. Total coal production in the United States today is only about 50% greater. If a modified in situ process were used, it would be necessary to dispose of only about 100 million tons of spent shale a year.

Environmental and ecological impacts

The western United States, where much of the oil shale is located, is a semiarid region of deep valleys, high plateaus, and generally spectacular scenery. Because of the necessity to move massive amounts of rock material and to dispose of enormous quantities of spent shale, as well as to process all the material, the potential environmental and ecological impacts are considerable. An oil shale industry may create problems with dust and sulfur dioxide emissions into the atmosphere, drainage from underground aquifers, consumptive use of water, increased salinity in the Colorado River, revegetation of disturbed sites and of spent shale deposits, disturbance of wildlife habitats, alteration of aesthetic and scenic values, and human sociological changes.

Ecology. The Piceance Creek basin is covered with a mixture of shrubs and trees. Forests dominate only a minor portion of the basin. Cottonwoods (*Populus sargentii* and *P. angustifolia*) and box elder (*Acer negundo* var. *interius*) forests grow along the stream beds, and aspen (*Populus tremuloides*) and Douglas fir (*Pseudotsuga menziesii*) forests grow on the northern and eastern slopes of the mesas at middle and upper elevations, where favorable moisture exists. A mountain shrub vegetation of Gambels oak (*Quercus gambelii*) and mountain mahogany (*Cercocarpus montanus*) dominates much of the upper elevations, particularly on north- and east-facing slopes. Big sagebrush (*Artemisia tridentata*) predominates in the drier habitats at all elevations. At lower elevations, where the soils are thin, there is a woodland type of vegetation dominated by pinyon pine (*Pinus edulis*) and juniper (*Juniperus osteosperma*). The deep-soiled valleys, once covered with grasses, are now occupied by big sagebrush, greasewood (*Sarcobatus vermiculatus*), and rabbitbrush (*Chrysothamnus nauseosus*). This change in vegetation type occurred following heavy grazing earlier in this century.

Large mammals include mule deer (*Odocoileus hemionus*), American elk (*Cervus canadensis*), black bear (*Ursus americanus*), mountain lion (*Felis concolor*), and ring-tailed cat (*Bassariscus astutus*). About 250 species of birds are found in the basin, including golden eagles (*Aquila chrysaetos*), bald eagles (*Haliaeetus leucocephalus*), and peregrine falcons (*Falco peregrinus*). Sage grouse (*Centrocercus urophasianus*) are found in the big sagebrush; chukars (*Alectoris graeca*) are found throughout the area from canyon floors to clifftops, and blue grouse (*Dendragapus obscurus*) are common on the plateau.

Parachute Creek contains cutthroat trout (*Salmo clarki*), and a tributary of this creek, Northwater Creek, has the subspecies *S. clarki pleuriticus*, which is an endangered species. DeBeque Canyon of the Colorado River contains the humpback sucker (*Xyrauchen texanus*), also endangered. Algae in Parachute Creek were dominated by a green alga (*Chlorotylium cataractum*) not found elsewhere in Colorado. The creek is a clean stream, but in the lower valley the stream has saline-tolerant species of algae.

There will be some impact on wildlife, particularly on the deer population. Most serious will be the number of vehicle collisions with deer. The Piceance basin has the largest deer herd in Colorado and one of the largest in North America. The valley of Parachute Creek, where the first oil shale plants are being established, is the winter feeding ground for this deer herd. When full-scale shale operations are in effect, vast numbers of deer will be displaced into neighboring regions and will become a nuisance to farmers and ranchers. Raptors will be disturbed because of loss of habitat and as the result of too much human activity in the area; the increase of human population in the region will increase the amount of hunting and off-road vehicle activity in the surrounding mountains and nearby canyon lands.

Revegetation of processed shale. Surface retorting of oil shale generates large volumes of waste material. This spent shale is a fine, powdery black material that remains after the oil has been removed. It has expanded and therefore occupies considerably more volume than the parent rock from which it came. Therefore, all of it cannot be put back into the mine, even if it were desirable to do so. Because of the underground water problem, spent shale should not be replaced in the mine: underground water percolating through the spent shale will leach arsenic and other toxic substances into the watershed. There is no question that a major oil shale industry in the western United States will leave behind enormous quantities of spent shale in the valleys of this arid region and greatly modify the ecology of the region.

For example, a processing plant producing 50,000 barrels of oil per day would use about 73,000 tons of shale per day and produce about 60,000 tons of spent shale per day. After 20 years of operation, this spent shale would cover more than 500 hectares to a depth of 60 m. The plant itself would occupy about 1700 hectares, in addition to the area required for the spent shale. If 10 oil shale plants of this size were in operation, a region representing about 3% of the total area of the Piceance basin would be disrupted. When oil shale operations finally shut down about 20 to 30 years later, the area would be completely revegetated.

Powdery spent shale is hauled by truck and dumped into a broad valley with a limited watershed and low stream-flow. The moistened spent shale compacts, in which state it is nearly impermeable to water. Therefore it allows little leaching or percolation of water into surrounding aquifers. The face of the pile of spent shale must have a slope less than 1 to 3 (vertical to horizontal) and should also have benches at regular intervals in order to minimize erosion.

Kilburn (1976) reports on the revegetation studies conducted by the Colony Development Operation in the Piceance Creek basin of western Colorado. Growth of plants on the processed shale reduces erosion and blowing dust, improves the appearance, and in general restores the area to a balanced ecosystem. Processed shale is almost devoid of nitrogen and phosphorus and often is low in potassium. Therefore, to grow vegetation on it, a high initial application of fertilizers, particularly of nitrogen, is necessary. The processed shale possesses a high pH and a high salt content, both of which are deleterious to the establishment of vegetation. Regular watering during the first year is needed to leach out some of the salts and reduce the pH, while at the same time providing an adequate water supply to the plants. Watering may need to continue the second year, but tests have shown that the vegetation requires no further watering after that. Nitrogen applications may need to continue for several years.

The first field tests of revegetation of processed shale were begun in 1968 by Colony Development and expanded during subsequent years. Studies compared the growth rates of 22 species of grass, 13 forbs, and 28 woody plants. Both native and exotic species were used. Successful growth was achieved for native species such as juniper (*Juniperus scopulorum*), skunkbush (*Rhus trilobata*), and four-winged saltbush (*Atriplex canescens*). Yellow sweet clover (*Melilotus officinalis*), a legume, was also very effective in establishing a cover on the spent shale, as were several grasses. Eventually any mound of spent shale would become covered with plants ecologically compatible with existing ecosystems. However, it is important to recognize the time frame involved for revegetation. Grasses may require 3 to 5 years to become well established; shrub and tree species may take 5 to 10 years to establish and from 50 to 100 years to grow into mature stands.

Water pollution. Water is a critical substance in the oil shale industry, a substance whose environmental impacts include effects on water quality, surface flow, and groundwater. An oil shale plant would use three to four barrels of water per barrel of oil produced. An operation producing 50,000 barrels of oil per day would require about 170,000 barrels of water per day, or about 7950 acre-feet per year. (An acre-foot is the amount of water to cover an area of one acre to a depth of one foot.) The water used would be totally consumed, with no planned discharges to surface streams or underground aquifers. According to Kilburn (1976), the rate of water use mentioned here is less than 1% of the water flowing from the Colorado River Cameo demand diversion near Grand Junction, water that is used for irrigation purposes. Although the amount of water required by one oil shale plant is reasonably small, the amount used by 5 to 10 such operations would be significant, and water could become a limiting resource for the oil shale industry. The allotment of water, based on the Upper Colorado River Basin Compact of 1948, is sufficient for considerable industrial activity (Kilburn, 1976). Amounts of water now available are 100 times greater than the amount consumed by a plant producing 50,000 barrels of oil per day.

If the use of water is almost entirely consumptive, its impact on the salinity of the Colorado River will be minimal. However, water used for the revegetation of the spent shale could leach salts from it. This problem could be avoided by building a catchment dam below the pile of compacted spent shale. The water in this pond, which would contain dissolved solids, would be pumped back to the plant area for moistening of process shale. Only at times of heavy precipitation would some of this contaminated water flow downstream. It is estimated that one oil shale plant producing 50,000 barrels of oil per day would contribute less than 0.02% of the present 730 mg/liter salinity of the Colorado River at Hoover Dam, Nevada. However, many processing plants operating simultaneously could increase salinity levels significantly in the Colorado River if they were not all operating in an exemplary manner. Some estimates place the increase in the Colorado River salinity resulting from extensive commercial development as high as 1.5%. However, Kilburn (1976) considers this estimate to be too high.

Not all of the area drains toward the Colorado River. Some of the region north of Parachute Creek drains into the White River watershed. Groundwater aquifers are extensive in this region. There is a high-quality freshwater aquifer above the Mahogany Zone and a lower-quality aquifer below this zone. Mining activities, such as dewatering in the mine itself, could cause a mixture of the waters in the two aquifers and degrade the higher-quality water.

In situ recovery of oil from shale would reduce the quantity of water used from 40 to 60%; however, serious water contamination may occur underground because underground fracturing, burning, and retorting produces many water-soluble products (e.g., hydrocarbons, dissolved solids, and particulates) that can get into the aquifers. How to handle this underground water when in situ processing is used is a serious, unsolved problem.

When spent shale, the by-product of surface retorting, is placed back in the mine, contamination of underground aquifers is a serious problem, just as it is for the in situ underground retorting scheme. Dewatering of the mine will be necessary, and the water pumped from the mine must be treated before it can be discharged to surface streams.

Air pollution. Air pollution resulting from commercial oil shale processing and refining plants, transportation, and associated industrial activities will become a serious problem if not very carefully controlled. This is a region of mountain valleys and canyons, where temperature inversions do occur and polluted air can concentrate.

Direct sources of emissions and their probable constituents are (1) particulates from mining and crushing operations; (2) CO, SO_2, H_2S, HC (hydrocarbons), NO_x, and particulates from retorting, from gas and oil recovery, and from burning off some of the gases; and (3) hydrocarbons from the storage of petroleum. Leakage of hydrocarbons and other gases through the ground to the surface also becomes a problem when in situ retorting is used.

Development of oil shale processing will bring secondary industries into the region, such as refineries, petrochemical plants, and electric power-generating plants. Probable emissions from these sources include CO, SO_2, H_2S, HC, NO_x, and particulates. Indirect sources of air pollution will come into the region with residential and commercial developments and associated transportation. Increased emissions of hydrocarbons, CO, SO_2, NO_x, and particulates, including fly ash from wood stoves and fireplaces, will occur.

These various emissions will behave very much like the Los Angeles-type smog and the Denver brown cloud, with secondary reactions generating ozone, peroxyacetylnitrate, aldehydes, and an abundance of photochemical aerosols. Photochemical air pollution results from the mixing and reaction of NO_x with hydrocarbons. There are many potential sources of these pollutants,

and they will be emitted in large amounts if the projected growth of the oil shale industry is realized. Impacts on human health, plants, and wildlife will be considerable if very careful controls are not instigated from the beginning. Western Colorado is a region of very clear skies, clean air, and vast vistas. It would be tragic in the extreme to see a significant deterioration of this magnificent environment.

Carbon dioxide release. Retorting of oil shales in the western United States will generate more carbon dioxide per unit of useable energy produced than any other synfuel development. Methods utilizing high-temperature retorting (above 600°C) will cause more CO_2 to be released through decomposition of carbonate rocks than will subsequently be generated by burning the oil produced (Sundquist and Miller, 1980). The amount of CO_2 produced by retorting oil shale varies greatly depending on differences in minerology of the rock, organic content, and retorting technique.

Shale oil produced by low-temperature (near 500°C) retorting yields about 30 kg of carbon as CO_2 for every million Btu of usable energy. This figure includes retorting and burning the oil and is significantly higher than the amount of CO_2 released when conventional fuels are burned (about 15 kg C/MBtu for natural gas, 21 kg C/MBtu for crude oil, and 25 kg C/MBtu for bituminous coal). Processing other synfuels and burning the products also generates about 30 kg C/MBtu. Retorting and burning shale oil by high temperature retorting could generate 70 kg C/MBtu for shale that yields 25 gallons of oil per ton and 110 kg C/MBtu for yields of 10 gallons of oil per ton. It is possible that some of these estimates are on the high side (particularly for high-temperature retorting) because they assume that all of the carbon in the carbonate rock is released at high temperature.

Undoubtedly, this is not true. However, there are no measured values with which to compare the estimates.

Because the global increase of atmospheric CO_2 appears to be the direct result of burning fossil fuels (see Chapter 9), it is of some interest to estimate how much CO_2 will be added to the atmosphere by shale oil production. An industry in the Piceance Creek basin producing 1 million barrels of oil per day would eject into the atmosphere from 0.06 to 0.17 $\times$ 10^9 tons of carbon per year. The present global rate of emission of carbon from conventional fossil fuels is 5 $\times$ 10^9 tons per year. Therefore, an oil shale industry would release only about 1 to 3% additional CO_2 to that already being released on a global basis by burning fossil fuels. If the production of shale oil substituted for other oil being produced for the world market, then the corresponding increase in CO_2 production would be 0.5 to 2.5% of the total world fossil fuel production today. Finally, if half of all the oil shale resources contained in the Green River formation (estimated at 3.30 $\times$ 10^{11} barrels from Table 2 in Chapter 4) were eventually recovered and burned, the total release to the atmosphere would be 55 $\times$ 10^9 tons of carbon. However, Sundquist and Miller (1980) use a much higher number for the oil shale reserves of the Green River formation (1.8 $\times$ 10^{12} barrels of oil), and they suggest that the total release to the atmosphere would be about 300 $\times$ 10^9 tons of carbon.

PIPELINES

Oil and gas are fluids that are readily passed through pipelines. The first gas pipeline was installed in Genoa, Italy in 1802 to carry gas for street lighting. Oil was first carried by pipeline in Pennsylvania over a distance of 10 km in 1861. By 1874, a pipeline 100 km long carried oil into Pittsburgh.

Early in this century, pipelines were of small diameter: usually 7 to 15 cm and eventually up to 30.5 cm. During World War II, the BIG-INCH PIPELINE of 60 cm was built to carry gas for 2000 km from Texas to Pennsylvania. The networks of natural gas pipelines throughout the world today are vast. In the United States alone, there are more than 300,000 km of gas pipelines. Diameters of 35 to 76 cm are commonplace; and recently pipelines of 90 to 122 cm have been built.

Major pipelines have been successfully constructed and operated in very hot regions of North Africa, the Middle East, and the United States and in extremely cold regions such as Siberia, Alaska, and the Canadian arctic. Someday pipelines may be built and operated in Antarctica. There are fundamental differences between gas and oil pipelines, and the interactions of the two types of pipelines with the surroundings require different design and construction techniques.

Permafrost

Oil and gas pipelines in polar regions must cope with the special properties of PERMAFROST and of a surface ACTIVE LAYER that is subject to much freezing and thawing. Permafrost is ground that remains frozen year in and year out. A state of being perennially frozen is more probable at a considerable depth in the soil than it is near the surface. At a depth of 15 m there is no detectable seasonal temperature change. At this depth a steady mean annual temperature prevails, and when it is 0°C or less, permafrost exists. If the temperature is very cold for much of the year, the permafrost layer will be close to the surface and the active layer will be quite shallow. Sometimes the depth of the permafrost layer varies a great deal.

Soils of cold regions are characterized by polygonal patterns of soil and rock, ice wedges, and pingos. A pingo is a hill 10 m or more high with an ice core.

When the active layer or the underlying permafrost is melted by the heat from a pipeline or in any other manner, the ground becomes soft, mushy, and unstable and pools of meltwater will form. Roads, houses, buildings, bridges, and pipelines must be isolated or insulated from the active surface layer and from the permafrost. Special construction techniques are required in the arctic. Roads and small buildings or homes must be built on a thick gravel mat and larger buildings must be placed on pilings sunk deep into the ground.

Alyeska oil pipeline

In 1968 geologic explorations proved the existence of a large oil field on the north coast of Alaska near Prudhoe Bay. As America approached oil shortages in the early 1970s, it became increasingly urgent to begin to move this oil out of the ground and deliver it to the lower 48 states. Because oil tankers can only reach Prudhoe Bay during the ice-free summer season—a few months of the year—it was necessary to build the Trans-Alaska or Alyeska pipeline connecting Prudhoe Bay with the ice-free port of Valdez on the south coast of Alaska. From here the oil is shipped by tanker to West Coast ports in the United States.

In 1977, an oil pipeline of diameter 122 cm was completed; it runs 1300 km from Prudhoe Bay to Valdez. This pipeline is claimed to be the largest engineering project in human history undertaken by private industry. Its cost was $7 billion. It carries nearly 2 million barrels of oil per day to Valdez. When pumped out of the ground the oil is at 80°C; and while it flows through the pipeline, it is between 60° and 65°C. When the oil cools to air temperature, it congeals and becomes a very sticky substance. Pumping stations lie along the route of the pipeline at frequent intervals. Each time the oil is compressed by the pumps and given a boost on its way along the pipeline, energy is added to it, so it maintains its warm temperature.

The Alyeska pipeline traverses a great variety of forest, swamp, mountain, and tundra terrain. At least three-quarters of the route from Prudhoe Bay to Valdez overlies permafrost. A warm pipeline laid on or in the ground would thaw soil to a depth of 10 m during the first year of operation. The thawing would be highly variable, depending on permafrost and ice conditions, and the pipeline would settle into the ground in a highly irregular and uncertain manner. Considerable disruption of the pipeline would occur, each event carrying the potential for breaks and serious oil spills into the ecosystem. Therefore, the pipeline had to be mounted aboveground on towers for much of its route; and when placed in the ground, it had to be insulated. The southernmost section of the pipeline passes through a major earthquake zone. However, pipelines have often been built through earthquake-prone regions, such as in the Middle East, and the technology is well developed for making them highly resistant to seismic damage.

The 122-cm diameter steel pipe of which the Alyeska pipeline is constructed has a wall thickness of only about 1 cm. The steel pipe is wrapped with insulation. Over much of the route, where the pipeline is elevated on vertical support members (known as VSMs), air passing beneath the pipe dissipates most of the heat and generally reduces the thermal impact on the active and permafrost layers. Problems of thermal construction and expansion of the pipe were overcome by allowing some pipe movement at the VSMs and by placing the pipeline in the trapezoidal zigzag as it traverses the landscape. In general, the pipe can move laterally through some 4 m, and in special situations more than this vertically—a scheme that also affords protection against earthquakes.

The original plan called for the Alyeska pipeline to pass under rivers. In fact, it does go 5 to 6 m beneath the Tonsina River. Here the pipeline was encased in 22 cm of concrete in order to weight it down. However, other large rivers are crossed by bridges to which the pipe is slung. Crossing the Tanana River requires a bridge and sling 360 m in length. Crossing rivers is particularly difficult because in the spring the upper 2 to 3 m of the riverbeds are periodically removed by scouring and erosion by the spring floods. Where the pipeline has been buried, there is a difficult problem with the permafrost near the river banks, where it is more prone to thaw. Crossing the rivers requires not only avoiding the scour problems, but also the preservation of conditions suitable for fish populations. Building bridges and slinging the pipeline from them was the best solution to these problems.

The frequent passage of men and equipment near the pipeline inevitably damages vegetation and initiates a thawing of the permafrost. Enormous quantities of gravel were laid down as pads about 1.5 m thick for work areas and for the road paralleling the pipeline. Often these were underlaid with a urethane plastic insulation. Gas and oil storage units at Prudhoe Bay and elsewhere were elevated on piles 2 to 2.5 m above the ground surface, wherever permafrost was present. The eight pumping stations operating at intervals along the Alyeska pipeline have advanced gas turbines that power centrifugal pumps and complex valve systems. At several of the pumping stations, the foundations of the buildings have refrigeration systems to maintain the underlying soil in a frozen condition.

Environmental and ecological impacts

The Alyeska pipeline and haul road cross the arctic tundra of the North Slope and forested areas in the mountains and south of the Brooks Range. The ecosystems traversed are highly varied and the impacts on these are extremely different.

Tundra. Webber et al. (1980) give an excellent discussion of the characteristics of arctic tundra and make the following statement: "Tundra is often viewed as easily disturbed or changed but it is quite stable and resilient to major environmental changes. It appears to be adapted to large, natural, often sudden environmental fluctuations." On the North Slope of Alaska, the coastal tundra undergoes various natural perturbations that are part of a THAW-LAKE CYCLE. Thousands of years are required for the cycle to return the ecosystem to its original alluvial state. The ion-exchange capacity of mosses and the nutrient uptake by all the plants and microorganisms removes the nutrients from snowmelt and runoff, thereby retaining nutrients in the soil surface.

In anticipation of the Alyeska pipeline carrying hot oil, an experiment was conducted in which a wet meadow substrate was heated in situ. A 10°C soil temperature increase for one month at Barrow, Alaska increased the thaw depth, the decomposition rate, the nitrogen availability, the plant nutrient absorption rates, and the primary production (Webber et al., 1980). Many years later very little of the disturbance effect could be detected. When, instead of short-term heating, soils were heated for one year, the result was much increased thaw depth, melting of ice in the permafrost, subsidence of the ground surface, and ponding of water. Rapid decomposition of organic matter depleted the oxygen concentration of the soil, and all the vegetation died within a year. Recolonization of the site did not occur over a 10-year span. On the other hand, experimental heating of an ice-free soil in the interior of Alaska throughout the year caused no subsidence, no die-off of vegetation, and, in fact, increased primary productivity. Detrimental effects associated with soil heating seem to relate to a series of events associated with melting ice, soil subsidence, and changing chemical and physical conditions.

There has been a long history of oil exploration across the arctic tundra of Alaska. Vehicle tracks are seen across the arctic tundra in many directions, some of which were created during World War II. The severity of vehicle impact upon tundra seems to depend very much on the nature of the plant community and the character of the underlying soil. The vehicle tracks may crush the vegetation, compact the surface, and result in water impoundment in deepening troughs through partial ice melting. Striking vegetation changes may occur along an old vehicle track within a few years. Vehicle damage is greater in the shrub tundra than it is in the meadow tundra because of greater breakage of shrub stems. Sites with low ice content are less susceptible to vehicular damage than are those containing more ice. When the soil surface is compressed below the shallow water table, particularly in poorly drained meadow soils, standing water develops and decreases the albedo to sunlight. The standing water absorbs more sunlight, raises the underlying soil temperature, and accelerates the thawing of permafrost. These changes result in increased nutrient availability and increased primary productivity. When vehicle tracks cross ice wedges, deep permanent ponds may form. When there is much slope to the surface, very deep erosion of the soil will occur. Recovery from some of these impacts may take hundreds or even thousands of years.

Alaskan crude oil is toxic to most vegetation. Oil spilled on the tundra produces a decrease on the surface albedo, hence increases in absorbed sunlight and heating of the soil. Vegetation growth may accelerate because of the warmer condition resulting from the oil spill, despite some toxicity to the plant leaves. When water fills the site, the oil is not as likely to penetrate the soil and the more toxic volatile fractions may have time to evaporate. On the other hand, it is known that oil may remain in the active layer of the tundra for as long as 30 years or more. These hydrocarbons in the soil may increase some microbial activity, thereby causing an increased demand for nutrients so that the nutrients become less available to vascular plants. Oil kills some mycorrhizal fungi, thereby further decreasing the ability of plants to assimilate nutrients. Oil is hydrophobic; therefore, once it penetrates the soil, it reduces water movement and reduces nutrient transport and availability to plants.

To test the sensitivity of tundra vegetation to oil spills, Walker et al. (1978) established a series of test plots within which crude oil or diesel fuel were spilled. The changes in the vegetation within these plots were observed and compared with control plots nearby. These experiments were done in the tundra at Prudhoe Bay. One year following a simulated crude oil spill, most plant species were dead. On dry sites almost all plant species (including *Dryas integrifolia*, the most important vascular species) and all lichens were killed. In more mesic (wetter) sites, many moss species and nearly all herbaceous dicotyledonous species were killed. A few plant species recovered a year after the spill. On a plot with standing water, total recovery of the vegetation occurred one year after the spill. The dry plots recovered very poorly, with the exception of willows (*Salix*) and sedge (*Carex rupestris*), which recovered well. The experiments using diesel fuel rather than crude oil on both wet and dry sites showed all species except an aquatic moss to be killed. Apparently, contact between diesel fuel and the plant leaves was sufficient to kill the plant and direct contact with the roots was not necessary. From these studies, augmented by other information, Walker et al. (1978) composed a sensitivity map of the tundra vegetation in the vicinity of Franklin Bluffs, where an 1800-barrel crude oil spill occurred on 20 July 1977. The spill created a gradient of oil that radiated out from a broken valve of the Alyeska pipeline. The oil apparently squirted vertically for 35 m and a strong north wind fanned it out over an area approximately 100 m long downwind. The soil was totally saturated with a thick layer of oil. Approximately 1400 barrels of oil were removed by cleanup procedures used on the area of heaviest impact. This left other areas covered with oil. About 400 barrels of oil remained on 8.3 ha. Of this, about 1.8 ha received a heavy oiling; a situation that was similar to the test plots at Prudhoe Bay. The Franklin Bluffs observations indicated that the dry areas away from the coast are more resilient than dry areas at the coast. It was also discovered that if the spilled oil is allowed to flow, it will tend to go to the wetter areas, where recovery is more probable. The oil may be more easily skimmed off a lake or off standing pools than off the land. A threat to waterfowl could exist, however, unless a means for frightening them can be devised.

Further information concerning the environmental and ecological impacts of petroleum exploration, development, and transport, as well as of other mineral resources, may be found in Rand (1982).

Haul road. Construction of the Alyeska oil pipeline required construction of a road adjacent to the pipeline over more than half its length. The first section (90 km in length) was constructed in 1969 and 1970. It connected Livengood, Alaska to the Yukon River. The 577-km second section between the Yukon River and Prudhoe Bay was built in five months in 1974. The road traverses highly variable permafrost conditions and diverse climatic, biotic, and geologic zones. The haul road includes 20 permanent bridges, more than 1000 culverts, and 135 gravel sites. There are 15 oil pipeline crossings of the road; nine of them are buried and six are aboveground. A fuel gas line (0.2 m diameter) crosses the road seven times and is buried along the edge of the road for 180 km of its 230 km length.

The haul road passes through 280 km of forest and 320 km of tundra vegetation. Mixed deciduous–coniferous forests characterize the vegetation between the Yukon River and the southern foothills of the Brooks Range. This is low forest of black spruce, with some white spruce; and, depending on the fire history, aspen and paper birch are mixed with the spruce. Above an elevation of 700 m, forest vegetation is taken over by tundra shrubs such as alder and dwarf birch. Tundra vegetation of the Arctic Slope is mostly sedge-tussock in the foothills and sedge-meadow on the coastal plain.

Large grazing animals along the haul road include huge herds of caribou, sheep, and moose. Black bears, grizzly bears, and arctic foxes are numerous. Grayling and slimy sculpins abound in streams along the entire length of the haul road; some round whitefish are also present. North of the Brooks Range, arctic grayling, ninespine stickleback, and arctic char are abundant.

Studies of the haul road and its environmental impacts show that the roadbed has subsided to some degree over nearly all of its length. Clogging of culverts by gravel pushed off the road by grading is a problem. Some gully erosion has occurred in areas where the flow of water has been concentrated through culverts and directed onto ice-rich soils. Dust from the haul road causes a problem on the adjacent tundra. Early snowmelt (two or three weeks early) occurs because the dust accumulates on the winter snow for distances up to 100 m on both sides of the road.

Government regulations require that erosion be controlled and that disturbed areas be revegetated along the haul road, along the oil pipeline, and along the fuel gas line. The road required many cuts and much fill. Material sites for gravel were opened in river terraces, floodplains, and uplands. These sites were connected to the haul road by access roads. A few of these sites were used as refuse disposal areas before being covered and revegetated, and some were left as gravel sites for maintenance. Access roads not in continuous use were blocked and revegetated. The fuel gas line (providing gas to heat buildings) produced a linear disturbance adjacent to the haul road.

Revegetation along the haul road involved mulching, fertilizing, and seeding with agronomic grass species. Seed was not available for widespread seeding of native grasses. In the future greater seed supplies of native grasses will be available because of planting and harvesting programs conducted by the oil companies. Many of the revegetated areas have 80 to 90% grass cover derived from the initial seed mix. Native species have been slowly invading revegetated areas. It is not yet known how long the introduced exotic plants will survive. Webber and Ives (1978) indicate that non-native grasses and herbs that are sown into disturbed areas are useful for stabilizing the surface for colonization by native plants, and there appears little likelihood that they will have significant effect after a few generations. Transplanting native trees and shrubs to areas along the haul road has been highly successful. A willow cutting program to replace lost wildlife browse has not been successful. The use of unrooted willow cuttings and tussock sodding may be successful in the future for restoring upland vegetation.

A very thorough description of the environmental and ecological studies made along the Yukon River-to-Prudhoe Bay haul road is given by Brown and Berg (1980).

Gas pipelines in the arctic

Gas and oil pipelines are very different in their design and modes of operation. A gas has a low volumetric heat capacity, warms quickly when compressed, and cools quickly when expanded. The quantity of gas transported is greater when the gas is compressed to a high density. However, the gas will get hot when passing through the compressor system. Between compressor stations some expansion and cooling will occur.

Deliberate cooling of the gas is necessary when it is transported through a pipeline buried in the permafrost. The gas in some pipelines is at a pressure as great as 12 megapascals. After compression, the hot gas must be cooled by refrigeration. A detailed discussion of gas pipeline construction and the environmental problems associated with its traversal of permafrost or of active soil layers is found in Williams (1979).

An application was made to the Canadian government in 1974 to build a gas pipeline up the Mackenzie River valley. The Canadian government decided not to allow the Mackenzie valley route, but agreed instead to build the Alcan gas pipeline parallel to the oil pipeline from Prudhoe Bay to Fairbanks, Alaska; this pipeline was to cross the southern Yukon and follow the Alcan highway to southern Canada, where the pipeline would branch out to California or to the upper Midwest. (This route avoids the northern Yukon coastal plain, where there is unique wildlife habitat and where the native rights of Eskimos are an important consideration.) However, because of a changing oil and gas market, this Alcan gas pipeline has not been built.

Other gas pipeline proposals are under consideration. Among these is a proposal for a pipeline to carry gas from sources in the Canadian Arctic Islands down the west side of Hudson Bay and into eastern Canada. This particular pipeline would need to be installed under the Arctic Ocean between the Boothia Peninsula, Bathurst Island, and Milville Island. The technical challenge of doing this will be enormous. Grooves in the seabed caused by the ploughing of icebergs can occur. Pressure ridges of winter sea ice may scrape the seabed, and floating ice islands may rake the bottom. The pipeline going from the seabed to shore encounters severe problems with waves and floating ice—problems more severe than those encountered with river crossings. However, these technical problems eventually will be solved, and we will have gas pipelines in the Arctic Ocean and in the Antarctic Ocean.

12

Electric Power Generation

Natural and man-made electricity come together as a lightning strike silhouettes electric power lines. Courtesy of the National Center for Atmospheric Research.

INTRODUCTION

World electricity demand will probably grow more rapidly the next few decades than will total energy demand for several reasons. First, a large part of electricity cost is due to capital expenses and therefore electricity prices are less sensitive to fuel costs. If primary fuel costs rise more than capital costs, electricity will become cheaper relative to other energy forms. Second, as societies become more affluent, they tend to prefer more convenient energy forms. Finally, if climate change from rising concentrations of atmospheric carbon dioxide are proved, there will be increased incentive for a nuclear power–hydrogen fuel economy. Hydrogen, when burned, forms only water and no carbon dioxide.

The current annual world energy consumption is about 276 quads per year. The total world electric energy production is about 10 trillion kWh/yr, or 34 quads. It is possible that by the year 2010 world electricity consumption will be three to five times greater than at present. It is possible that nuclear energy would supply up to two-thirds or more of this increase in electricity production. However, many doubt that nuclear generating capacity will be allowed to expand so rapidly, and the alternative is increased use of coal.

Electric power growth in the United States

Americans have enjoyed the privilege of low-cost electrical energy and have added to their homes a large number of energy-intensive electrical appliances, such as air conditioning, dishwashers, and clothes driers. The total U.S. energy consumption is currently about 80 quads per year. Electric power generation consumes approximately 30% of this energy, or 24 quads per year. However, the electrical energy output is 9% of the total U.S.

energy consumption or 7.2 quads per year. This is equivalent to 2100 billion kWh/yr. There are approximately 3069 electric power plants in the United States. Demand for electricity grew at 6.8% per year between 1960 and 1973, and 3.1% per year between 1973 and 1980. It is expected to increase by only 2.5% per year between 1980 and 1990, and 1.9% per year between 1990 and 2000.

The sources of U.S. electric utility generation from 1960 to the present and for the future, to the year 2000, are given in Figure 1. Coal supplies about 49% of the electric utility fuel in the United States today. Coal use for utilities more than doubled between 1964 and 1980, but is expected to remain near 50% for the remainder of this century. Gas consumption will decline from 11% of the total energy supply for utilities today to less than 2% by the end of the century, while oil will go from 12% to less than 2%. Nuclear energy supply is expected to go from about 15% of the total fuel supply for utilities today to 27% in 1990, and 34% by the year 2000. Hydroelectric is about 13% of the total present supply and will remain nearly constant for the remainder of the century. It is evident that about 72% of our electric power generation today in the United States is from fossil fuels.

Projections of energy consumption are very uncertain because of the many factors affecting energy demand. To project electric power consumption far into the future is also highly uncertain. However, one can place upper limits on the amount of electric power generation that can occur before the year 2000, because any power plant in operation then must either be in operation today, about to be in operation, or be in the planning stage with construction to begin in the near future. Power plant construction has a very long lead time from the original planning and site selection to the day when it comes online, generating electrical energy for the consumer. A fossil fuel plant has a lead time of 10 to 12 years and a nuclear plant from 14 to 16 years.

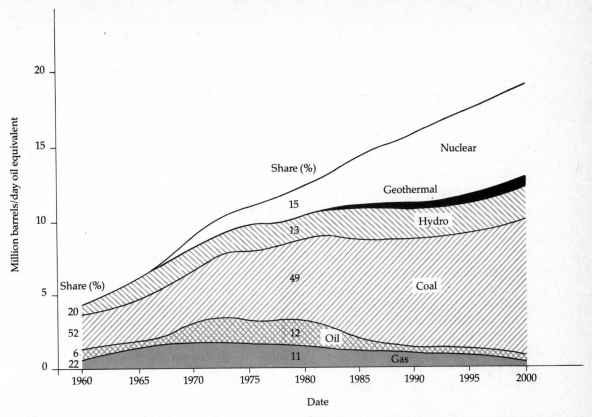

FIGURE 1. Sources of U.S. electric utility generation from 1960 to the present and the anticipated distribution of sources from the present until the year 2000, in millions of barrels of oil equivalent per day and in percentage of the total. (Courtesy of Exxon, 1979.)

Decisions concerning future fuel requirements for electric power generation, or the strategy necessary to avert too much of a buildup of atmospheric carbon dioxide, must be made in the near future in order to be effective by the year 2000 or shortly thereafter. At the present time, no new nuclear power-generating plants are on order in the United States.

Population and power consumption

To get a perspective on the size of electric power-generating plants needed as related to the size of the community consuming the power, consider the case of Ann Arbor, Michigan. How large a city may be served by a 1000-MWe generating plant? Much depends upon whether the community is largely residential, commercial, or industrial. Ann Arbor is a university town with a good mixture of all three types of consumers. It has a population of about 110,000. A peak electric power demand of 141 MWe occurred for Ann Arbor on 9 July 1981. Of this, 36 MWe went to residential customers, 43 MWe to commercial, and 62 MWe to industrial. On this particular day the Detroit Edison Corporation, which supplies electric energy to all of southeastern lower Michigan and the city of Detroit, produced 7171 MWe. A 1000-MWe plant would clearly provide about seven times the peak demand of Ann Arbor, but only one-seventh the peak demand of all of southeastern Michigan. The Ann Arbor population had a peak demand of 1318 We/person. At this per capita rate, a 1000-MWe plant would serve a community of 760,000 persons having the same mix of consumer types as Ann Arbor. The U.S. Council on Environmental Quality (1973) indicates that a 1000-MWe plant operating at 75% capacity would generate sufficient energy to meet the demands of 900,000 people, or 6.57 billion kWh annually.

The demand for electrical power depends on the time of day, day of the week, and season. On a hot humid summer day in the eastern half of the United States, for example, the PEAK POWER DEMAND can be more than twice the minimum demand on the same day. Utilities meet this fluctuation in demand by generating power in three ways. That part of the total power needed throughout the day and night—called the BASE LOAD—is generated by large steam turbines powered by fossil fuels, water turbines (hydroelectric), or nuclear steam turbines. The power consumed only during the daylight hours (dawn to midnight) is referred to as INTERMEDIATE POWER. It is supplied by so-called cycling units, usually older and smaller fossil fuel-fired plants. At times of exceptional or peak power demand, additional steam units, gas turbines, and hydroelectric or pumped-storage generators are used.

POWER PLANTS

An electric power generating plant transforms the chemical energy of coal, oil, or gas, or the nuclear energy of uranium or thorium, into electrical energy by a sequence of three conversion processes. First, the fuel is burned, heat generated, and the heat trans-

ferred to water to make steam. Second, the hot, high-pressure steam expands, impacts the blades of a turbine, and turns the turbine to create mechanical energy. Third, the rotating shaft of the turbine turns the shaft of a generator. When the rotor of the generator is turned, electric current is produced. The rotor of an electric generator is simply an array of conducting wires that are rotating relative to a magnetic field. When a conducting wire, usually of copper, moves across a magnetic field, an electric potential is inducted in it and a current flows. An electric generator has an energy conversion efficiency from 96 to 99%.

The elements of an electric power generating system are shown in Figure 2. A power plant is made up of many components in addition to the boiler, turbine, and generator units. There must be roads and usually a rail spur or barge terminal on the property. Space is needed to stockpile coal or oil for a fossil fuel plant or to store fuel rods for a nuclear plant. A place is needed to dispose of fly ash and coal ashes. A chemical treatment facility is needed for cleaning the waste water from coal scrubbers. Conveyors, shutes, and hoppers are needed to get the coal to the power house. Boilers, steam pipes, ducts for flue gases, chimney stacks, electrostatic precipitators, desulfurization units, and other auxiliary equipment are needed. A control room, electrical distribution system, and transmission lines also are necessary parts of the electrical system.

Major electric power-generating plants typically have a rated capacity of 650 to 1300 MWe. Coal-burning power plants have thermal efficiencies of about 38%. This means that a plant with a rated capacity of 1000 MWe must burn 2632 MW of fuel equivalent and will lose as waste heat about 1632 MW. A 1000-MWe nuclear power plant has a thermal efficiency of about 33% and loses 2000 MW of waste heat.

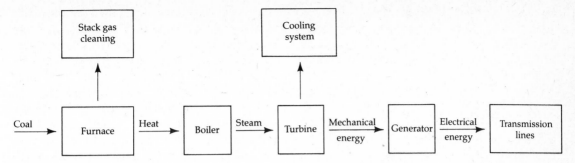

FIGURE 2. The basic elements of an electric power generating system, from fuel and furnace to turbine to generator to electric transmission lines.

Furnaces

Heat is produced in a furnace by burning coal, oil, gas, uranium, or thorium. Each fuel type requires a somewhat different furnace design. The arrangement used for generating heat from nuclear fuel rods is described in Chapter 13. In oil-fired furnaces, oil is sprayed into a combustion chamber, along with jets of air, and the mixture is ignited. A gas-fired furnace operates essentially the same way. Both oil and gas are relatively clean burning and leave little residue in the furnace. Coal, being a solid material, is more difficult to handle and after burning leaves behind a substantial residue of ash.

In coal-fired power plants, furnaces are of three basic types: STOKER-FIRED, CYCLONE FUR-NACE-FIRED, and PULVERIZED COAL-FIRED. Stoker-fired furnaces are only used in small power plants of less than 40 MWe. Cyclone furnaces burn crushed coal in a horizontal cylinder into which is injected a high-velocity stream of air. These furnaces are efficient but do create a high concentration of nitrogen oxides. With a pulverized coal system, the coal is pulverized to a powder, mixed with air, and blown into the furnace. Most new coal-fired plants are of the pulverized coal-fired type.

Boilers

A steam boiler is actually quite complicated. To maximize thermal efficiency, the boiler contains a superheater, reheater, economizers, and air preheater. These consist of various tubes and chambers in which saturated steam is heated by the combustion gases. The economizer transfers heat from the combustion gases (after the superheater and reheater) to the boiler feedwater. The air preheater extracts additional heat from the flue gases and passes it to the combustion air before it is fed into the furnace. The steam produced will be at a pressure of about 24×10^6 pascals (3500 pounds per square inch) and will be at a temperature of about 538 °C.

Condensers and waste heat

In order to have as high an efficiency as possible for the steam turbine, it is necessary to expand the steam against the turbine blades from as great a temperature differential as possible. The spent steam emerges at a much lower temperature and pressure than the high-pressure, high-temperature steam coming from the boiler to the turbine. It is condensed into water, and the water is returned to the boiler to complete the cycle. The heat extracted by the condenser water is ejected to a lake, river, or pond, or to the atmosphere by means of cooling towers or other methods. (Cooling systems are described in another section of this chapter.)

Electrical energy efficiency

If we could use efficiently all of the energy contained in our fuel reserves, we would extend their lifetimes considerably. As it turns out there is a great deal of energy wasted between the source and the consumer. A 1000-MWe power generating plant must burn approximately 2.30 million metric tons of coal annually, or 6300 metric tons per day. (See the section on coal requirement.)

Let us follow a ton of coal from the mine to the electric power-generating plant and to the home or factory where it is eventually consumed. At an underground mine, 43% of the coal in the ground is either lost in the mining process or is too difficult or costly to remove. By comparison, only 20% of the coal is lost in strip-mining. This is one reason strip-mining is preferred by the coal companies. From a metric ton of coal in the ground, only 570 kg is removed from a deep mine and 800 kg from a strip mine.

Processing the coal for burning loses another 8%, and about 1% is lost during transport from the mine to the electric power plant. By the time it reaches the plant, only 520 kg of coal from the deep mine and 725 kg from the strip mine are available for burning. The electric power generating plant can extract only about 38% of the energy in the coal because of the basic thermodynamic inefficiency of the steam–electric system. Much energy is lost during the burning process and in the conversion of heat to electricity because heat must be rejected in the steam condenser, some of the electrical energy generated must be used to power auxiliary equipment within the plants, heat is lost up the stack, heat is lost from the boiler walls by radiation and convection, and energy losses occur in the electric generators. Of the ton of coal with which we started, we have been able to convert the energy equivalent of only 200 to 275 kg. This energy has not yet reached its destination. As the electrical energy travels through transmission lines, another 10% or more of the energy is lost. This takes the energy available down to the equivalent of 180 to 248 kg of coal. At this point the overall efficiency is between 18 and 25%. For oil, the system efficiency is 10% for onshore oil and 13% for offshore oil, for natural gas it is 24%, and for nuclear energy, 16%. At its destination, if the electricity is used for lighting, less than 10% of the energy is converted to light, or the fuel equivalent of between 18 and 25 kg of coal. The overall efficiency of such a process is between about 2 and 2.5%. The situation is only a little better with the use of oil, gas, or nuclear fuel.

Energy production and consumption have some potential for improved energy efficiencies and conservation. Only about 30% of the oil in a reservoir is being extracted from onshore wells today. Secondary recovery techniques may allow the recovery of more oil. For electric power systems, a major source of inefficiency is the power plant itself. A number of promising techniques, including the use of magnetohydrodynamics and combined cycles, may increase power plant efficiency from 38 to 50%. When power plants are located close to industrial sites, the waste heat may be used for industrial processes, a method referred to as cogeneration.

Coal requirement

The daily coal requirement (DCR) for an electric power generating plant may be determined from the energy content of the coal used (B_c, in kJ/kg), the rated capacity of the power plant (C, in MWe), the plant capacity factor (P, in %), and the plant efficiency (E, in %):

$$DCR = 86,400 \frac{PC}{EB_c} \text{ metric ton/day} \quad (1)$$

See the section in Chapter 7 concerning the use of biomass residues for fueling an electric power-generating plant for other details. The quantity of biomass needed was calculated on an annual basis; DCR is calculated on a daily basis. In addition, the biomass calculations were done for a 100-MWe plant and the coal calculations are for a 1000-MWe plant.

An electric power plant operates only about 70% of the time throughout the year, because the plant must be shut down for maintenance occasionally. Hence, $P = 70\%$. The plant efficiency $E = 38\%$. For this example, the rated plant capacity $C = 1000$ MWe. When anthracite coal or very good bituminous coal from Appalachia is burned with an energy content $B_c = 30,000$ kJ/kg, the daily coal requirement is

$$DCR = 86,400 \frac{70 \times 1,000}{38 \times 30,000} = 5305 \text{ mt/d}$$

When, instead of burning Appalachian coal of high energy content, the power plant burns Illinois coal of $B_c = 26,500$ kJ/kg or western coal from the Powder River basin of $B_c = 19,000$ kJ/kg, the daily coal requirement becomes 6006 mt/d and 8376 mt/d, respectively.

Coal storage

There are two types of coal storage at a coal-burning power plant: LIVE STORAGE and RESERVE STORAGE. The live-storage pile contains the coal needed on a daily basis. Actually the quantity held in live storage must be greater than the daily requirement to allow for any short-term variation of supply. The reserve storage contains at least a 100-day supply of coal.

It is undesirable to put any more coal than necessary into live storage because oxidation of this loosely piled coal produces heat, generates a risk of fire, and reduces the heat value of the coal as fuel. The density of coal is about 1.1 mt/m³. When coal is placed in reserve storage, it is compacted to a density of about 1.3 mt/m³ to reduce the air spaces within the coal pile.

If the quantity of coal required by a 1000-MWe plant were from 5305 to 8376 mt/d, the minimum size of live-storage piles would be about 6000 to 9500 m³. If the coal were piled 5 m high, these piles would cover areas approximately 35 × 35 m and 45 × 45 m. Reserve storage requiring a 100-day supply must contain 530,500 to 837,600 mt of coal. These reserve storage piles would contain compacted coal volumes of 410,000 and 645,000 m³, in round numbers. If the coal were piled 10 m high, these piles would cover areas 202 × 202 m and 254 × 254 m, respectively.

A freighter unloading coal at the Saint Clair, Michigan power plant is shown in Figure 3.

Land use

Electric power generation requires a substantial amount of land. A 1000-MWe plant would include the following on-site land: for coal delivery, handling, and storage (5.6 ha), land for the power house (1.1 ha), land for cooling towers (9.0 ha), and land for ash and sludge ponds (4.5 ha), for a total of 20.2 ha. Additional land for roads, parking areas, switchyards, landscaping, and other activities would consume about 120 ha—for a grand total of 140 ha.

A deep underground mine requires about 3640 ha of land to produce coal to keep a 1000-MWe plant operating for a year; and a surface strip mine requires about 5670 ha. A rail transportation corridor 1000 km long and

FIGURE 3. A freighter unloading coal at the power plant in Saint Clair, Michigan. (Courtesy of Detroit Edison Co.)

15 m wide would occupy 1500 ha. Transmission line corridors may take as much as 4000 ha. Altogether, a 1000-MWe coal-fired power plant requires about 11,000 ha of land.

Environmental and ecological impacts

The operation of a fossil fuel electric power generating plant creates many potential environmental and ecological impacts. These cannot be totally avoided, but every effort can be made to minimize them. Coal, our most abundant fossil energy resources, is the worst polluter. An oil-fired power plant of the same size is not as harmful to the environment as is the coal-burning plant, and natural gas-fired power plants cause less environmental disruption than do either coal- or oil-fired plants.

Many environmental and ecological impacts occur away from the power plant, such as at the coal mining site or oil well, or along the transportation corridor. Carbon dioxide, sulfur dioxide, oxides of nitrogen, and fly ash escaping from a coal-burning power plant chimney can cause environmental and ecological damage both near and far from the power plant. Without emission controls, a 1000-MWe plant burning 2.30 million metric tons of coal per year produces about 350,000 metric tons of potential pollutants each year. Two-thirds of the potential pollutants are fly ash, but more than 91,000 metric tons are sulfur dioxide, carbon monoxide, and oxides of nitrogen. About 1.0 million metric tons of carbon as carbon dioxide are released each year from a 1000-MWe power plant. Heavy metal trace elements with very low levels of radioactivity may also be released. Flue gases and particulates emitted from the power plant stack may impact biota near the plant. They may do so most adversely under special meteorological conditions. When the bottom of an inversion layer rises above the top of the stack, as sometimes happens after sunrise, stack effluents mix downward, rapidly fumigating the ground nearby. Upward mixing is limited by the inversion (Figure 4). The highest concentrations of pollutants are usually reached within a few kilometers of

the plant. At night the ground surface is an effective radiator to the clear sky, and the air near the ground becomes colder than air at the height of 10 to 100 m above the ground. With the coldest air near the ground surface, a stable temperature inversion exists, and any pollutants entering this cold air become trapped and do not rise. When the sun warms the surface and the air just above the surface, this warm air will tend to rise and carry away the pollutants. However, a temperature inversion may exist higher up, and if it is above the height of the stack, the effluent and pollutants will be trapped beneath this layer or within it.

The height of power plant stacks has increased steadily since the mid-1960s when the solution to the problem of localized pollution was to disperse the flue gases at higher levels in the atmosphere. The attitude was that the solution to pollution is dilution. However, more recently it has been recognized that this technique has simply transferred the problem elsewhere and may have contributed significantly to an increase of acid rain. The average stack height of fossil

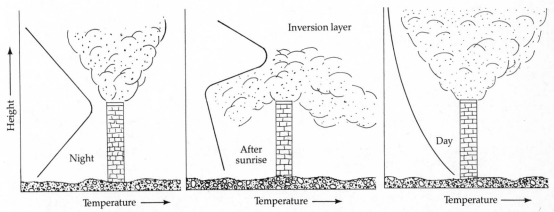

FIGURE 4. Characteristic atmospheric temperature profiles and their effects on the dispersion of emissions from a power plant stack. The inversion layer, which often occurs soon after sunrise, can cause stack effluents to mix rapidly downward, fumigating the ground near the plant.

fuel plants in the United States is about 200 m, and the tallest stacks are more than 300 m (Patrick et al., 1981).

At the plant site, the fuel must be stored and handled. Coal piles produce runoff of silt, acids, and dissolved materials. Ash and fly ash removed from the stack gases must be disposed of in huge settling ponds before the residue can be transported by truck or pipeline to a permanent disposal site. Seepage of salts and various dissolved or suspended materials from the settling basins into surface or underground waters is always probable. In addition to water seepage and runoff, dust blows off coal piles and may be a neighborhood nuisance. Power companies control dust by keeping the coal surface moist.

AIR POLLUTION FROM COAL

When coal is burned, three undesirable substances may be emitted into the atmosphere. These are particulates, sulfur oxides, and nitrogen oxides known as ROCKS, SOX, and NOX, respectively. These pollutants act as lung irritants to humans and are associated with acute and chronic respiratory problems, including asthma. In addition, these pollutants may, in association with polycyclic organic material such as benzo pyrene, contribute to lung cancer. These emissions may also relate to cardiac problems in some people.

Fly ash

Fly ash, a fine particulate from the burning of coal, is emitted into the atmosphere and is a serious problem to human health, ecosystems, and climate. Most trace elements become concentrated in fly ash. Typically, about 10% of the weight of coal goes off as fly ash when the coal is burned; however, about 70% of the total ash produced at coal-burning power plants starts up the stack as fly ash. Bag filters or electrostatic precipitators on power plant stacks can catch up to 99% of the particles; however, only the heavier particles may be removed efficiently and at least 50% of the finer particles of diameters less than 5 μm will remain in the flue gases. About 95% of the major coal-burning plants in the United States have installed precipitators. A big problem in the future is how to dispose of the filtered material. More than 56 million metric tons of fly ash are collected each year in the United States. Of the particles that stay in the atmosphere, those of diameter 1.0 μm or less are a special problem because they remain airborne and become deposited in the lungs of humans or animals. They also persist in the atmosphere to produce potential climate effects.

Fly ash is siltlike in composition and has a pH value from 6 to 11. Its chemical composition depends on the particular coal being burned, that is, fly ash from lignite differs from that of bituminous coal. Lignite fly ash has larger particles and a higher lime content that tend to make it set into a hard mass when combined with water. The chemical constituents going up the stack during the burning of coal may be classified into several groups.

Group I. Elements that are not volatilized by combustion and form a uniform melt in the fly ash. These elements are Al, Ba, Ca, Ce, Co, Eu, Fe, Hf, K, La, Mg, Mn, Rb, Sc, Si, Sm, Sr, Ta, Th, and Ti.

Group II. Elements that are volatilized by combustion and absorbed on the fly ash as the flue gas cools. These elements include As, Cd, Cu, Ga, Pb, Sb, Se, and Zn. These elements concentrate in the fly ash and leave the slag depleted of them.

Group III. Elements that remain in the gas phase but associate with the fly ash. They include Hg, Cl, and Br.

Group IV. Radioactive elements, including uranium, thorium, and radon isotopes.

Not only does fly ash contain all the concentrated trace elements normally recognized in coal, but it also contains a family of compounds called chlorinated dioxins that form during the combustion process. Coal normally has a chlorine content of about 1300 ppm and water mixed with coal when used to clean it may add chlorine. Combustion processes must be more than 99.9% efficient to ensure a reduction of the concentration of chlorinated dioxins from 1 ppm to 1 ppb. Most common combustion processes are not adequate to do this. These chlorinated dioxins are known to be toxic to humans, and, in particular, they are carcinogenic.

SO_x and NO_x

The two most abundant primary pollutants emitted by coal-fired electric power-generating plants are oxides of sulfur and nitrogen. The sulfur and nitrogen cycles of the biosphere are described in Chapter 1. Sulfur is introduced into the atmosphere as hydrogen sulfide from bacterial decay, as sulfates from sea salt, and as sulfur dioxide from both natural and human sources. Nitrogen is released to the atmosphere as ammonia from organic decay and fertilizers and as oxides of nitrogen from denitrification by soil organisms, by natural fires, and by high-temperature combustion. On a global basis, most NO_x is produced by bacteria. The total estimated NO_x emissions in the United States by human sources is about 20 million metric tons per year, and, of this, about 19% is due to electric power-generating plants. About 86% of the atmospheric SO_x in the continental United States is from human sources. Of this, 62% is related to coal combustion, 11% to metal smelters, 10% to other industrial sources, 2% to transportation, and 1% to miscellaneous sources. Coal combustion in power plants accounts for 40% of the total SO_x emissions in the United States.

Once the NO_x or SO_x goes into the atmosphere, they undergo chemical transformation into other compounds such as sulfate, aerosols, ozone (O_3), and peroxyacetal nitrate (known as PAN). These secondary pollutants have major sources other than power plants, for example, transportation. They generally form at considerable distance from the sources because of the slow reaction time and a dependence on other factors, such as catalysts, sunlight, water vapor, and temperature.

Air quality standards

In 1963 the U.S Clean Air Act was passed by Congress. In 1967 the Air Quality Act was passed; this legislation set goals for state and federal standards and zoned the nation into air quality control regions. By the late 1960s, enormous public concern for a clean environment forced Congress to enact stricter amendments to the Clean Air Act. These were signed into law 31 December 1970. These amendments established the standards given in Table 1 and required the states to devise implementation plans for enforcing the air quality standards for stationary sources, including incinerators, factories, and power plants.

Two sets of air quality standards were established for ambient air. PRIMARY AIR QUALITY STANDARDS are the exposure levels to a pollutant that can be tolerated by humans without ill health effects. SECONDARY AIR QUALITY STANDARDS are the exposure levels tolerable without serious impact on materials, crops, visibility, personal comfort, and climate. In addition to the five pollutants listed in Table 1, there are strict controls on other hazardous substances such as asbestos, beryllium, mercury, and vinyl chloride. Lead levels are also controlled and must not exceed 1.5 $\mu g/m^3$. Additional amendments to the Clean Air Act made in 1977 restrict the expansion of indus-

TABLE 1. National U.S. ambient air quality standards

Pollutant	Averaging period	Primary standard[a] ($\mu g/m^3$)	Secondary standard[a] ($\mu g/m^3$)
Particulates	1 year	75	60
	24 hours	260	150
Sulfur oxides	1 year	80 (0.03)	—
	24 hours	365 (0.14)	—
Carbon monoxide	8 hours	10 (9)	10 (9)
	1 hour	40 (35)	40 (35)
Nitrogen dioxide	1 year	100 (0.05)	100 (0.05)
Ozone	1 hour	235 (0.12)	235 (0.12)
Hydrocarbons	3 hours (6 to 9 p.m.)	160 (0.24)	160 (0.24)

[a]Numbers in parentheses are expressed as parts per million (ppm).

tries in regions where air pollution is already near the limit. The Environmental Protection Agency is permitted to allow the establishment of new industrial sources of pollution if the NEW SOURCE can achieve an OFFSET by an equivalent reduction in emissions from other sources, such as by the closure of another plant or the modernization of an existing plant with the use of better pollution control equipment.

Emission standards from power plants have been set by the U.S. Environmental Protection Agency to protect public health and welfare. NEW SOURCE PERFORMANCE STANDARDS (NSPS) set limits for the quantity of pollutants emitted by a power plant. The federal emission standards for new fossil fuel-fired electric power plants are given in Table 2. The emissions listed are the maximum allowable over a 2-hour averaging period.

TABLE 2. Federal emission standards for new fossil fuel-fired steam generators

Pollutant	Emissions ($lb/10^6$ Btu)	Fuel type
Particulates	0.1	All
Sulfur dioxide	0.8	Oil
	1.2	Coal
Nitrogen oxides	0.2	Gas
	0.3	Oil
	0.7	Coal

Impacts on biota

The effects of gaseous pollutants on plants and animals are characterized as either acute, chronic, or long-term. An excellent review of these is to be found in Dvorak (1978). ACUTE

EFFECTS result from relatively short (minutes, hours, months) exposure to high concentrations of pollutants. CHRONIC EFFECTS occur when organisms are exposed for many months or even years to relatively low levels of pollutants. LONG-TERM EFFECTS include subtle physiological alterations in organisms or abnormal shifts in ecosystems resulting from low-level exposures for decades or longer. Acute and chronic effects are caused by the direct action of the pollutant on the organism, whereas long-term changes may be secondary ones, such as those resulting from a change of soil pH due to acid rain, which in turn is formed by SO_x and NO_x emitted into the atmosphere. Very few studies of the long-term effects of pollutants exist. It is easiest to do fumigation experiments in greenhouses or animal houses and assess the short-term acute effects on the biota.

Impacts on vegetation. Acute injury to vegetation is usually associated with short-term exposure to high SO_2 levels, usually at concentrations in the range of 131 to 4000 $\mu g/m^3$ for exposures of 3 to 24 hours for sensitive plants. Table 1 gives the primary standard for SO_2 on an annual basis as 80 $\mu g/m^3$ (0.03 ppm) and for a 24-hour period as 365 $\mu g/m^3$ (0.14 ppm)—concentrations that are generally below the threshold sensitivities for most plants.

Plant species and varieties exhibit an enormous range of sensitivities to SO_2 as a result of complex interactions among microclimate factors, such as sunlight, temperature, and moisture, and morphological and genetic factors that influence plant response. A concentration of SO_2 that will kill one species may not affect another species. Injury to a plant always depends on the combination of pollutant concentration and length of exposure. For example, plants that are highly sensitive to SO_2 will likely be damaged when exposed for two hours to a SO_2 concentration of 4240 $\mu g/m^3$ (2.0 ppm) but will require at least an eight-hour exposure to be damaged at a concentration of 1310 $\mu g/m^3$ (0.5 ppm). In other words, concentration times exposure time equals a constant. For highly sensitive plants, the constant is 10,480 $\mu g/m^{-3}$ hr^{-1}. For plants that are highly resistant, the constant is 41,920 $\mu g/m^{-3}$ hr^{-1}. Sensitive plants require only one-fourth of the exposure time to the same SO_2 concentration as do resistant plants. Native plant species are generally less sensitive to SO_2 than introduced species, such as ornamentals and crop species. Hill et al. (1974) exposed 87 species of plants that are native to the Arizona desert to concentrations of SO_2 from 1300 to 26,000 $\mu g/m^3$ for two hours and found that at least 5200 $\mu g/m^3$ of SO_2 was needed to damage most species. Only one species was damaged by 1300 $\mu g/m^3$. The relatively short exposure time may be one of the reasons for the apparently high resistance of these plants.

According to Jones et al. (1973), soybeans, which are SO_2-sensitive plants, had no loss of yield when exposed to SO_2 fumes from a large coal-fired electric power-generating plant at concentrations that caused some severe leaf injury at an early leaf stage. SO_2-induced leaf injury was not a factor in accounting for variations in yield for the soybeans on 110 fields. Factors contributing to poor yields were low soil fertility, soybean cyst nematode infestation, continuous soybean cropping, and late planting, rather than SO_2 exposure. Apparently, most of the plants whose leaves were affected by SO_2 exposure attained near-normal growth. In several other studies on soybeans, it was found that when exposure to SO_2 occurred during the flowering period and when leaf chlorosis (yellowing) was greater than 5%, some yield reductions were obtained (Muller et al., 1979).

Lichens are highly sensitive to SO_2 pollution and are good indicators of the quality of the air within urban areas. The extreme sensitivity of lichens to SO_2 appears to be due to destruction of the alga within the fungus.

Grain, vegetable, pasture, and forage crops are mostly in the SO_2-sensitive group of plants. Corn, however, is one of the most resistant of crop species and has rarely been injured by SO_2 under field conditions. Where SO_2 fumigation from power plants is known to be a problem, more resistant crops can be planted.

Impacts on animals. The effect of sulfur oxides on animals is highly variable, as it is with plants. Sulfur oxides usually irritate or injure the respiratory passages or mucous tissues. Animals with the highest ventilation rates for a given body size appear to be the most sensitive. From epidemiological studies with humans it is concluded that chronic exposure (several months) to concentrations of SO_2 of about 100 $\mu g/m^3$ can have adverse effects. Chronic exposures to concentrations of 13,000 $\mu g/m^3$ have produced pulmonary dysfunction in dogs. Acute exposures (two to three hours) of some laboratory animals to SO_2 have resulted in deleterious pulmonary action at concentrations down to 18,000 $\mu g/m^3$. Mortality in laboratory animals does not occur during exposure to SO_2 concentrations from 65,000 to 3.1 million $\mu g/m^3$ for less than 90 hours.

Less seems to be known concerning the response of plants or animals to NO_x concentrations. Effects on guinea pigs begin to show up at 1000 $\mu g/m^3$ and more pronounced with exposure to above 2000 $\mu g/m^3$ of NO_2 for several hours. Humans appear to be more sensitive to NO_2 than are most other animals. One wonders why this should be so, or whether it is a communications problem. Epidemiological studies of human populations suggest that chronic exposure to NO_2 concentrations of about 100 $\mu g/m^3$ may be deleterious.

Modeling emissions

Modeling of emissions from a 700-MWe power plant shows that maximum short-term (three hours) concentrations of SO_2 near the ground surface within a radius of 1 km from the plant may exceed 1800 $\mu g/m^3$ if no scrubber is used when burning western coal. Burning eastern interior coal, using an 85% scrubber efficiency, would produce a SO_2 concentration of about 1400 $\mu g/m^3$ near the ground within 1 km of the plant. SO_2 concentrations in the vicinity of a 750-MWe coal-burning power plant are estimated to be far below the levels at which laboratory animals are adversely affected but can exceed the levels for which adverse effects on human health are indicated.

The predicted NO_x levels within 1 km of a 700-MWe coal burning power plant are 171 $\mu g/m^3$ during a 24-hour period and levels as high as 600 $\mu g/m^3$ over a 3-hour period. Acute and chronic threshold levels for plants and animals exposed to NO_x are much higher than any doses predicted in the vicinity of the model plants.

Trace elements and heavy metals

Trace elements are emitted into the atmosphere from a coal-fired power plant as vapor, adsorbed on the surface of fly ash, or as a combination of both. Elements emitted as vapor are mercury, selenium, or beryllium, but generally these are in such small quantities as vapor that they represent no real hazard in the environment. Of most concern are the trace elements adsorbed onto fly ash particulates that become airborne, such as Be, Pb, As, Se, F, Cd, B, and Ni. The height of the stack determines to a considerable degree the amount of particulate or gaseous deposition on the soil or biota in the vicinity of the plant. Once the plume is widely dispersed, trace element concentrations are negligible. Trace elements also get into the environment as leachate from coal piles or ash.

Terrestrial impacts. There have been relatively few studies of the flow of trace elements from coal-fired power plants through terrestrial ecosystems. As a general rule, it is found that the level of trace elements deposited on leaves or on the soil would have no adverse effect on vegetation.

Klein and Russell (1973) sampled soils, native grasses, maple leaves, and pine needles at sites near a 650-MWe power plant on the shore of Lake Michigan near Holland, Michigan. The plant burns Ohio coal. The stack is 122 m high and is equipped with a 90% efficient electrostatic precipitator. The top two centimeters of soil near the plant had concentrations of Ag, Cd, Co, Cr, Cu, Fe, Hg, Ni, Ti, and Zn that were higher than concentrations in similar soil farther away from the plant. Concentrations of Cd, Fe, Ni, and Zn were higher in the vegetation near the plant than in similar vegetation on the control sites. Levels of heavy metals were far below toxic levels. Measurements made near the 870-MWe Thomas A. Allen power plant near Memphis, Tennessee on the flood plain of the Mississippi River showed no significant uptake by cotton or soybeans, nor any adverse effects on this vegetation.

Studies were made of trace element concentrations in the tissues of wildlife living near two lignite-fired plants near Stanton, North Dakota. One plant had a capacity of 215 MWe and the other 530 MWe. Tissue concentrations of Ca, Na, K, Sr, Zn, and Cu were obtained from white-tailed deer, deer mice, redback vole, and meadow vole in two areas near the plants. For all species examined, no significant differences were found between animals in experimental and control areas for each of the trace elements in bone, kidney, liver, or fresh antler tissues. Further studies of deer mice showed no bioaccumulation of Hg, Cd, Pb, Se, and Mo for both liver and skin tissue. This may have been a result of the relatively short lifetime, about 6 to 12 months, of the mice, a time too short for much bioaccumulation to occur.

Extreme effects of long-term accumulation of trace elements on biota may be found in the vicinity of smelters. The nearly total destruction of the forested ecosystems surrounding the huge nickel smelter at Sudbury, Ontario is a classic example of the devastation generated by heavy metal concentrations that are much above toxic levels. It would be incorrect to use such examples as illustrations of the effect of coal-fired power plants on ecosystems.

Radioactive emissions

Coal contains several radioactive elements; and although coal has been underground since primordial times, it still retains some radioactivity. Radioactive elements include uranium 238 and thorium 232. The concentration of these in coal is highly variable but can range from 0.2 to 47 ppm. Most of the radioisotopes will be left in the slag and ash when coal is burned, where their concentrations are higher than in the coal. The concentration of uranium and thorium in fly ash will be about 16 and 12 times greater than that in coal, respectively. Emission rates of release for uranium, thorium, and radon from coal-fired plants have been estimated for plants of various sizes. The dose rate to humans for emissions from a 1000-MWe coal-fired plant is estimated to be 1.9 mrem/yr for the whole body and lungs, compared to 80 and 180 mrem/yr, respectively, from natural background. For bone, the coal-fired plant will deliver a dose of 18.2 mrem/yr versus 120 mrem/yr from the natural environment. Further information may be found in McBride et al., 1977.)

Coal cleaning

The best way to reduce sulfur emissions is to burn coal of low sulfur content. This means burning mostly western coal rather than eastern coal, although there are some exceptions. Coal can be cleaned prior to burning to reduce emissions, but the flue gases also need to have pollutants removed. Recently it was proposed that all coal should be scrubbed to remove 90% of its sulfur emissions. Western coal companies protested, so a new Environmental Protection Agency proposal was made to require 90% removal of sulfur from high sulfur coal and 50% removal from low sulfur coal. Low sulfur·coal from the western United States now must be scrubbed.

Coal is cleaned to reduce the sulfur content and the content of unwanted noncombustible materials that become fly ash during the burning of coal. Cleaning increases the uniformity of the chemical and physical properties of coal. Cleaning reduces the quantity of particulates in flue gases, increases the heating values of the coal, and reduces its sulfur content. Most deep-mined coal is cleaned because mining machines fail to discriminate between coal and roof or bottom slate. All anthracite coal is cleaned, most bituminous coal is cleaned; only some subbituminous coal is cleaned, and no lignite is cleaned. Cleaning requires crushing and screening into sized fractions which are then sent to different cleaning units. The crushed material is washed, and either gravity or centrifugal forces are used to separate slate, rock, and sulfur-containing pyrites from the coal.

Trace elements in coal

The trace elements of heavy metals in coal are serious, potential pollutants of air and water. The major trace elements found in coal are arsenic (1 to 16 ppm), barium (31 to 380 ppm), beryllium (0.6 to 2.4 ppm), boron (15 to 85 ppm), lead (4.0 to 6.0 ppm except 0.6 in Powder River basin and 33 in Illinois), mercury (0.05 to 0.20 ppm), molybdenum (2.0 to 9.8 ppm), vanadium (12 to 30 ppm), and zinc (15 to 140 ppm). In addition, coal contains concentrations of uranium (1.3 to 2.5 ppm) and thorium (3.3 to 5.4 ppm), both radioactive elements. Cadmium, antimony, and selenium may also be found in coal in concentrations greater than average in the Earth's crust and are of concern to human health. All of the above elements can be toxic to plants and animals.

STACK GAS CLEANING

The U.S. Environmental Protection Agency has developed New Source Performance Standards (NSPS) that place limitations on particulates and oxides of sulfur and nitrogen emitted from power plants. The hot gases leaving the stack of a power house carry particulates, the uncombustible residue known as fly ash. For a pulverized coal-fired plant, 80% of the ash becomes entrained in the flue gas. The remaining 20% settles in hoppers at the base of the boiler, from which it is removed for disposal. The fly ash generated by pulverized coal burning is much lighter than that produced in either stoker or cyclone furnace-fired systems because pulverized coal is burned as a suspension in air. With cyclone furnace firing, only about 20 to 30% of the total ash is carried up the stack as fly ash.

There are four methods used for removing fly ash from stack gases: electrostatic precipitators, wet scrubbers, fabric filters, and mechanical collectors. Figure 5 illustrates an electric power plant flue gas cleaning system showing the electrostatic precipitator and desulfurization units.

Electrostatic precipitators

An ELECTROSTATIC PRECIPITATOR is made up of a series of flat parallel collector plates spaced from 15 to 30 cm apart with a wire discharge electrode mounted between each pair of plates. A strong electric field is put across the wire and plate. The particulate-laden flue gas passes between the plates. An electric corona discharge is formed, and this ionizes the flue gases such as O_2, CO_2, and SO_2. These ions then charge the fly ash negatively. The particles of fly ash are pulled by the electrostatic field forces to the positive collecting electrode, which is periodically shaken to dislodge the accumulated fly ash. The fly ash falls into a hopper beneath the electrodes,

from which it is removed and transported to a land fill. Figure 6 shows the electrostatic precipitators of the River Rouge, Michigan coal-burning electric power plant. Collection efficiencies of 99% or greater may be achieved. Efficiency increases with the time the flue gas spends in the precipitator and with the strength of the electric field. Efficiency declines with a reduction of the sulfur content and with lower temperatures. When burning low-sulfur coal, it is necessary to place the precipitator upstream of an air heater. Electrostatic precipitators are reasonably compact and have low maintenance cost and low down time. The fly ash escaping from a precipitator into the atmosphere has fly ash particles smaller than about 5 μm.

Wet scrubbers

A WET SCRUBBER removes fly ash from the flue gas by mixing the particles with water. There are two basic types of wet scrubbers, the ven-

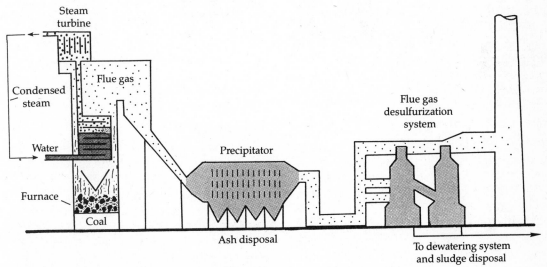

FIGURE 5. Schematic illustration of a flue gas-cleaning system for a coal-burning electric power generating plant.

FIGURE 6. The coal-burning electric power generating plant at River Rouge, Michigan. The electrostatic precipitators are in the foreground. (Courtesy of Detroit Edison Co.)

turi and the moving-bed scrubber. A venturi scrubber utilizes a venturi throat, which is a constricted region in the flue gas stack that forces the flue gas to move at a high velocity (60 to 120 m/sec). Water is injected into this high-velocity stream, and impact between the water and particulates causes the fly ash to be removed in the water stream. Venturi scrubbers have efficiencies of about 99%.

A MOVING-BED SCRUBBER is made of a bed of loosely packed plastic or glass spheres that are resting on a perforated plate; the spheres and plate are mounted within a cylinder. Flue gas passes upward through the sphere bed as water flows downward through it. In addition to extracting fly ash particles, moving-bed scrubbers are effective in also removing SO_2 because of the long residence time of the flue gases in the scrubber. By contrast, the venturi scrubber provides a short residence time and a poor removal efficiency of SO_2. Fly ash collection efficiency is high for moving-bed scrubbers, ranging from 98.7 to 99.9%.

Fabric filters

Very large fabric bags through which stack gases flow are used to remove fly ash. The particles are simply trapped by the fine threads of the fabric material. Curiously, as fly ash accumulates on the filter, its efficiency increases. The bags must be cleaned periodically, after which their efficiency is reduced for a while. Fabric filters can only be used where humidity is not excessive. Often they cannot compete with electrostatic precipitators for efficiency.

Mechanical collectors

Flue gas enters a rapidly spinning centrifuge in the mechanical collector, and the fly ash particles are thrown against an outside wall from which they are removed. Mechanical separators only work well for very large particles. The removal efficiency is less than 90% for particles greater than 10 μm in diameter.

Nitrogen oxide reductions

Oxides of nitrogen are produced when nitrogen and oxygen combine at high temperatures, for example, when coal burns in air. The nitrogen may originate from the air (78.8% N_2) or from nitrogen contained in coal. The formation of nitrogen oxides depends on combustion conditions in the primary flame zone of the boiler. Oxides of nitrogen are recognized as undesirable contaminants of atmospheric pollution because they contribute to smog and acid rain. Four techniques are available for controlling NO_x emissions from coal-fired power plants: modification of combustion conditions, modification of the coal, extraction of NO_x from flue gas, and use of low NO_x-producing combustion processes such as catalytic combustion or fluidized-bed combustion. Modification of combustion conditions appears to be the best

developed and most effective technique of the four for reducing NO_x emissions. Burning with low excess quantities of air is the most widely used combustion modification strategy. However, when the excess air level becomes too low, smoke and increased amounts of carbon monoxide are produced. This modification usually results in NO_x emission reductions of less than 50%.

Another modified combustion process sometimes used is the staged combustion system. Here the furnace is fired so as to make the primary flame zone rich in fuel. Sufficient air is introduced above the flame zone to complete combustion, while the amount of air in the high temperature flame zone is minimized. Nitrogen oxide formation proceeds rapidly at temperatures above 1650°C. By good radiative transfer, the primary flame zone is maintained in a cool, oxygen-poor state. The use of staged combustion results in NO_x emission reductions of 50 to 65%.

Sulfur oxide reductions

Several methods are available for reducing sulfur oxides from stack gases: lime/limestone scrubbing, double alkali scrubbing, magnesia scrubbing, and the Wellman-Lord process.

Lime/limestone scrubbing. The lime/limestone scrubbing process removes SO_2 from the flue gas by reacting it with lime ($Ca(OH)_2$) or limestone ($CaCO_3$) in a water slurry to form a precipitate. Lime or limestone is mixed with water and pulverized within a giant cylinder that rotates and tumbles steel balls (about 75 cm in diameter) with the mixture. The pulverized lime or limestone goes to a classifier, where the remaining oversized stone is returned to the pulverizer. The chemical reactions involved in the scrubbing process are

$$SO_2 + Ca(OH)_2 \rightarrow H_2O + CaSO_3$$
$$SO_2 + CaCO_3 \rightarrow CO_2 + CaSO_3$$
$$2CaSO_3 + O_2 \rightarrow 2CaSO_4$$

The flue gas comes in contact with the slurry in the scrubber, and it is there that the first two reactions occur. The reacted slurry then flows to the reaction tank, where additional lime or limestone is added to cause precipitation of calcium sulfite ($CaSO_3$) and calcium sulfate ($CaSO_4$) in hydrated form. The remaining slurry is sent to a separator and then to a holding pond. A sludge (the precipitate), which is made up of hydrated calcium sulfite and calcium sulfate, unreacted lime and limestone, and some fly ash, is removed for disposal in a land fill. Large quantities of sludge are produced by a lime/limestone scrubbing system.

After scrubbing, the flue gas is reheated to above its dew point temperature and released up the stack. SO_2 removal efficiencies can be as high as 80%. This is the most widely used method in the power plant industry.

Although the chemical reactions shown above indicate that only one atom of calcium is required for each sulfur atom removed, there must be additional lime or limestone added to the process because some of it does not react. A 1000-MWe coal-burning plant requires about 195 to 375 mt of lime and 435 to 836 mt of limestone per day. These amounts are for burning northern Appalachian coal (that is, Pittsburgh and West Virginia coal) and for burning Illinois coal, respectively. When burning western low-sulfur coal, scrubbing is not required to meet the NSPS requirements.

Double-alkali scrubbing. The double-alkali scrubbing process uses a sodium hydroxide (NaOH) or sodium sulfite (Na_2SO_3) solution to absorb the SO_2 from the flue gas, and lime or limestone is used to regenerate the sodium compounds. A solid waste of calcium sulfite or calcium sulfate is generated. Double-alkali

scrubbers do not collect scale very rapidly and are more efficient than lime/limestone scrubbers. SO_2 removal efficiencies as high as 90 to 97% have been reported.

Magnesia scrubbing. A magnesium sulfite ($MgSO_3$) slurry is reacted with the flue gas:

$$SO_2 + MgSO_3 + H_2O \rightarrow Mg(HSO_3)_2$$

Magnesium oxide (MgO) is then used to regenerate magnesium sulfite:

$$Mg(HSO_3)_2 + MgO \rightarrow 2MgSO_3 + H_2O$$

The precipitated $MgSO_3$ is concentrated by a separator, dried, and sent to a sulfuric acid plant. There the magnesium sulfite is heated to drive off the magnesium oxide, leaving SO_2, which is combined with hydrogen to yield sulfuric acid (H_2SO_4).

Magnesia scrubbing is regenerative, producing no waste sludge and not requiring massive amounts of lime or limestone. It has an SO_2 removal efficiency of 90% or more and a useful end product, sulfuric acid.

Wellman-Lord process. The flue gas is reacted with sodium sulfite in the Wellman-Lord SO_2 removal process:

$$SO_2 + Na_2SO_3 + H_2O \rightarrow 2NaHSO_3$$

The $NaHSO_3$ is heated in an evaporator, which simply reverses the above process. The products go to a condenser and separator units to condense and separate the H_2O and SO_2. The SO_2 can then be converted to sulfuric acid. The condensed water is mixed with the regenerated sulfite slurry, some makeup alkali, and makeup water to provide the scrubber feed solution. SO_2 removal efficiency by the Wellman-Lord process is over 90%.

Water requirement

Water is required in the lime/limestone scrubbing processes, in the sluicing of bottom ash and captured fly ash, and in the other processes described earlier. As a general rule the amount of water required for these purposes is from 16 to 30% of the cooling water requirement. A 1000-MWe plant requires about 7 million gallons of makeup water per day, depending on the cooling process used; scrubbing processes require an additional 0.9 to 2.1 million gallons per day. Therefore, a 1000-MWe coal-fired power plant will use between 8 and 9 million gallons of water per day. A 500-MWe plant has just half the water requirement of a 1000-MWe plant. A primary siting consideration is location: a power plant must be near an ample supply of water. Therefore, most electric power generating plants are near rivers, lakes, or estuaries.

Fly ash and bottom ash disposal

Considerable amounts of fly ash and bottom ash are generated at a coal-burning electric power generating plant. A 1000-MWe coal plant would create about 470, 400, and 270 mt/d of fly ash when burning Illinois, northern Appalachian, and western coal, respectively. Bottom ash production rates are 120, 99, and 69 mt/d when burning the same three coal types in that order.

The disposal of fly ash from the electrostatic precipitator is by wet sluicing into an ash pond. Effluent from the ash pond has previously gone directly to natural surface waters. New Source Performance Standards now require that there be no direct discharge of suspended solids or of oil and grease. Settling ponds must be used, and dewatering by natural evaporation may leave a residue of sludge that is then hauled to a landfill. Often the fly ash slurry is mixed with that containing bottom ash, allowed to dewater by evaporation or by withdrawal of the clear surface water, and then transported to a landfill. Sometimes bottom ash is collected separately from fly ash and placed on top of the compacted fly ash at the disposal site. When placed as a cover on compacted fly ash, it reduces runoff of rain because it is more permeable to water than is the fly ash. Sometimes ash disposal is in abandoned mines rather than in a landfill site. Often when flue gas desulfurization with lime or limestone slurry is used, the sludge is combined with the fly ash slurry for disposal.

Scrubber sludge disposal

The scrubber sludge flows to large reaction tanks, where it is aerated by bubbling air through it. This process oxidizes the calcium sulfite to calcium sulfate, which forms large particles that settle out rapidly. The calcium sulfate is then precipitated by the addition of gypsum. The slurry containing them goes to a clarifier, where these waste solids settle out; the clear surface water is returned to the cycle. The sludge is pumped to a settling pond and from there conveyed by truck to a landfill. Here it will be compacted for permanent storage. As mentioned earlier, the scrubber sludge is often combined with the fly ash slurry and, after some treatment, transported to a landfill. Combining fly or bottom ash with flue gas desulfurization sludge is one method of improving the physical and chemical properties of the sludge for land disposal. The ash acts as a chemical hardener and neutralizes the acidity of the sludge. Lime may be added to the mixture to promote hardening.

Aquatic impacts

Sludge from scrubbers contains almost all of the trace elements found in the coal utilized by the power plant. These trace elements enter the groundwater by runoff from sludge ponds or landfills. The effects of trace elements on aquatic ecosystems are similar to the effects of heavy metals in streams and rivers receiving acid mine drainage (see Chapter 8). In aquatic ecosystems, trace elements are mainly associated with the sediments that act as both a sink and a reservoir. Very small amounts are found dissolved in the water. Trace element concentrations in water tend to decrease as pH is increased. The pH of ash-pond water is usually so high (i.e. between 6 and 11) that the trace elements precipitate in hydroxides of iron and aluminum. Trace elements are taken up from the sediments by rooted plants and benthic (bottom) invertebrates. Grazers and some lower-order consumers seem to concentrate trace elements the most. The greatest bioaccumulations are found among the sediment or detrital feeders. Higher-order consumers receive trace elements through the food chain. Curiously, trace element concentrations are usually higher in the prey than among the predators, with certain exceptions.

Microorganisms are known to play an important role in the biogeochemical cycles of carbon, oxygen, sulfur, nitrogen, phosphorus, iron, zinc, and manganese. Some microbes are able to convert elemental mercury to methylmercury, in which form it enters the aquatic food chain. Methylmercury is readily absorbed by animal tissue, passes through all membranes, and is carried by red blood cells to the brains of animals, including humans. Fish absorb methylmercury from water 100 times more rapidly than they absorb elemental mercury. People eat the fish and are poisoned by the methylmercury. The classic examples of mercury poisoning of human populations were at Minamata and Niigata, Japan in the 1950s and 1960s. These were situations in which large amounts of mercury entered the marine environment from industrial processes; they were not the result of power plant operations.

Bacteria pick up heavy metals from the water, concentrate them, and pass them on to other organisms in the food chain. Aquatic plants, both phytoplankton and higher plants, are known to accumulate trace elements. Some synergistic effects are known as well. In the alga *Scenedesmus*, a high concentration of copper enhances nickel uptake.

Accumulation and bioconcentration of heavy metals in duckweed (*Lemna minor*) lead to their uptake by waterfowl because duckweed is a major food source for them. Trace element uptake in invertebrates, such as oysters or clams, is very species specific. The main uptake of heavy metals by fish is through the food chain and not from the water directly. Seasonal changes of temperature appear to affect the uptake rate in fish. Within a given species, resistance to a toxicant, such as a heavy metal, can vary with age, sex, growth rate, physiological condition, and life cycle stage. Fish eggs appear to be more tolerant to trace elements than are larvae, fry, or adults. For chinook salmon and bluegills, larvae were more susceptible to copper concentrations than were older fish.

Guthrie and Cherry (1976) studied the biota of a stream receiving the effluent from a fly ash settling basin. They measured concentrations for 29 elements and found that fishes had the lowest concentrations, invertebrates always more than fishes and plants, with some exceptions, more than invertebrates. However, fishes had the highest concentrations of calcium and selenium. The major role of each plant or animal species in the cycling of specific trace elements is highly varied. Cattails (*Typha latifolia*) concentrated over 50% of the manganese present, duckweed more than 40% of the titanium, tadpoles more than 40% of the chlorine, midges 38% of the chromium, and crayfish 31% of the calcium. Plants were generally more efficient concentrators of manganese and potassium than were animals.

As mentioned earlier, the pH of the water

is an important factor with respect to the fate of heavy metals and other trace elements in the aquatic ecosystem. Sludge from power plant scrubbers tends to be acidic (pH 5 to 5.5) and so is often neutralized by adding fly or coal ash to it. Usually the pH of most ash-pond and scrubber-pond leachate will be neutral or alkaline. This prevents trace elements from becoming soluble in the water. When the surrounding soils are acidic, the leachate flowing through them may release trace elements into the ecosystem.

COOLING SYSTEMS

Water is always used for carrying away the heat from the spent steam in the condenser unit of the power plant. Enormous quantities of water must be withdrawn from a river, lake, or estuary, passed through the condenser, and returned to the environment. Its temperature may rise as much as 22°C when passing through the condenser. This waste heat can either be used (a process known as cogeneration) to heat buildings and greenhouses or as industrial process heat, or be discharged into the environment. Four methods used to disperse the waste heat into the environment are once-through cooling, wet-tower cooling, dry-tower cooling, or spray-pond cooling (see following sections).

The amount of water used by an electric power plant depends upon the type of cooling system used. A 1000-MWe coal-burning plant rejects about 1632 MW of heat. When the water used to carry off this heat is raised in temperature by 10°C, nearly 900 million gallons of water per day must be used in a once-through cooling system. When cooling towers are used, the amount of water required is greatly reduced. A coal-burning electric power plant using cooling towers requires 0.4 gallons of water per kilowatt-hour of electrical output. A 1000-MWe coal plant operating at 70% capacity uses 8 or 9 million gallons of water per day. A 1000-MWe nuclear plant must reject more heat and would use 10 to 11 million gallons of water per day with cooling towers.

Once-through cooling

The ONCE-THROUGH, or OPEN-CYCLE COOLING SYSTEM, pumps water from a river, pond, lake, or ocean through condenser pipes across which the hot steam from the turbine is passing. A simplified drawing of such a system is shown in Figure 7. The condenser water is heated in the process and the steam from the turbine is cooled and condensed to water. Very little water is actually consumed, but the water is returned to the original body of water warmer than when it was withdrawn. The condensed steam is returned to the boiler as water. The cooling water never comes in direct contact with the steam but only receives heat from the steam through a heat exchanger. A survey of 52 electric power-generating plants showed that the temperature gradient between condenser and receiving water was from 2° to 16°C.

A lake is the most vulnerable of the various types of receiving waters to heat input. It is estimated that an offshore discharge from a 1000-MWe nuclear plant with a temperature rise in the condenser of 11°C will raise ambient lake surface temperatures by about 1°C over nearly half a square kilometer of a large lake. For a small lake, the temperature rise could be considerably more. At Lake Julian, North Carolina, a 200-MWe power plant discharged its cooling water into the epilimnion (top layer) of the 1.3-km^2 lake and raised its temperature about 5°C. The effect on estuaries may be every bit as great or greater than the effect on lakes. However, in estuaries there can be an additional impact by changing the salinity of the water as well as the

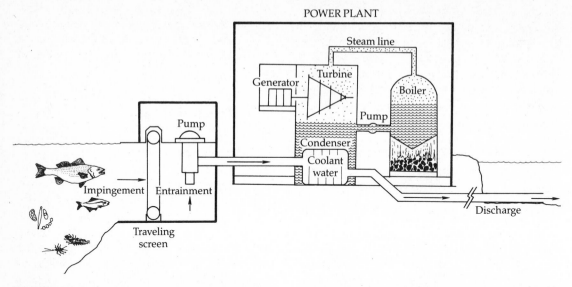

FIGURE 7. A once-through cooling system. The coolant water causes the steam from the turbine to condense; this condensation is then pumped into the boiler. Although the coolant water is not in direct contact with the steam, it does absorb heat and is returned to the environment at a heightened temperature.

temperature. On rivers, the flow of water tends to disperse the heat downstream. Figure 8 shows the thermal effluent entering Lake Erie from an electric power plant on the shore.

The once-through cooling system has the greatest ecological impact on aquatic ecosystems of any of the cooling processes, but from the economics of the various methods, it may be the most attractive.

Cooling towers

One sees cooling towers near every large city, along many major rivers, near lakes, and even far removed from surface waters. They are usually massive concrete structures, often rising 100 meters or more above the landscape. Figure 9 shows a typical power plant with its accompanying cooling towers. In contrast to the once-through or open-cycle cooling system, cooling towers make use of a partial closed-cycle system, whereby water is recycled between the cooling tower and the condenser.

There are basically two types of cooling towers, the dry tower and the wet tower. The mechanical-draft evaporative cooling tower is basically a wet tower. Usually a cooling tower must reduce the temperature of the condenser water from 6° to 22 °C.

A dry tower uses less water than a wet tower. Because the wet tower operates as an evaporative process, a good deal of water loss occurs. In addition to evaporative water loss, there is DRIFT, that is, fine droplets carried away with the air blast leaving the tower. (Additional information concerning the design of cooling towers can be found in Woodson, 1971.)

Dry tower. The dry tower illustrated in Figure 10 is more conservative in the use of water than is the wet tower because the heat in

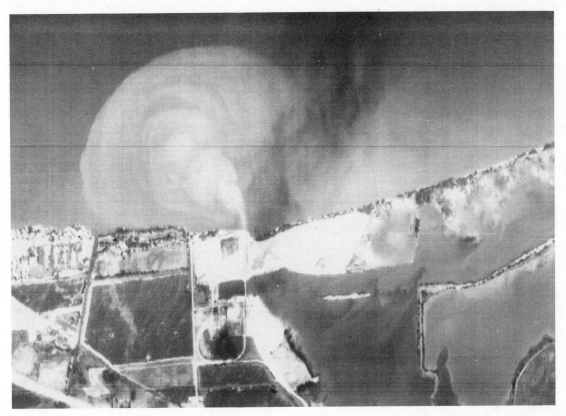

FIGURE 8. Aerial photo of the thermal effluent into Lake Erie from the Whiting power plant, Monroe, Michigan on 29 August 1972. The warm, circular plume extended 900 m from shore and had a temperature of 29°C. Surface water temperature beyond the plume was 24°C.

the condenser water is indirectly released to the atmosphere. The cooled condenser water recycles to the condenser; only a small amount of makeup water needs to be added occasionally. Some water loss in incurred by the dry tower operation because of the BLOW-DOWN process necessary for cleaning scale out of the condenser pipes.

Air flows across the pipes carrying the hot condenser water and, by forced ventilation, picks up the heat and carries it off and out the top of the tower. Air is a very good coolant, but very large volumes of it must be pulled through the tower and quite elaborate heat-exchange surfaces must be provided.

Large fans in the tower force the flow of air. (Electricity is required to keep the fans running, a process consuming considerable energy.)

Very few dry towers are built, partly because they are the most expensive of the cooling processes. Because the dry towers operate with minimal loss of water, they are particularly suited to regions where water supplies are limited.

Wet tower. The wet-tower cooling system is the design most frequently used (Figure 11). Warm water from the condenser is released at the top of a heat exchanger that is made up of

FIGURE 9. The Belle River power plant and cooling towers. (Courtesy of Detroit Edison Co.)

a large series of louvers or slats. The water cascades down across these louvered vents, which have the appearance of a gigantic venetian blind. Air flowing across the wet vents evaporates some of the water and cools it. The cooled water collects at the bottom and is pumped back to the condenser. The heat exchanger is only in the skirt or ring around the bottom of the tower. The massive hyperbolic concrete structure above this is the chimney and is designed to create the greatest natural draft of air possible. Moist air is less dense and more buoyant than dry air. The density difference between the warm, moist air inside the tower and the cooler, drier air outside creates a massive flow of air upward.

Localized climate changes from the use of wet cooling towers can occur in the form of increased humidity, ground-level fogs, cloud formation, snowfall and rain, and thermal effects. Legionnaires' disease may be transmitted by wet-tower cooling. Various pathogens

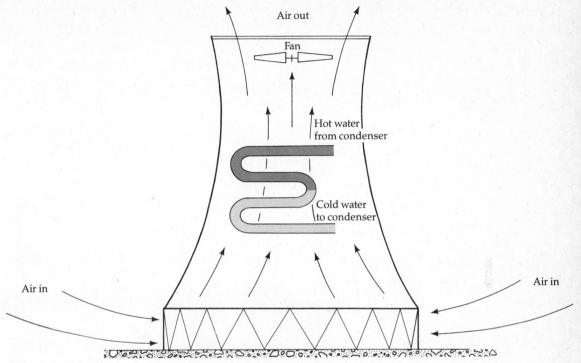

Air out

Fan

Hot water from condenser

Cold water to condenser

Air in

Air in

FIGURE 10. A dry-tower cooling system with forced ventilation. The air cools the heated water, which is recycled back into the condenser.

(fungi, in particular) may be harbored in the tower. Chemicals, such as chlorine, must be used to clean the tower periodically to reduce biofouling in the tower and the pipes. The chlorine can poison organisms in the receiving water—a stream, lake, or ocean. In addition, water intake results in entrainment or impingement of fish and other organisms.

Mechanical-draft evaporative cooling tower. Mechanical-draft evaporative cooling towers have been in use for many years but are now defined as state-of-the-art technology by the EPA. They have a fan at the top of the tower to draw air through the baffles, across the falling water, and out the stack (Figure 12). A power plant will have a series of many such towers.

Because a mechanical-draft evaporative cooling tower uses a fan, there is an energy penalty at a power station of about 2 to 4% of the electrical output, depending upon the specific design. Some plants have a penalty as high as 10 or 12%. For a 1000-MWe plant, an energy loss of 20 to 40 MW occurs. Because it takes an additional 1500 to 2000 MW of fuel for each 1000-MWe produced, the actual energy penalty as fuel is 2.5 to 3.0 times 20 to 40 MW, or from 50 to 120 MW. There is also an energy penalty for building a hyperbolic wet-tower cooling system because a substantial amount of energy is involved with its manufacture. The same would be true for a dry-tower system.

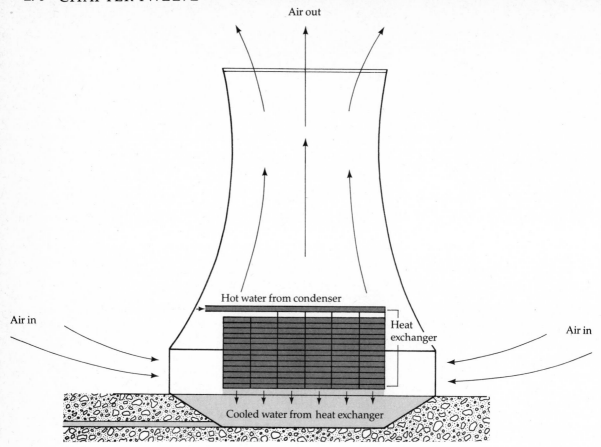

FIGURE 11. A wet-tower cooling system. Hot water from the condenser is air-cooled as it flows across the heat exchanger. (No fan is used in this system.)

Spray-pond cooling

The pond system is a semiclosed system of evaporative cooling. Warm water from the condenser is pumped into a pond, where evaporative cooling occurs by repeatedly spraying it into the air. Then the cooled water is returned to the condenser. Fresh water must continually replace the evaporative losses. A pond system is in use at the Greenwood Detroit Edison plant (Figure 13).

Cogeneration

Instead of throwing away the waste heat of the condenser by putting it into a lake, stream, or estuary, or into the atmosphere, it can be used to heat buildings or as process heat. Usually the hot water ejected by an electric generating plant is at about 40°C—to maximize the amount of electricity produced. Although this heat is useable as is, it would be much more useable if it were at a higher

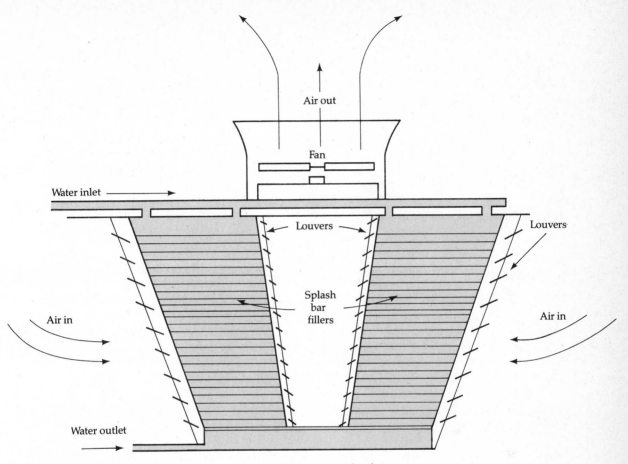

FIGURE 12. A mechanical-draft evaporative cooling tower with a fan.

temperature. Then it could be used to do mechanical work in addition to providing warmth. It is often worthwhile to reduce the output of electricity somewhat in order to reject the waste heat at a higher temperature. Using the rejected heat for useful processes is known as COGENERATION.

Cogeneration has been used by central power stations in Europe much more than in the United States. However, it is now being given careful consideration here. The use of cogeneration may reduce fuel inputs by 30%.

Advantages to cogeneration include (1) reduced capital costs for generating equipment; (2) reduced cooling water requirements; and (3) possible reduced costs for energy to industrial and commercial customers. Major applications for cogeneration are district heating of residential and commercial buildings and process heat for industries such as food, textiles, pulp and paper, chemical, petroleum refining, and steel. It is more difficult for central power stations to use cogeneration than it is for industrial power plants to use it. This is

FIGURE 13. A pond spraying cooling system at the Greenwood power plant north of Detroit, Michigan. (Courtesy of Detroit Edison Co.)

partly because of the distances from a central power station to the points of application. On the other hand, within a particular industry, it is effective to use cogeneration.

When a gas turbine is used for electric power generation, the exhaust gases are still very hot (300° to 550°C) after they have passed through the turbine, and are very useful for further work. The exhaust gases can generate steam in a waste boiler, and the steam then can be used for useful work. Schurr et al. (1979) discuss cogeneration in considerable detail and point out that the fuel savings achieved using cogeneration are quite substantial. For a steam turbine operating as a cogeneration system, the net fuel consumption per kilowatt-hour of electricity produced is less than half that required for conventional central power plant generation.

ECOLOGICAL IMPACTS OF COOLANT WATER

There are many environmental and ecological impacts associated with the construction and use of a coolant water system for an electric power generating plant. Some of these impacts are purely physical, such as the modification of a body of water by dredging and the laying of the intake system, or the entrainment and the impactment of fishes during the operation of the coolant system. Other impacts include thermal and chemical effects on the biota of the body of water from and into which the coolant water flows. Coutant (1978) gives a list of the many possibilities for environmental and ecological impacts. The following list is a modification of the Coutant list:

1. Change in the body of water by dredging, filling, or other construction at the intake and discharge points.
2. Current changes near and far from intake and discharge points, which will affect siltation, thermal stratification, and even salinity in estuaries.
3. Entrapment and impingement of larger organisms, principally fishes, on intake screens.
4. Entrainment in the intake of plankton, larvae and juveniles of fishes, small adult fishes, and invertebrates. Once in the coolant water, these organisms become exposed to mechanical shock involving pumps, pressure changes, and shear forces; temperature shock; and chemical shock by chlorine or radionuclide activation products.
5. Plume entrainment in the discharge stream, where organisms receive mechanical or chemical shock.
6. Temperature elevation for organisms near the discharge area.
7. Rapid temperature changes in the vicinity of the discharge area as a result of plant operations, such as shutdown for maintenance or variations in operating power levels.
8. Changes in dissolved gas concentrations

in the intake and discharge areas as a result of increased biochemical oxygen demand of warmed waters, pumping of oxygen-poor waters, or gas saturation of discharge waters in the winter.

9. Chemical changes in the receiving water from blowdown and cleaning operations. Substances used in the coolant loop include chromates, zinc, and organophosphates to prevent corrosion and chlorine to reduce biofouling.

Impingement and entrainment

When water is withdrawn from a lake, river, stream, or estuary, many organisms, including fishes, are entrained with it. Screens of various designs, including traveling screens, are placed across the intake to the pump house as shown in Figure 7. The screens have openings about 1 cm in diameter. Every attempt is made by power plants to minimize the impingement and entrainment of fishes, but it is nearly impossible to avoid some fish losses. The water current will sweep them against the screens where, if they are not killed outright, they may become exhausted or damaged while swimming against the current. They may drown from an inability to ventilate the gills or be physically damaged by moving screens or by high-pressure screenwash, designed to keep the screens open. Impingement acts selectively on fish populations, affecting large size classes much more than small ones.

During entrainment, any organism not capable of swimming against the induced current near the intake will be drawn into the cooling water stream. This includes phytoplankton, zooplankton, invertebrates, larval fishes, and many smaller adults. In the cooling circuit the organisms receive a series of stresses in rapid succession. They first experience an abrupt pressure drop; then may impact the pump impeller or encounter shear stress near it; then rapid pressurization downstream of the pump followed by shear stress wherever the hundreds of condenser tubes divide off of the main flow; then they encounter an abrupt temperature rise as heat is transferred through the condenser tubes; they remain in an environment at 8° or 10°C above ambient until finally they are disgorged into the receiving body of water, where they encounter turbulent mixing, abrupt cooling, and decreasing pressure. Many organisms cannot survive such a dizzying ride. In addition, when the pipes need to be cleaned with a biocide, such as chlorine, the entrained organisms get a lethal toxic shock.

Entrainment of fishes at electric power-generating plants became an issue of great concern in the early 1970s for two reasons. First, to reduce the thermal impacts on the receiving body of water, increased flow rates in the coolant stream were used, and this caused increased entrainment. Second, increased use by large nuclear plants of estuarine cooling water threatened the large numbers of eggs and larvae of fishes and invertebrates inhabiting these waters.

There are about 90 thermal electric generating stations using once-through cooling on the shores of the Great Lakes. An additional nine plants used closed-cycle cooling. Reports from individual power plants indicated that in excess of 40 million fishes were lost annually by impingement on power plant screens. Clupeids, including alewives (*Pomolobus* sp.) and gizzard shade (*Dorosoma cepedianum*), were most frequently impinged, followed by smelt (*Osmerus mordax*). Fish valued by sport and commercial fisheries, such as smelt, percids, and salmonids, amounted to only about 15% of the total number of impinged fish. Kelso and Milburn (1979) estimated that the total annual impingement by existing and proposed power plants on the Great Lakes would be approximately 100 million fishes, equivalent to 7500 metric tons per year. The annual harvest for all fish species in the Great Lakes by Canada and the United States is estimated at 50,000 metric tons.

Temperature effects on fishes

Fishes are generally quite temperature sensitive. A few examples of the preferred and lethal temperature for fishes are shown in Figure 14 (Clark, 1969). It is evident that most fishes have rather cool preferred temperatures, in the range from 12° to about 23°C. Lethal temperatures for some are as low as 25°C, but for many other fishes, the lethal temperature lies between 30° and 35°C; and

for the goldfish (*Carassius auratus*), it is 42°C. It should be noted that largemouth bass (*Micropterus salmoides*) have a preferred temperature around 28°C, which is quite warm. This much sought-after game fish finds the thermal effluent of power plants an attractive habitat. Some thermal fish kills occur naturally in nature, but not often. Fishes usually are able to move about with sufficient speed to find a suitable habitat temperature for survival.

It is not always the lethal temperature with which we need to be concerned but those temperatures that are unfavorable to the fishes in a variety of ways at each stage of their life histories. The temperature re-

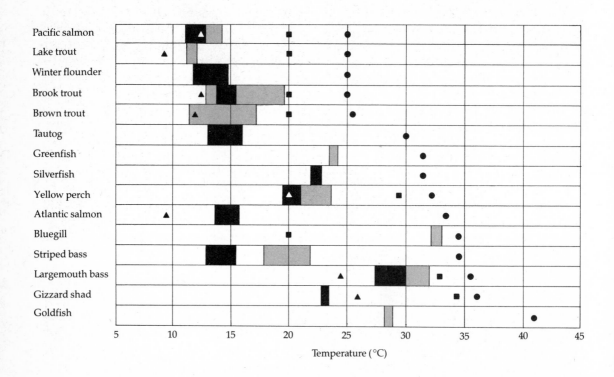

FIGURE 14. Preferred temperature ranges are shown here for some fish species as measured in the field (light gray boxes) and for younger fish as measured in the laboratory (dark gray boxes). The chart also indicates the upper lethal limit (●) and the upper limits for satisfactory growth (■) and for satisfactory spawning (▲). (Redrawn from Clark, 1969.)

sponses of eggs, juveniles, and adults are different. Temperature levels that adversely affect metabolism, feeding, growth, reproduction, and other vital functions may be as harmful to a fish population as thermal death. Generally the preferred temperature for most adult fishes is about 7°C below the lethal temperature, but for young fishes it is only 5°C below the lethal temperature.

The growth rate of fishes is strongly temperature dependent. Figure 15 shows the growth rates of juvenile sockeye salmon (*Oncorhynchus nerka*) as a function of the water temperature when fed at saturation and when fed on a daily basis an amount measured as a percentage of body weight. A well-fed salmon has optimum growth at about 15°C, and above this temperature the growth rate diminishes rapidly. The optimum growth rate temperature decreases with a reduced food supply. When the fish gets a daily ration of 1.5% of its body weight, the optimum temperature is about 5°C. Temperatures above the optimum for maximum growth rate cause increased metabolic demand and perhaps decreased food conversion efficiency—that is, the growth rate decreased. The data shown here come from laboratory experiments by Brett et al. (1969). However, field observations seem to confirm this growth-versus-temperature response. The lethal temperature for the juvenile sockeye salmon is around 25°C.

Hatching success is strongly temperature dependent. Fertilized eggs of the Atlantic salmon (*Salmo salar*) will hatch in 114 days in water at 2°C and in 90 days at 70°C; herring eggs in 47 days in water at 0°C and in 8 days at 14.5°C; trout eggs hatch in 165 days at 3°C and in 32 days at 12°C. Higher temperatures will decrease hatching success and finally will be disastrous to it. There always is a maximum temperature above which fishes will not reproduce. At a temperature above 22°C, the banded sunfish fails to develop eggs. For carp, temperatures in the range of 20° to 24°C

prevent cell division in the eggs. The possum shrimp (*Neomysis*), an inhabitant of estuaries, cannot lay eggs when the temperature rises above 7°C.

Organisms often adapt to the temperature of the habitat in which they are living. They undergo acclimation to a warmer or cooler temperature with time. Plants growing in cool habitats have low photosynthetic temperature optima and those of warm habitats have higher optima. Thus, when a stream or lake becomes warmer from a power plant effluent, two things are likely to happen. There may be some acclimation to the warmer temperatures by the plants present in the stream or lake, or those plants best suited to a warmer habitat may dominate this region of the stream and other species will decrease.

When the eggs of largemouth bass (*Micropterus salmoides*) are suddenly transferred

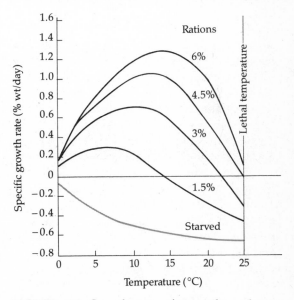

FIGURE 15. Growth rate of juvenile sockeye salmon for various levels of feeding as a function of the water temperature. Daily feeding levels (rations) are given as a percentage of the body weight. (Redrawn from Brett et al., 1969.)

from water at 18°C–21°C into water at 30°C, 95% of the eggs are killed, but when the eggs are acclimated by gradually raising the temperature to 30°C over a period of 30 to 40 hours, 80% of the eggs will survive. For the possum shrimp, its lethal temperature can be raised by as much as 13°C, to a high of 34°C, by acclimating it slowly to higher temperatures. As a general rule, aquatic animals can acclimate to increased temperatures more rapidly than they can to lowered temperatures. Hence, the manner by which a power plant effluent is changed over time makes a considerable difference as to how and to what degree organisms are affected by it. A sudden shutdown from full operating conditions can deliver a thermal shock from which many organisms will not recover.

The optimal temperature of a habitat for a specific species of fish not only must be suitable to the fish itself but also to the organisms upon which the fish feeds or in other ways depends upon. A food chain in the ecosystem is a very delicate and intricate web; and if some part of it is broken, other parts will also be affected. The algae, crustaceans, and insects in the water must be in healthy balance with the fishes that feed upon them. Estuarine eelgrass does not reproduce above 20°C, and most benthic (bottom-dwelling) organisms in rivers are damaged by temperatures above 32°C.

There is some evidence of a variation of thermal tolerance within species among geographically distributed populations. Largemouth bass had differences in thermal responses between populations in Florida and Ohio according to Hart (1952). Gibbons et al. (1980) reported genetic adaptation of bluegills (*Lepomis macrochirus*) to heated effluent in Pond C from nuclear reactors at the Savannah River plant compared with the same species found in the cooler Par Pond nearby. (See the discussion in this chapter entitled the Savannah River reactor effluent.)

Thermal effluent does not always have serious deleterious effects on a body of water. Fishermen are finding that some game fishes are thriving under the influence of warmer water near the outfall of electric power generating plants. This is often true for largemouth bass, which have preferred temperatures of 30°C or higher. Fish will move about seasonally, often seeking deeper, colder water in summer and warmer, thermal effluent water near the surface in winter.

The effects of heated water discharges on a variety of organisms are reported by Cairns (1956, 1972, 1976) and by Cairns et al. (1978).

Thermal cycles for reproduction

Reproduction of temperate zone aquatic organisms is tied to annual thermal cycles acting in synchrony with the seasonal changes in daylength. Thermal alterations of lakes, rivers, or estuaries may disturb normal reproductive cycles. In many cases, power station operations have advanced the spawning date of fishes in discharge areas. However, there is no evidence that this has caused a great decline of fish populations. Reproduction by fishes occurs in a very species-specific manner within a certain temperature range and usually during a time of rising or falling temperatures. Reproduction must have been preceded by certain physiological changes in response to habitat conditions that ready the gametes (reproductive cells) for release and fertilization. Although most fish species spawn only once a year, a few species are reproductively active throughout the summer or have multiple spawnings. Natural weather changes may advance or retard spawning by several weeks. Warm water effluent from electric power-generating plants may actually act as a stabilizing influence for reproduction.

Some fish species seem to require a winter cold period in order to spawn in the spring. According to Foltz and Norden (1977), smelt require a cold winter period with reduced feeding for the development of gametes. Temperature requirements during egg incubation are critical for successful hatching, as also are the upper and lower thermal tolerance limits for embryos. Temperatures during embryo development are critical. Too rapid a development during the winter can lead to premature hatching under unfavorable spring conditions of cold and of limited food supply. (See Wrenn, 1980.)

Manatees

The manatee (*Trichechus manatus*), also known as the seacow, is a large aquatic herbivore and a mammal that evolved along the same ancestral line as the elephant—their closest living relative. The West Indian manatee lives in the coastal waters of northern South America, the Caribbean, Florida, and as far north as Georgia. Other manatee species are found in the Indo-Pacific coastal waters and in West Africa. Surveys show that only about 800 to 1000 animals remain in Florida's coastal waters today. This slow-moving mammal can remain submerged for as long as 20 minutes, can swim and feed in both saltwater and fresh water, and consumes 30 to 45 kg of aquatic vegetation in a day. Manatees weigh close to a metric ton as adults and are up to 4 m or more in length. The manatee has no known natural enemies, except cold water. It has a diet of shallow water plants, such as the water hyacinth (*Eichhornia crassipes*), and it needs to live in water of temperatures above about 16°C for adults and 20°C for the young. Their greatest enemy has been the human. Manatees are regularly, and increasingly, sliced by motorboat propellers, crushed by barges, drowned in flood control structures, entangled in discarded fishing gear, harassed by divers, and often killed deliberately. Unfortunately, they are slow to mature, requiring 7 or 8 years for

the female and 9 or 10 years for the males. A seacow may bear a calf only once in 2.5 or 3 years. Of nine warm water sanctuaries in Florida used by manatees, only six are natural; the others border power plants or factories. Five of these power plants are owned by Florida Power and Light. Scientists with this utility have studied intensively this rare, unique, and endangered mammal. Manatees are on the federal endangered species list and they have been protected by the Marine Mammal Protection Act since 1972. However, it is estimated that from 60 to 80 animals die each year in Florida's coastal waters.

Warm water effluents from the power plants of Florida Power and Light Company (FPL) act as winter refuges for the endangered manatee. It would appear that the artificial warm water habitat is indeed affording protection during the winter. However, it is not yet clear just how beneficial this is. At the FPL Fort Meyers plant during the 1978–1979 winter season, there were 265 manatees. To quote Ross Wilcox of FPL, "During and after the passage of cold fronts, where air temperature may be as low as 0°C, the animals are literally stacked up like cordwood in certain stretches of a river receiving our warm-water effluent." At all five FPL plants there were a total of 568 manatees. Other endangered species enjoying FPL plant effluent are the American crocodile (*Crocodilus americanus*), southern bald eagle (*Haliaeetus leucocephalus*), green turtle (*Chelonia mydas*), leatherback turtle (*Sphargis coriacea*), and everglade kite (*Rostrhamus sociabilis*).

Plankton populations

In general, phytoplankton and zooplankton, as well as many invertebrates, are more resistant to thermal damage than are fishes. They seem better able to withstand mechanical, chemical, and thermal shocks, such as those delivered with entrainment into a power plant coolant system. The continuous flow of warm water effluent into a lake, stream, or estuary may have considerable influence on the plankton community by changing the species composition. Phytoplankton seem to be better able to withstand thermal shock than zooplankton.

Lake phytoplankton are classified into three major groups: the diatoms (*Bacillariophyta*), green algae (*Chlorophyta*), and blue-green algae (*Cyanophyta*). Each of these groups has a distinct water-temperature preference (Figure 16): diatoms (15° to 28°C), green algae (30° to 35°C), and blue-green algae (35° to 45°C). All of these algae are primary producers, but the diatoms and green algae are preferred food by fishes and other organisms. The blue-green algae do not seem to be eaten as much and also tend to be toxic to some aquatic life. When a lake or stream becomes warmer because of thermal effluent, the algal population shifts from diatoms to green algae to blue-green algae. There is, of course, often a natural shift in algal populations that is seasonal, from winter to summer

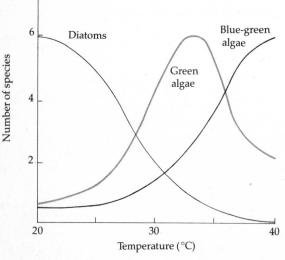

FIGURE 16. Temperature preferences for diatoms, green algae, and blue-green algae.

and back to winter, with the blue-green algae dominating in the summer months. However, with thermal effluent affecting the temperature of stretches of river, the blue-green algae may be abundant the year round. Lake Julian, North Carolina, a lake of 1.3 km², receives warm water into the epilimnion from a 200-MWe power plant that raises its temperature about 5°C. This lake has experienced growth of mats of the blue-green alga *Oscillatoria* in thicknesses increasing with temperature. In Polish lakes receiving warm effluent, there was a replacement of diatoms with blue-green algal taxa. At Biscayne Bay, blue-green algae have replaced red (*Rhodophyta*), brown (*Phaeophyta*), and green algae.

Clark and Brownell (1973) state that losses of zooplankton in power plant cooling systems may range from 15 to 100%, but 30% may be fairly representative. The percentage will vary enormously, depending on the manner of plant operation, season of the year, water temperature, and the design of the plant coolant system. In the estuarine food chain, some of the zooplankton of most value to fish appear to be most susceptible to damage by once-through cooling. These are gammarids, tiny shrimplike amphipods that live on estuarine bottoms, but whose young are suspended in the water column; mysids, also small shrimplike organisms; and copepods, crustacean zooplankton. Clark and Brownell state that during the summer and early autumn, copepod kills are nearly 100% at the Northport power plant on Long Island Sound, because of the high temperature of the thermal effluent. The discharge temperature of this plant was about 34°C when Long Island Sound temperatures reached 19°C. Copepod kills dropped off to 33% in late autumn and to 4% in the winter. At the Brayton Point power plant on Mt. Hope Bay, Rhode Island, copepod kills of 36 to 71% occurred with summer effluent temperatures of 26° to 30°C.

Gammarids, essential food for juvenile fishes, do not survive a passage through a cooling system as well as copepods. Some species are killed by temperatures above 26°C and most species are killed by temperatures above 32°C.

Mysids are even more sensitive to thermal shock than are gammarids or copepods. A 100% kill of mysids occurred when temperatures were above 26°C for long periods, and above 31°C for short periods of four to six minutes.

The warmer the water intake temperature, the warmer the effluent temperature in a once-through cooling system. It is for this reason that once-through cooling systems have their greatest impact on zooplankton during summer months and more often for power plants located in warmer, year-round climates than for those in cool climates. Because plankton are so vital to the food chain of fish and other organisms, it is extremely important that these impacts be minimized.

Higher plants

Among higher aquatic plants, there is a shift in populations as a result of warm water effluent. At the 730-MWe fossil-fueled Chalk Point power plant on the Patuxent River estuary in Maryland, the higher plant *Ruppia maritima* was replaced by *Potamogeton perfoliatus* in the area near the effluent discharge. More robust growth of the salt marsh cordgrass (*Spartina alterniflora*) occurred near the mouth of the effluent canal. Here fishes moved away from the discharge area in summer and were attracted to it in winter.

Dissolved oxygen

Oxygen concentration in lake water goes down as water temperature increases (naturally or as a result of mixing with power plant effluent). On the other hand, there are examples of increased oxygen levels in the water because of oxygen entrainment during the cooling process. The cold water in the bottom of the lake, the hypolimnion, is naturally low in oxygen concentration because of organic decomposition and lack of mixing. There are advantages to using the deeper water for once-through cooling because the temperature rise in the condenser simply brings it up to the temperature of the epilimnion. However, because this water is poor in oxygen it can cause a substantial shock to the organisms living near the surface. Fish kills below the Fort Loudon Dam due to the low oxygen content of the hypolimnion water have been reported.

For many bodies of water receiving nutrients and oxygen simultaneously, there may be all sorts of synergistic effects. In 1956 fish mortality in the Grand River below Lansing, Michigan was traced to organic waste (sewage) loading aggravated by a 5°C temperature increase from power plant effluent. The reduced oxygen concentration of the water after passing through the power plant turbines combined with the oxygen demand of the organic waste simply did not leave sufficient oxygen for the fish. A similar situation occurred on the Coosa River in Georgia, where a paper mill of the Georgia Kraft Company was surrounded by a peaking hydroelectric plant upstream, an adjacent steam electric plant, and an impoundment downstream. The paper mill had to reduce its organic waste output in order not to overload the impoundment, whose temperature had increased.

Savannah River reactor effluent

The U.S. Department of Energy has operated nuclear reactors at its Savannah River plant near Aiken, South Carolina for more than 30 years and released high-temperature cooling waters into the aquatic habitats there. Five plutonium-production reactors were built on the site in 1952. Three of the reactors continue to release hot water at temperatures above 70°C into reservoir systems, canals, or major streams that are tributaries of the Savannah River about 19 km away. Many different thermal gradients exist in the stream and lake habitats of the area, and as a result a diversity of aquatic ecosystems are present. This affords a unique opportunity for studying ecosystem impacts of thermal effluent. Scientists at the University of Georgia's Savannah River ecology laboratory have availed themselves of this opportunity and have reported results of studies conducted over a period of many years (see Gibbons et al., 1980; Gibbons and Sharitz, 1981; Sharitz and Gibbons, 1981).

Figure 17 shows the location of the reactors of the Savannah River plant. The 1100-ha Par Pond reservoir was constructed in 1958 to serve as a closed-system cooling reservoir for heated effluents from the production reactors. The reactor effluent goes directly into the middle arm of the reservoir, labeled Pond C, where the water temperature is often above 50°C, because it is receiving 70°C water. From Pond C, the water is released through the hot dam, where temperatures exceed 35°C. The warm water then disperses throughout the central part of the reservoir. The north and west arms of Par Pond are at temperatures only slightly above normal for the area. The two tributaries of the Savannah River—Pen Branch and Four Mile Creek—have received water at temperatures above 70°C since the early 1950s. This water cools as it flows down these tributaries. When the water enters the swamp along the Savannah River, it is still at a temperature above 40°C.

Biotic responses

Turtles. Temperature has a strong influence on the metabolism of animals and results in noticeable effects on growth rates and body size. Slider turtles (*Pseudemys scripta*) show higher juvenile growth rates and larger body sizes when living in waters warmer than the natural water temperature. Two factors—temperature and diet—may play a role here. The warmer water may allow the turtle to extend its activity period during the year and may allow it to feed and digest at a more rapid rate. According to Gibbons and Sharitz (1981), "Turtles residing in the thermally influenced reservoirs have diets enriched by fish and high-protein seeds, whereas diets of turtles in most natural areas are composed primarily of aquatic vegetation and insects." Female turtles in the warmer parts of Par Pond not only are larger but also have larger clutches than those living in natural areas.

Fishes. Juvenile largemouth bass (*Micropterus salmoides*) are larger when living in the heated areas of Par Pond than they are in nonheated areas. This is also true for other fish species. The warmer water temperature may be increasing the rates of primary productivity and these in turn enhance the food base for the fishes. This can be a positive factor when the fish can feed in a warm area and then retreat to cooler water below the thermocline or to a cool water habitat nearby to avoid the constant high metabolic cost at elevated temperatures. However, very often, the consequences of high metabolic cost are apparent in bass. Adult bass in the heated areas of Par Pond are often emaciated because fat reserves are depleted as a result of an increased metabolic rate. There is also an increase in red-sore disease and in numbers of internal parasites; bass living at elevated temperatures have poorer health than bass living at normal temperature. During the winter, bass are attracted to the effluent heated waters that they avoid during the summer. This behavior is also noticed at power plants in other parts of the country.

Some fishes acclimate to higher water

temperatures and others appear not to. For example, the thermal tolerance of mosquito-fish (*Gambusia affinis*) from natural areas is similar to that of mosquito fish from heated areas, but bluegills (*Lepomis macrochirus*) living in warm water have a much higher thermal tolerance than those from cool water. Mosquitofish normally occupy shallow surface waters warmed by solar radiation, and natural selection may have resulted in a higher temperature tolerance. Bluegills, on the other hand, normally live in deeper,

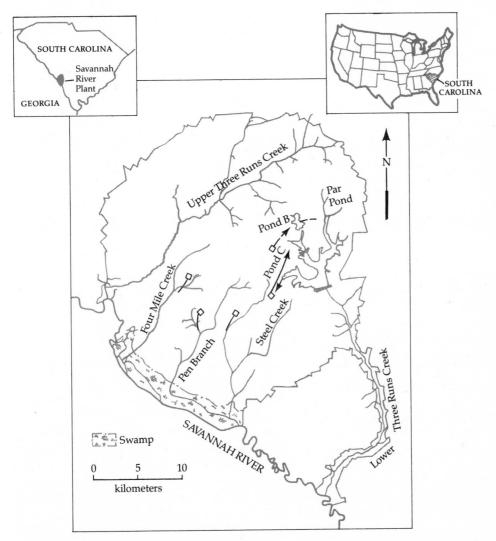

FIGURE 17. The main physical features of the Savannah River Atomic Energy plant, showing the locations of the reactors on the various bodies of water. A map of the United States shows the location of South Carolina, and a map of South Carolina shows the location of the Savannah River plant.

cooler water and may undergo genetic selection for higher thermal tolerance when occupying the warm water areas of Par Pond.

Alligators. Many adult male alligators (*Alligator mississippiensis*) congregate in the heated areas of Par Pond during the winter, rather than going into dormancy in the cooler areas nearby. The females remain in the cooler water and go into dormancy. As a result, in the spring the male alligators are out of breeding synchrony with the females by several weeks.

Primary producers. Par Pond has developed into a moderately productive reservoir compared with other lakes nearby, largely the result of higher water temperatures and increased nutrient concentrations. The latter is the consequence of having makeup water pumped directly from the Savannah River to Par Pond and the progressive evaporative concentration of the recycled cooling water.

Tilly (1975) reported that measurements of plankton made between 1965 and 1973 showed a sixfold increase in primary productivity during that time, but no increase in the standing biomass. Tilly proposed that this increase in productivity, but not in the standing crop, may be the result of more detritus going to the bottom and an increased consumption by primary consumers of all kinds. Other studies of this area are inconclusive as to whether or not primary consumption increased significantly.

Submerged higher plants, the macrophytes, began to grow in shallow waters, 0.5 to 1.0 m deep, at temperatures of 33° to 34°C. No higher plants were found in Par Pond at the hot dam. The Eurasian water milfoil (*Myriophyllum spicatum*) and spike-needle rush (*Eleocharis acicularis*) constituted 92% of the total biomass of submerged vegetation in the unheated areas of the reservoir, but only 64%

in water that was 5°C or more above normal temperature. The inability of *M. spicatum* to grow at temperatures higher than 32°C was attributed to increased respiration rates. Several other plant species, such as bushy pondweed (*Najas guadalupensis*) and pondweed (*Potamogeton* sp.) has a greater biomass in heated than in unheated water. Although these species do not usually outcompete *Myriophyllum* and *Eleocharis* at normal temperatures, at somewhat higher temperatures they seem to compete successfully.

Two species of cattails—the common cattail (*Typha latifolia*) and the giant cattail (*Typha domingensis*)—show very different responses to elevated water temperatures. *Typha latifolia* will survive temperatures above 35°C, but *T. domingensis* does not grow where year-round water temperatures exceed 30°C. The thermal stability of certain metabolic enzymes in these plants seems to be the differentiating factor because those of *T. domingensis* denature readily by heating and several in *T. latifolia* are stable above 50°C. Typically, these cattail species occur in adjacent stands, but seldom as mixed stands. *Typha latifolia* grows in shallow water of 0 to 1 m and *T. domingensis* in deeper water up to 2.5 m. In Par Pond water temperatures along the shoreline become warmer as the hot dam is approached. Near the hot dam *T. domingensis* disappears entirely from the shoreline community, but *T. latifolia* continues to grow in water temperatures up to 35°C.

The Savannah River tributary and swamp aquatic ecosystems are ideal for studying plant adaptation to environmental stresses. Localized races or ecotypes of a plant species may be clearly arrayed along a thermal gradient. There is a major difference in vegetation composition between the natural and thermally disturbed swamps that receive thermal effluent from the Savannah River plant reactors (Christy and Sharitz, 1980). Populations of two species of semiaquatic plants, the swamp primrose (*Ludwigia leptocarpa*) and the swamp loosestrife (*Ammannia coccinea*) were studied in heated and unheated sections of the swamp. Both species flowered earlier and produced more fruits and seeds when growing in heated areas than in the natural temperature habitats. Seedlings of the swamp primrose grown from plants in the warmer water showed distinctly higher growth rates at elevated laboratory temperatures than did seedlings from normal temperature populations. Christy and Sharitz (1980) and Gibbons and Sharitz (1981) suggest that evolution in a population of plants can occur at a rapid rate in response to thermal effluent.

Many of the waterways on which power plants are located in the southeastern United States pass through swamps or have flood plains with tree species about the same as those along the Savannah River in Georgia and South Carolina. The undisturbed ecosystems here are dominated by bald cypress (*Taxodium distichum*), tupelo gum (*Nyssa aquatica*), red maple (*Acer rubrum*), water ash (*Fraxinus caroliniana*), and other hardwood species. In 1951 a closed forest canopy extended over the entire 3020-ha swamp. As warm reactor effluent entered the swamp by way of the tributary streams, trees died in approximately two-thirds of the swamp. When water temperatures were 25°C above normal, total tree kill resulted. Not only did high temperatures kill trees, but also elevated water levels from the reactors, flooding, and siltation stunted or killed trees. The bald cypress and tupelo gum were often able to grow where hardwoods, such as red maple and water ash, previously grew in the flood plain. Successional recovery of the swamp ecosystem following removal of the thermal stress was slow. Following tree kill, the woodpeckers constituted more than 40% of the nonflocking birds in two thermally affected swamp areas, but made up only 14% in the undisturbed swamp.

Species diversity. As a general rule the diversity and evenness of aquatic species decreased in areas receiving thermal effluent from the reactors compared with the natural ecosystems. For example, 22 species of insects were present in a thermal area as compared with 45 species in a postthermal area and 54 species in a natural aquatic habitat nearby. The number of ostracod species was much lower in the thermal areas of Par Pond than the number in the parts at normal temperatures. This may have been the result of both temperature and food supply, because there was also a decrease in the density of vascular plants upon which the ostracods feed. The abundance and diversity of fish species was also much lower in the heated waters of several of the streams receiving reactor effluent than in the unheated waters, as also were the frog and toad species along the adjacent shorelines. The densities of larval forms of amphibians was inversely proportional to the temperature of the water.

ELECTRIC POWER TRANSMISSION LINES

Electric utilities in the United States now operate more than 483,000 km of overhead transmission lines involving more than 1.6 million hectares of right-of-way. The amount of land in these easements is more than the areas of Delaware and Maryland combined. The length of electric power lines has increased steadily, commensurate with increased demand for electricity in the United States. Not only have the number of kilometers of line increased, but the transmission voltage has gone up significantly during the last decade. One of the reasons for the increase in voltage is that high-voltage lines can transmit electricity over longer distances more efficiently and at less cost than can low-voltage lines. A single 765-kV line can carry as much power as thirty 138-kV lines. Furthermore, much less land is used. A 765-kV line requires only one-thirteenth the land area per kilowatt of capacity as its equivalent 138-kV line. The construction cost per kilowatt is ten times less for the high-voltage line than for the low-voltage line.

Land use

It is clear that enormous amounts of land are occupied by electric transmission lines. Land areas utilized are long and narrow, usually 35 to 65 m wide, but as much as 150 to 300 m wide in the case of multiline corridors. These corridors may cross highly variant terrain—from swamps to mountains, may be in remote rural regions, or may be in high-population urban centers. Rights-of-way must be easily accessible to work crews for inspection, maintenance, and repair. The vegetation along the corridor must be kept low, a condition often requiring the use of herbicides or fire. Trees must be kept away from the power line because contact between the vegetation and wires can cause electrical grounding or wire breakage.

Whenever a right-of-way corridor is established, it changes the wildlife habitat of a considerable area. Clear-cutting a corridor through forests is followed by the establishment of successional species of plants. What was once a climax forest may now be an open field of grasses, herbs, and shrubs. The animals associated with the open corridor will be completely different from those occupying the climax forest. When the right-of-way is across farmland, the habitat changes in the corridor are minimal, and often farming can continue on the strip of land. Lands altered by human activities will be occupied by those species of animals capable of living in a dis-

turbed ecosystem. Game species such as the ruffed grouse (*Bonasa umbellus*), mourning dove (*Zenaidura macroura*), bobwhite (*Colinus virginianus*), and cottontail rabbit (*Lepus floridanus mallurus*) are all representative of animals occupying the open corridors.

Clear-cutting of a forest to establish a power line corridor may cause a loss of nutrients from the soil, as demonstrated at the Hubbard Brook Experimental Forest in New Hampshire by Bormann et al. (1968); or it may cause no loss of nutrients, as demonstrated by Richardson and Lund (1975) in the sandy soils of northern Michigan. Nutrient loss was negligible in northern Michigan because of the rapid regrowth of aspen sprouts on the clear-cut sites. Clear-cutting may result in increased runoff or stream flow, particularly on steeper slopes at times of heavy rainfall. Increased sedimentation in streams may also occur, and this can result in an alteration or destruction of bottom organisms in a stream.

Clever management of a power line corridor can lead to a stabilized ecosystem and at the same time can keep out tree seedlings that would eventually interfere with the power line. This can be done by utilizing native species of grasses and shrubs. A grassland strip in the center of the corridor can be bordered with low shrubs such as viburnum, dogwood, azaleas, and low forms of juniper, which in turn can be bordered on the outside by willows, alders, amelanchiers, and redbud. This, of course, presupposes that these species can grow in the particular region. Elsewhere, other plants will need to be considered.

Herbicides

Controlling vegetation with herbicides is the most frequently used method on utility rights-of-way. Treatment is usually most effective when large quantities of solution are used. The solution contains active chemical ingredients, oil, emulsifiers, and surfactants. One of the most serious problems associated with the chemical treatment of rights-of-way occurs when the spray drifts onto crops and woodlands adjacent to the treated areas. Not only are the chemicals transported through the atmosphere by wind, but they are washed out by rainfall onto adjacent areas or water courses. Herbicides absorbed on soil particles may then be carried by soil erosion and deposited into stream beds or lake bottoms.

The herbicide 2,4-D is quite toxic to a wide variety of plants. Grasses seem to be the most resistant and broad-leafed plants the most sensitive. The herbicide 2,4,5-T is similar to 2,4-D in its manner of action, but it is more effective on woody plants. The half-life for these herbicides in the environment under moist conditions is a few weeks. Other herbicides often used include diphenamid, which is used prior to seedling emergence; propanil, a postemergence herbicide; and trifluralin, a preemergence chemical. A group of herbicides known as triazines are photosynthetic inhibitors. Picloram is an active growth-regulating chemical that is effective against a large spectrum of plants. It continues to persist in soils where extensive leaching does not occur, but it is highly mobile in water. Dichlobenil inhibits plant germination. It is effective against germinating seeds, rhizomes, tubers, and young seedlings.

Any chemical used in the environment, whether herbicide or insecticide, is almost certainly going to have some direct effect on organisms other than the target ones. The use of insecticides along power line corridors generally is unnecessary and therefore the biological effects of these chemicals are not described here. Many herbicides have been considered not to be particularly toxic to animals, but this certainly is not categorically

true. Recently there has been some evidence that dioxin, a by-product of 2,4,5-T, is toxic to animals. Toxicity tests of 2,4-D, 2,4,5-T, and picloram with cattle and sheep showed no effect. Laboratory rats injected with amitrole developed thyroid carcinomas.

Crustaceans form an important part of the aquatic food chain, being essential food for some fish. The herbicide dichlone was most toxic and 2,4,5-D least toxic to aquatic crustaceans. Daphnia was generally the most sensitive to these herbicides, followed by seed shrimp and grass shrimp, sowbug, scud, and crayfish; there was no effect on brown shrimp. Concentrations of herbicide in water required to produce a 50% mortality within 48 hours of continuous exposure by daphnia through crayfish (in the above order) were 50 to 100 ppm for dichlobenil, 55 to 100 ppm for diphenamid, and 0.25 to 2.0 ppm (except 50 ppm for crayfish) for trifluralin. In the case of 2,4-D, the concentrations to produce 50% mortality were highly variable, depending on the particular 2,4-D compound—of which there are several. Daphnia were affected only when concentrations exceeded 100 ppm, but scud were affected at 3 ppm for the acid form of 2,4-D. When the herbicide sodium arsenate gets into bottom waters, there is usually a reduction of benthic organisms. Among zooplankton in a farm pond, the most sensitive to sodium arsenate were the rotifers, followed by copepods and cladocerans.

Aquatic organisms take up chemicals either directly from the water or by the ingestion of contaminated food. Some species concentrate chemicals in the food chain, a process that is well documented for insecticides, but herbicides may also be concentrated biologically. The effects of herbicides on fish are highly variable. For example, it only takes a concentration of 0.011 ppm in water of trifluralin to produce a 50% mortality of rainbow trout (*Salmo gairdnerii*) and 0.019 ppm for bluegill (*Lepomis macrochirus*), but it takes from 200 to 2400 ppm for various forms of 2,4-D to produce a 50% mortality on rainbow trout. (See Johnson, 1968 for further references and details.)

Ozone

Ozone (O_3) is produced by electrical coronal discharge into air by high-voltage electrical transmission lines. Concentrations of ozone directly beneath transmission lines may approach 1 to 2 ppb (parts per billion), but no direct effect on plants growing in the corridor or nearby has been demonstrated. Ozone damage to plants has been much studied in California where it is a significant photochemical constituent of smog. Toxicity by ozone to plants is manifested by lesions, discoloration, and reduced photosynthetic rates. Injury occurred to tobacco plants when they were exposed to ozone of 2.5 ppb concentration for six hours. Various plant species were exposed by Hill et al. (1961) for two hours to ozone concentrations ranging from 1.3 to 7.2 ppb. Alfalfa, spinach, and tobacco were the most sensitive species, with injury noticeable at concentrations of 2.0, 2.3, and 2.4 ppb, respectively. Exposure for eight hours at ozone concentrations from 1 to 10 ppb by Ledbetter et al. (1959) showed injury at 1.0 ppb by tomato, bean, spinach, tobacco, potato, and smartweed.

Electric fields

An electric field extends outward from a transmission line carrying an electric current. The highest voltage transmission lines in the United States are nominally rated at 765 kV. Under a 765-kV line at its closest proximity to the ground, the electric field intensity is about 8 kV/m at a point about 1.5 m above the ground. Because all of our alternating current appliances operate at a frequency of 60 Hz (60 cycles per second), electric power transmission lines are carrying power at this frequency. Any alternating electric field has a wavelength affiliated with it. A 60-Hz frequency has a wavelength of 5000 km. The amount of electrical energy absorbed by an object is a function of the wavelength of the inducing field. The energy absorbed by an object becomes progressively smaller as the wavelength of the alternating electric field becomes longer. The energy absorbed from a 60-Hz power line field will be one-trillionth of that from a 60-MHz (wavelength, 5 m) television transmitter field of the same intensity. Microwaves from 300-MHz to 300-GHz sources (wavelengths of 1 m to 1 mm) are much more dangerous to organisms than are the microwaves from a 60-Hz electric power line.

There is much concern with the effects of electric fields near high-voltage electric transmission lines on human health and the wellbeing of plants and animals. A person standing beneath a 765-kV, 60-Hz line will have a current of about 120 microamperes flowing through the body. This is about 1/10,000 of the current flowing through a 100-W light bulb. Leakage current from household appliances will be about 20 to 300 microamperes through the body of the person operating them.

There are reported cases of humans affected by high-voltage fields, but they have been difficult to verify. A few people have reported a tingling sensation in the skin of their forearm when standing beneath a 765-kV line. Workers in electrical railroad switchyards in the Soviet Union have reported loss of appetite, listlessness, and diminished sex drive when exposed to high-voltage gradients for long periods of time. The Soviet government has now set time limits for exposure to electric fields greater than 5 kV/m. In Spain, five of eight switchyard workers, transferred to a new 500-kV power station, reported headaches, fatigue, and loss of appetite. In the United States, where carefully conducted experiments have been done over a nine-year period with 10 utility linemen who worked on energized lines, none showed any changes in physical, mental, or emotional characteristics. Comparisons were made between two groups of French people of the same social class. One group was living within 25 m of a corridor containing a 200-kV and a 400-kV transmission line, and the other group lived at least 120 m distant from the lines. No differences were discernible between the two groups in their medical care, drug use, or frequency of medical examinations. In Germany, no physiological or biochemical changes were noted with volunteer subjects exposed to fields of 20 kV/m for periods of three hours under carefully controlled conditions. In Sweden, where similar experiments were conducted, there were no significant psychological effects noted. However, three-hour exposures are too short to produce much of an effect. No other studies have been able to verify the Soviet and Spanish results.

Although farmers have occasionally reported receiving slight electric shocks when standing under a 765-kV line, most farmers interviewed had received no shocks (Miller and Kaufman, 1978). Alternating electric and magnetic fields might induce some interference into certain types of cardiac pacemakers. However, it is fairly certain that for electric transmission lines less than 345 kV, pacemakers will not be perturbed. Only when a person with a highly sensitive pacemaker is under the most powerful high-voltage line is there a possibility for an effect. When the person is in an automobile, there can be no influence of the power line on any pacemaker, because the automobile acts like a Faraday cage and shields anything inside the car against all electric fields outside. Many other sources besides electric transmission lines are as likely, or more so, to interfere with a pacemaker's signal. Such sources include automobile ignitions, radar installations, microwave ovens, and electric shavers. (Many additional references to the biological effects of electric transmission lines may be found in Miller and Kaufman, 1978.)

Many experiments concerning the response of organisms to electric fields have been conducted in connection with the planning of Project Seafarer, also called Sanguine, the U.S. Navy's proposed submarine communication system that is to be buried in the Upper Peninsula of Michigan. This system comprises several hundred kilometers of buried cable operating at a frequency of 30 to 90 Hz. The voltage gradients are much lower (0.01 to 0.02 kV/m) than those from high-voltage transmission lines, whereas the magnetic fields are higher (2 to 3 gauss). In approximately 40 experiments many different organisms, from soil microorganisms to fruit flies, birds, cows, and humans, have been subjected to simulated Seafarer electric fields. Most of the comparisons showed no detectable effect of the magnetic fields on the organisms, except for an apparent disturbance to bird navigation (Southern, 1975, 1977). In addition, a rise in the serum triglycerides in several human subjects was observed 24 to 48 hours after exposure; but the rise was followed by a return to normal (Kornberg, 1977). Experiments at Pennsylvania State University on high-voltage effects on vegetation showed that leaf tip damage occurs for pointed leaves that are subjected to electric fields between 20 and 22 kV/m. Only about 1% of the leaf is damaged and photosynthesis is not affected. Extended exposure of rounded leaves to field gradients up to 50 kV/m showed no damage and no effect on plant growth.

Polychlorinated biphenyls

Polychlorinated biphenyl (PCB) compounds were a chemist's dream when discovered, and they are some of the most widely used compounds ever to come out of the chemistry laboratory. As with DDT, they are colorless, odorless liquids with a consistency about like molasses. However, because they are fire-resistant, good heat conductors, and poor electrical conductors, they are ideally suited for uses in transformers, in capacitors, and in various hydraulic systems. PCBs are also widely used as plasticizers in inks and dyes. Between 1930 and 1977, 570 million kilograms of PCBs were manufactured in the United States alone. PCBs are extremely resistant to physical, chemical, or biological degradation. The fact that they do not break down readily made them ideal for industrial use.

Then why all the fuss about PCBs? It turns out that they are highly toxic to organisms, including humans. PCBs interfere with reproduction in rodents, fishes, birds, and primates. Animals fed low levels of PCBs exhibited loss of hair, enlarged livers, gastrointestinal lesions, and abnormalities of the lymphatic system. Rodents fed diets containing 100 ppm of PCBs developed tumors. People working with PCBs have complained of dermatitis, nausea, dizziness, eye irritation, and asthmatic bronchitis. PCBs are sometimes considered to be carcinogenic, teratogenic, and mutagenic. However, one must be careful not to overextrapolate from this conclusion, for many public health experts consider that there is little evidence to support this level of toxicity in humans. PCBs in the environment are far more stable than DDT and the other chlorinated pesticides. As with DDT, the PCBs concentrate at upper trophic levels of food chains. They are particularly soluble in fatty tissues. Their highest concentrations are found in various predators, including birds of prey, fishes, and humans. But unlike many organic contaminants, they are not broken down by bacteria or by various metabolic processes. Because of their toxicity and persistence, the U.S. Food and Drug Administration has adopted a safety level of 2 ppm for PCBs in food.

PCBs are found everywhere in the world today, including in the flesh of polar bears and seals north of the Arctic Circle and in fishes throughout much of the world. PCBs vaporize into the atmosphere from transformers and capacitors, are disposed of in landfills, are dumped into sewers, and generally escape to the environment in about every conceivable manner. When any material containing PCBs is burned, the PCBs enter the atmosphere and become attached to dust particles. The dust then settles throughout the landscape on land or water, or gets washed down with precipitation; and the PCBs end up in the watersheds of the world. It is estimated that 50 to 85% of the PCBs entering Lake Michigan come through the atmosphere. The average surface water contamination of Lake Erie and Lake Ontario by PCBs is about 30 parts per trillion (ppt) by volume. Many highly polluted areas are much worse. The Milwaukee River carries up to 260 ppt, and the water in Green Bay, Wisconsin contains up to 450 ppt of PCBs. Highly localized sewage outfall will have between 40 ppt and 2.5 parts per billion (ppb).

The major link in the Great Lakes food web for PCBs is phytoplankton, the freefloating algae. The algae take up PCBs very quickly and efficiently, and then they suffer toxicity effects. Some algal species are more susceptible than others to PCBs. There are reductions in the numbers of algal cells and in their chlorophyll content. Smaller-celled algae seem to be more sensitive to PCBs than are larger-celled species. Among zooplankton, the rotifers are highly sensitive to PCBs. At concentrations above 20 ppb, their reproductive rate is reduced 50%. Above 50 ppb, half die within a few hours. The freshwater shrimp (*Mysis relicta*) also picks up PCBs readily. Fishes feed on plankton and shrimp, and the PCBs become concentrated in their fatty tissues. Fishes also pick up PCBs directly from the water. Mink, eating fishes, also become contaminated with PCBs and eventually lose their ability to reproduce. Although in general Great Lakes fishes have PCB concentrations less than the federal food limit of 2 ppm, it is recommended that humans not consume fishes from the Great Lakes more frequently than once per week.

PUMPED STORAGE

The only economical system for the storage of electrical energy on a large scale is pumped hydroelectric storage. Water to operate the turbines must first be pumped uphill and stored in a reservoir before it can be used to generate electricity during periods of high demand. When the water runs back downhill, the plant operates like any hydroelectric power plant. Figure 18 illustrates in schematic form the principle involved in a pumped-storage system. It requires electrical energy to pump the water into the reservoir during off-peak hours, that is, about 2.7 watt-hours of energy are required to pump the water uphill for every 2 watt-hours of electrical energy generated by the water running back down. Hence, the efficiency is about 75%. The cost of electricity generated in this way is often less than the cost of using gas turbines or using the older coal- or oil-burning power plants during times of peak demand. The first pumped-storage hydroelectric plant in the United States was built in western Connecticut in the 1930s and had a power output of 32 MWe.

Although the idea of using pumped hydroelectric storage is a good one, there are serious environmental and ecological difficulties associated with their locations. There are only a limited number of sites available that have the elevation and size needed for large storage of energy. The potential energy of the water in the reservoir depends on the total amount of water stored and its height above a lower reservoir, lake, or river into which it will be discharged.

Ludington, Michigan plant

The Detroit Edison Company and Consumers Power Company of Michigan completed a pumped-storage system at Lud-

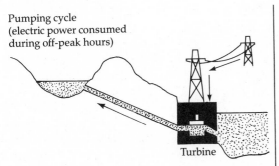

Pumping cycle
(electric power consumed during off-peak hours)

Turbine

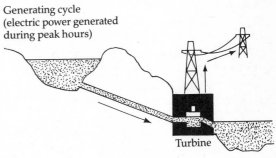

Generating cycle
(electric power generated during peak hours)

Turbine

FIGURE 18. A pumped storage system. Water is pumped from a lower reservoir, lake, or river to a higher reservoir. Pumping uses electrical energy during off-peak hours. The water is released from the upper reservoir and passes through the turbines to generate electricity during peak demand periods.

ington in 1973; the facility has a peak power delivery of 1800 MWe. The total cost was $315 million. It was a massive project requiring 10 years of planning and 5 years of construction. A view of the Ludington reservoir and plant are shown in Figure 19. The reservoir is nearly 120 m above Lake Michigan and has a surface area of 337 ha (1.3 mi²). It is located on top of a high sand dune bluff on the lakeshore. The reservoir, when full, has a capacity of 27 billion gallons of water. The earth-filled embankment has a circumference of 9 km and required the excavation of nearly 21 million m³ of sand and silt. Because of the pervious nature of the silty sand material, the reservoir had to be lined with asphalt and clay. The reservoir, when full, contains 15,000 MW-hr of stored energy. The maximum rate of flow is more than 33 million gallons per minute. The six penstocks at the Ludington plant are 350 m long and 8 m in diameter.

The Ludington plant serves as an integral part of Michigan's energy system, where it plays a complementary role to the coal, oil, and nuclear plants. Other advantages are that it improves the efficiency factor of operating large steam plants, requires a small operating maintenance staff of about one-fifth the size of that for a steam plant, offers increased reliability, and provides rapid generation of power as needed.

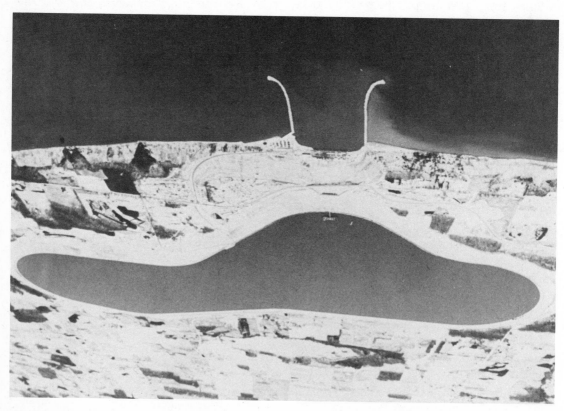

FIGURE 19. An aerial view of the pumped storage reservoir and power plant at Ludington, Michigan. (Courtesy of Detroit Edison Co.)

Environmental impacts include fish entrainment and impactment and some temperature effects on the receiving body of water. Ecologically the reservoir will behave in the same manner as most hydroelectric reservoirs. (See Chapter 14 for a discussion of these.)

The Storm King controversy

A massive amount of electric power is consumed by the megalopolis of the greater New York region. Electric power-generating stations for this region of the country are situated along coastal estuaries and the Hudson River. The lower Hudson River is estuarine in character with ocean tides influencing it up to 246 km inland from the sea. The river is used as a spawning and nursery area by many resident and anadromous species of fish. (An anadromous fish is one that lives in the sea but spawns upstream in the rivers.) The fish species having the greatest potential to be adversely affected in the Hudson River are striped bass (*Morone saxatilis*), tomcod (*Microgadus tomcod*) alewife (*Pomolobus* sp.), blueback herring (*Alosa aestivalis*), and anchovy (*Anchovia brownii*). The striped bass is an important sport and commercial fish. Therefore, when Consolidated Edison Company of New York proposed in the early 1960s to build a 2000-MWe pumped-storage electric generating plant at Storm King on the Hudson River near Cornwall, New York, a violent controversy erupted. The main spawning area for the striped bass is exactly the region of the river where the pumped-storage plant was to be located. The major nursery area for the striped bass is just downstream from Storm King. Therefore, the juvenile striped bass move from the spawning area to the nursery area by traveling past the Storm King part of the river. A report by Barnthouse et al. (1984) shows that entrainment by the power plant would have reduced the striped bass population from 11 to 18%. Impingement would cause an additional reduction from 1 to 9%.

A parallel controversy arose in the early 1970s in connection with Con Edison's Indian Point Unit 2, a large nuclear generating station nearing completion on the Hudson River. The Nuclear Regulatory Commission, after exhaustive, bitterly contested, highly technical hearings required Unit 2 to have a closed-cycle cooling system, and this meant building a cooling tower. This system would reduce the volume of water withdrawn from the river by 90% or more, greatly reduce entrainment and impingement of fishes, and minimize thermal discharges. In 1975, a requirement for a cooling tower at Indian Point Unit 3 was written into the operating license, as was done at the oil-burning electric generating stations at Bowline Point and Roseton, operated by Orange and Rockland Utilities, Inc. and by the Central Hudson Gas and Electric Corporation.

On 20 December 1980 the *New York Times* reported that an agreement had been reached between the Federal Environmental Protection Agency and the Consolidated Edison Company of New York, Inc. that the *Times* headlined "Peace Treaty for the Hudson." This agreement ended 17 years of controversy over the impact of electric power generation on the fish populations of the Hudson River. The EPA agreed not to require Con Ed and other Hudson River utilities to build cooling towers at the Indian Point, Bowline, and Roseton generating stations in return for abandonment of plans to build the Storm King pumped-storage facility. The utilities also agreed to implement a number of mitigating measures intended to reduce the number of striped bass and other fishes killed at the intakes for these plants.

13

Nuclear Power

A nuclear reactor vessel for a boiling water reactor. Courtesy of the Detroit Edison Company.

INTRODUCTION

No source of energy today is more controversial than nuclear power, nor may any source be more needed for the advancement of civilization than energy derived from the atomic nucleus. According to Schurr et al. (1979), "No technology shows more potential than nuclear power for long-run supplies of electricity at reasonable costs, nor does any other technology pose issues of public policy that are more complex and unyielding to solutions with wide public acceptance."

There are only two energy sources of very large, concentrated supply in the United States. One is coal and the other is nuclear energy. Nuclear energy has been seen by some people as the savior of mankind and by others as the ultimate destroyer of civilization. But, for good or bad, it is with us and if our political systems allow it, nuclear energy will give human society energy for the future. Before we describe how it is that nuclear energy can do work, we need to understand something about the basic physics of the process.

Radioactivity

The discovery of RADIOACTIVITY at the end of the nineteenth century and the investigations of subatomic particles spontaneously emitted from disintegrating atomic nuclei ushered in a period of scientific discovery unmatched in challenge and excitement throughout human history. The names of Becquerel, Röntgen, Thomson, Marie and Pierre Curie, Rutherford, Planck, Einstein, Bohr, Schroedinger, Fermi, Compton, and many others stand along the path of understanding in atomic physics through the first decades of the twentieth century.

Throughout this early period of research on atomic physics, all nuclear reactions studied involved the ejection of relatively light particles, such as electrons, protons, or neutrons, from the atomic nucleus. In 1896 Becquerel discovered that uranium spontaneously gave out rays that would penetrate several thicknesses of black paper and fog photographic plates wrapped inside them. The element uranium was identified as radioactive. It was only two years later that Pierre and Marie Curie isolated the radioactive elements polonium and radium. Research by Ernest Rutherford in England showed that three kinds of rays, called alpha, beta, and gamma rays, were emitted from the nucleus of radioactive elements, and that these rays could ionize atoms along their paths of travel. ALPHA RAYS turned out to be equivalent to a helium nucleus with two protons and two neutrons and, hence, with a double positive charge. BETA RAYS were found to be electrons with a single negative charge. GAMMA RAYS turned out to be light waves with about the same wavelength as X rays, with a neutral charge.

It was Rutherford who discovered that when one radioactive atom disintegrates by ejecting an alpha or beta particle, the remaining atom is still radioactive and will in turn disintegrate to become a still different atom. This process continues through a series of atoms until a final stable element is reached that is not radioactive. Nearly all natural radioactive elements disintegrate to lead as the final element.

The atom of any element is made up of a heavy nucleus of positive charge, containing protons and neutrons, surrounded by electrons of equal negative charge. Atomic elements are classified by two properties: an atomic number and an atomic mass number. The ATOMIC NUMBER represents the number of positive charges on the nucleus. This is the same as the number of protons in the nucleus. The ATOMIC MASS NUMBER is the number of protons plus the number of neutrons in the atomic nucleus. The neutron and proton have approximately the same mass but are 1837 times heavier than the electron. Because the atomic nucleus contains all of the protons and neutrons, most of the mass of an atom is located in the nucleus. The hydrogen atom is the simplest element of the periodic table, having a nucleus comprising one proton (and no neutrons) surrounded by one electron. The atomic number for hydrogen is 1 and its atomic mass is 1. The helium atom has two protons and two neutrons in the nucleus. Its atomic number is 2 and its atomic mass is 4. Uranium has an atomic number of 92 and an atomic mass of 238. It happens that every element has several forms and that these differ in the number of neutrons contained within the atomic nucleus. These various forms are known as ISOTOPES of an element. Uranium has several unstable isotopes of atomic masses 233, 234, 235, and 236, but each has an atomic number of 92.

Fission

In 1938 Otto Hahn and Fritz Strassmann in Germany bombarded uranium, atomic number 92, with neutrons and obtained barium, atomic number 56, and krypton, atomic number 36—two elements in the center of the periodic table. They had actually been trying to make elements heavier than uranium (the transuranic elements) when they discovered NUCLEAR FISSION. Measurements showed that an enormous amount of energy—200 million electron volts, the equivalent of 600,000 kilowatt-hours—is released from the fission of one nucleus of uranium 235. Enrico Fermi, an Italian physicist, then suggested that neutrons were being ejected from the uranium nucleus during fission and that they in turn might collide with other uranium nuclei to produce more fission, thereby leading to a CHAIN REACTION (Figure 1). When uranium 235 undergoes fission, it ejects two neutrons and sometimes even three neutrons. The average number of neutrons released is 2.07.

These two or more neutrons may strike other uranium 235 nuclei and release four or more additional neutrons. These then strike other uranium nuclei and release eight or more additional neutrons. This chain reaction proceeds in the same way as any other exponential growth function. Here we would write $(2.07)^n$ equals the number of fissions. n does not need to be extremely large before we get a very large number. If $n = 100$, the number of fissions would be 3.95×10^{31}. If hundreds or thousands of such reactions occur in succession, an enormous amount of energy would be released and would be available for work.

When neutrons bombard nuclei of uranium 235, not only are more neutrons released but the nucleus breaks into two parts, each having about half of the atomic weight of the uranium atom. These fission fragments, as they are called, are like projectiles shooting out in all directions. When they collide with other atoms, their kinetic energy is converted into heat energy. This heat energy can be used to heat water to steam and the

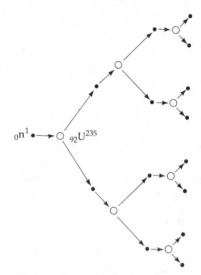

FIGURE 1. Chain reaction. Slow neutrons ($\bullet$) emitted from a uranium nucleus during fission strike more uranium nuclei, creating a multiplier effect.

steam used to drive a turbine and electric generator. The vessel containing the uranium fuel rods is called a REACTOR. The reactor vessel has the same function as the furnace in which coal, oil, or gas is burned. Reactors that consume more fissionable material than they produce are known are BURNER REACTORS. Those that produce more fissionable material than they use are known as BREEDER REACTORS.

If the chain reaction just described were to occur sufficiently rapidly, it would produce an uncontrolled nuclear explosion. However, a controlled chain reaction releases heat energy that can be used for beneficial purposes. The world's first controlled nuclear chain reaction was produced in an atomic pile set up by Enrico Fermi and Arthur Compton under the athletic stadium at the University of Chicago in 1942. There they witnessed the birth of nuclear power and realized that they had unleashed a new force of enormous potential to the human race.

Uranium 235 fuel

The isotopes of uranium with atomic mass numbers of 234, 235, and 238 occur naturally with abundances of 0.006, 0.711, and 99.283%, respectively. In fact, uranium 235, the isotope of uranium that undergoes fission, is present only as 1 part in 141 parts (0.711%) of natural uranium. The importance of uranium 235 is signified by the fact that of the several hundred naturally occurring isotopes, it is the only one that is spontaneously fissionable by the capture of slow, thermal neutrons. One gram of pure uranium 235 contains 2.56×10^{21} atoms. Because each fission event releases a quantity of energy equivalent to about 200 million electron volts or 3.20×10^{-11} joules, the total energy released by fissioning one gram of uranium 235 is 8.19×10^{10} joules. This is equivalent to the heat of combustion of 2.7 metric tons of coal or 13.7 barrels of oil.

Uranium oxide ore as taken from a mine must be refined and the concentration of uranium 235 must be increased from 0.711% to about 3% in order for it to be useful as nuclear fuel. This process of increasing the concentration of uranium 235 is known as ENRICHMENT. A kilogram of uranium oxide, when enriched to 3% uranium 235, will have an equivalent fuel value of about 16 metric tons of coal. Following an operating cycle of about three years in a nuclear reactor, the fuel mixture will contain 1% uranium 235 and 1% plutonium 239, each valuable as fissionable material. These can be separated from the spent fuel and refabricated into fuel elements for reloading in the reactor. This is called the PLUTONIUM CYCLE and is described in the next section. Plutonium is generated in a burner reactor because the enriched uranium ore still contains a high concentration of uranium 238.

When uranium 235 undergoes fission it breaks down into several fission products and most of these are isotopes of elements near the center of the periodic table. When the uranium 235 nucleus captures a neutron, it is destabilized and splits into large fragments. The fissioning of uranium 235 is a random process; that is, the nucleus never splits in a predictable manner. The particular isotopes ejected vary with each fission. Uranium 235 nuclei are known to fission in more than 30 different ways, forming over 450 types of fragments, most of which are radioactive isotopes.

A typical reaction is

$$^{235}_{92}U + ^{1}_{0}n \rightarrow ^{140}_{54}Xe + ^{94}_{38}Sr + 2^{1}_{0}n + 200\,MeV$$

where the fission products, xenon 140 and strontium 94, are unstable because they have too many neutrons in their nuclei. They are, therefore, radioactive and decay successively through other unstable isotopes to stable elements, in this example to xenon 129 and strontium 88. However, strontium 90 also is formed during this decay process. Radioactive strontium is always a problem in the environment because it behaves chemically like calcium. (Strontium is classified as an alkaline earth element in the periodic table, along with beryllium, magnesium, and barium. These elements all behave chemically in the same manner.) Radioactive strontium is incorporated into human and animal bones and teeth. Strontium 90 has a half-life of 28 years and, therefore, once released to the environment, will persist for hundreds of years. Cesium 137, another radioactive product of fission, has a half-life of 30 years and will also persist in the environment.

Conversion

Although uranium 235 releases an enormous amount of energy per gram of material, the total energy available from this resource is limited because of the scarcity of uranium 235 in the Earth's crust. Fortunately, it turns out it is possible to convert both nonfissionable uranium 238 (constituting 99.28% of natural uranium) and thorium 232 into FISSILE (fissionable) material. The reaction for converting uranium 238 is

$$\underset{92}{238}U + \underset{0}{1}n \rightarrow \underset{92}{239}U \underset{\beta-}{\rightarrow} \underset{93}{239}Np \underset{\beta-}{\rightarrow} \underset{94}{239}Pu$$

where Np is the transuranic element neptunium and Pu is another transuranic element, plutonium, about which a great deal of concern is expressed. During the reaction two beta particles, which are electrons, are ejected from the nuclei.

Fissile material is made from thorium 232 by the following reaction:

$$\underset{90}{232}Th + \underset{0}{1}n \rightarrow \underset{90}{233}Th \rightarrow \underset{91}{233}Pa \rightarrow \underset{92}{233}U$$

where Pa is protactinium.

The isotopes uranium 238 and thorium 232, which are not by themselves fissionable, are known as FERTILE MATERIAL. The process of converting fertile to fissile material is known in nuclear engineering as CONVERSION.

If Q_0 is the initial quantity of fissile material and Q is the quantity of fissile material remaining after one fuel cycle, during which Q_0 was consumed, then the CONVERSION RATIO K is defined by

$$K = Q/Q_0$$

If $K = 0$, the reactor is a pure burner; if $0 < K < 1$, the reactor is a converter; and if $K > 1$, the reactor is a breeder. The Soviet Union has breeder reactors operating at $K = 1.2$ and 1.5; the French at 1.0, 1.16, and 1.24; the British at 1.2; the Japanese at 1.01 and 1.2; and the United States at 1.01 and 1.27.

A report by the National Academy of Sciences (1969) has this to say about nuclear energy: "The energy potentially obtainable by breeder reactors from rocks occurring at minable depths in the United States and containing 50 grams or more of uranium and thorium combined per metric ton is hundreds or thousands of times larger than that of all of the fossil fuels combined. It is clear, therefore, that by the transition to a complete breeder-reactor program before the initial supply of uranium 235 is exhausted, very much larger supplies of energy can be made available than now exist. Failure to make this transition would constitute one of the major disasters in human history."

REACTORS

The reactor is the place where energy is released from fissile material and where fertile material may be converted to fissile material. The nuclear reactor is designed to produce heat that can then be used as part of a steam cycle to generate electric power. The steam expands against the blades of a turbine and the turbine rotates under the impulse of the steam. The turbine turns an electric generator, which produces an electric current.

Fuel rods and assemblies

The fuel in a nuclear reactor is in the form of small pellets of uranium dioxide (UO_2). These pellets are about 5 mm in diameter and 10 mm long. They are inserted into thin-walled stainless steel or zirconium alloy tubes about 3.5 m long. This metal tube that surrounds the fuel pellets is referred to as CLADDING. It prevents the radioactive fission products from getting into the coolant water and also provides support for the fuel. Bundles of about 200 of the fuel rods are constructed. These FUEL ASSEMBLIES are about 20 cm wide and are 3.5 m long. They weigh about 545 kg.

A REACTOR CORE will contain about 180 of the fuel assemblies, very closely packed but with space for CONTROL RODS and for tubes containing the coolant liquid. A boiling water reactor vessel, fuel assemblies, and coolant system in the reactor vessel are shown in Figure 2. Coolant water flows down around the outside of the core and is forced up past the fuel rods, removing the heat of fission and keeping the rods at a reasonable temperature. It is this heat that is transported by the coolant to a heat exchanger, where it generates steam at 300° to 400°C and at 13.8 million pascals (2000 lb/in^2).

The reactor core and all the associated components and radiation shields are contained in a reactor pressure vessel that is designed to withstand the enormous pressures of the coolant and to isolate the steam supply system from the core. The reactor pressure vessel is completely enclosed in a concrete containment structure to prevent the release of any radioactivity in the case of a failure of the coolant system.

Moderator

It is necessary to slow down the fast fission neutrons that are emitted by uranium 235 or by other fissile material to enhance the proba-bility that they will induce other fissions in a controlled reaction. The material used to slow down the neutrons is called a MODERATOR. Often water or graphite is used as the moderator material. The fuel concentration of fissile material must be such as to achieve a critical chain reaction. There is a certain critical mass of fuel that depends on its particular composition and geometry as well as on the amount of nonfuel material in the core. It is necessary to have not only a moderator material interspersed among the fuel rods, but also a coolant flowing through the core to carry off the heat generated. When water is used as the coolant, it can also act as a moderator to slow down the neutrons.

Usually a nuclear reactor is loaded with much more fuel than that necessary just to achieve a self-sustaining chain reaction. The extra fuel is included to compensate for fuel burnup over time and to accommodate a variable power demand. In order to compensate for the excess fuel and to balance the production and loss of neutrons, the moderator material is introduced into the system as CONTROL RODS. The control rods may be used to adjust the criticality of the reactor. Control rods may be withdrawn to speed up the chain reaction and to make it slightly super-critical. When the power level has reached some predetermined level, the control rods can be reinserted to achieve a critical rate, or they may be further inserted to shut down the reactor.

Negative feedback

All power reactors are made with negative temperature feedback so that increasing power, and therefore temperature, breaks the chain reaction and decreases the power production. Duderstadt and Kikuchi (1979) write "Such inherent feedback mechanisms, coupled with the dilute nature of the reactor fuel, eliminate any possibility of a runaway chain reaction and remove the concern about possible nuclear accidents involving the

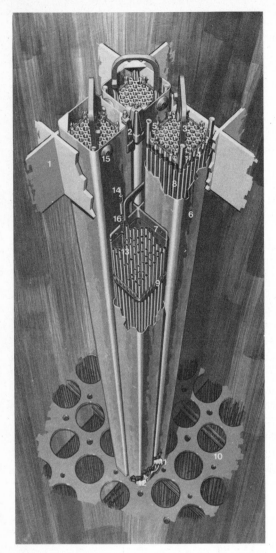

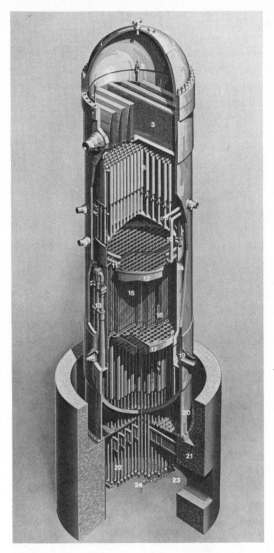

(A) FUEL ASSEMBLY   1. Top fuel guide, 2. Channel,
3. Upper tie plate, 4. Expansion spring, 5. Locking tab,
6. Channel, 7. Control rod, 8. Fuel rod, 9. Spacer,
10. Core plate assembly, 11. Lower tie plate, 12. Fuel
support piece, 13. Fuel pellets, 14. End plug, 15. Channel
spacer, 16. Plenum spring.

(B) REACTOR ASSEMBLY   1. Vent and head spray,
2. Steam dryer lifting lug, 3. Steam dryer assembly,
4. Steam outlet, 5. Core spray inlet, 6. Steam separator
assembly, 7. Feedwater inlet, 8. Feedwater sparger,
9. Low pressure coolant, injection inlet, 10. Core spray
line, 11. Core spray sparger, 12. Top guide, 13. Jet pump
assembly, 14. Core shroud, 15. Fuel assemblies,
16. Control blade, 17. Core plate, 18. Jet pump/recircula-
tion, water inlet, 19. Recirculation water outlet, 20. Vessel
support skirt, 21. Shield wall, 22. Control rod drives,
23. Control rod drive, hydraulic lines, 24. In-core flux
monitor.

FIGURE 2. (A) Reactor fuel assembly and control
rod module. (B) Boiling water reactor. (Courtesy of
General Electric.)

chain reaction itself from nuclear power reactor design and operation." A nuclear reactor cannot and will not explode like a bomb. A reactor core meltdown will not produce an explosion but could create the hazard of radioactivity released to the environment.

Burner reactors

The burner reactor operates off of fissile material. Although, ostensibly there is an enormous amount of energy available in uranium, the burner reactors in commercial service utilize less than 1% of the energy contained in the uranium. They consume the fissionable uranium 235 isotope while converting only small amounts of the more plentiful uranium 238 into fissionable plutonium. As a result, the burner reactor uses inordinate quantities of uranium, a fuel resource that is quite limited. Advocates of nuclear energy consider burner reactors to be only the near-term solution to human energy needs until breeder reactors and, ultimately, fusion reactors can take over.

Nuclear reactors are characterized by the type of coolant they use, such as light water, heavy water, gas, or liquid metal. The primary coolant liquid or gas is used to transfer the fission energy to the steam generating system, or the steam may be produced from coolant water directly inside the reactor itself.

The most common coolant in power reactors is ordinary water, called LIGHT WATER in nuclear terminology. Light water acts as a moderator as well as a coolant. There are two major types of light water reactors (LWR): PRESSURIZED WATER REACTORS (PWR) and BOILING WATER REACTORS (BWR). In the PWR, shown schematically in Figure 3, the heat picked up by the water in the reactor core primary coolant loop is circulated through a heat exchanger, where it converts water in the secondary loop into steam; the steam drives a turbine. The BWR, shown in Figure 4, is called a direct-cycle system because no heat exchanger is used. In the BWR, the primary coolant water serves as a coolant, moderator, and working fluid in the steam cycle. The steam is generated in the reactor itself at a pressure of about 7 million pascals (70 atmospheres), whereas in the PWR the primary coolant water is at about 15.2 million pascals (150 atmospheres) to permit the use of high coolant temperatures of about 300 °C without steam formation. In the case of a BWR, the primary coolant water becomes radioactive. This means that the steam turbine unit and building must be heavily shielded. The BWR is the most frequently used reactor in the United States.

Another type of reactor is the HEAVY WATER– (deuterium) cooled system. One such system is the Canadian deuterium–uranium, pressurized, heavy water reactor (CANDU-PHW).

GAS-COOLED NUCLEAR REACTORS usually use graphite as a moderator. They were developed in Great Britain using carbon dioxide gas as the coolant fluid. More recently, the HIGH-TEMPERATURE GAS-COOLED REACTOR (HTGR), using helium as a coolant fluid, was developed. The HTGR operates with a two-loop steam thermal cycle similar to that of the PWR. The helium gas circulates in pipes totally enclosed in the reactor vessel. They pass through a heat exchanger, where the energy produced by the reactor is transferred to the water–steam cycle that is in another set of pipes. Only the steam pipes leave the reactor vessel to carry steam to the turbine and to transport the condensed steam (as water) back to the heat exchangers. Coolant cannot be lost outside the reactor vessel because the coolant-carrying pipes do not leave it. The reactor core and steam generators are enclosed in a massive prestressed concrete reactor vessel. The thermal efficiency of the PWR is only about 32% compared with 38.5% for the HTGR. Great Britain has more than 40 HTGRs and France has 7. (Further informa-

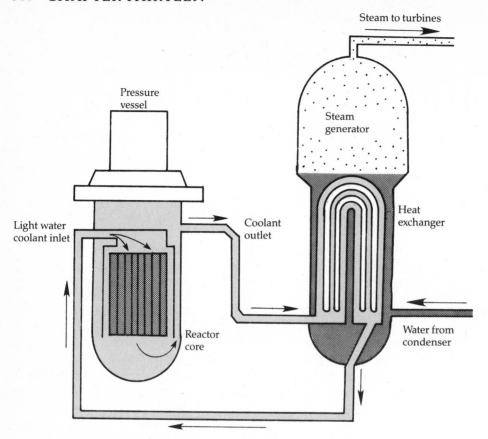

FIGURE 3. Schematic drawing of a pressurized water reactor and the heat exchanger for the steam generator.

tion may be found in Agnew, 1981.)

Not only are the HTGRs more thermally efficient than PWRs, they are much safer because of the massive graphite core. If something goes wrong, there is much more time to take corrective action than with PWRs; and higher temperatures can be sustained without difficulty. At Three Mile Island in Pennsylvania, the accident that disabled one of the reactors was caused by an interruption of the normal flow of coolant water. Operator error resulted in a delay in the emergency water flow to cool the reactor core. In light water re-

actors such as this one, there is simply not much time to respond, and operator error can be catastrophic.

Breeder reactors

The purpose of a BREEDER REACTOR is to generate more fissile material than it burns, while at the same time generating electricity. Clearly the excess fissile fuel (plutonium) produced can be used to fuel a burner reactor. The time required for a breeder reactor to produce enough excess plutonium to fuel a

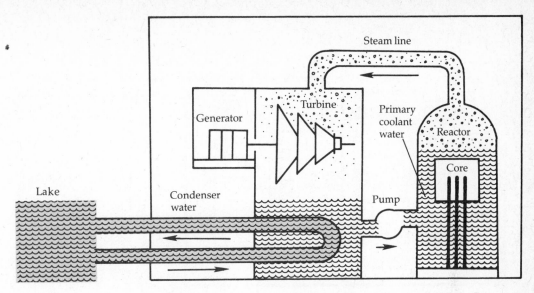

FIGURE 4. Schematic drawing of a boiling water reactor. In this system, the steam is generated in the reactor and the coolant water becomes radioactive. (Courtesy of General Electric.)

second reactor is called the doubling time. Most breeder reactors are designed to have doubling times of between 10 and 20 years.

Although breeder reactors may be fueled with either uranium 238 or thorium 232, it is more effective to use a combined uranium 238 and plutonium 239 fuel. The fuel elements consist of 85% uranium and 15% plutonium. The core is then surrounded with a blanket of natural uranium 238. The blanket can capture neutrons leaking from the core and convert still more uranium to plutonium. For a reactor to operate at a conversion ratio greater than unity (that is, to produce more fissile material than it burns), it needs to use high-energy, or fast, neutrons rather than slow, thermal neutrons. But for fission reactions, slow neutrons are most effective. Both fast neutrons and slow neutrons are needed in the breeder reactor. Liquid sodium as a coolant slows down some of the fast neutrons for the fission process while allowing other fast neutrons to participate in the breeder process.

The LIQUID METAL-COOLED FAST BREEDER REACTOR (LMFBR) uses liquid sodium as a coolant (Figure 5). Sodium has excellent heat transfer properties and also has a high boiling point (882 °C), which allows the primary coolant loop to operate at atmospheric pressure. Sodium, exposed to the intense neutron flux of the core, becomes radioactive. Therefore, the steam generator loop is isolated from the primary sodium coolant loop by an intermediate sodium coolant loop. Sodium and water react violently to produce hydrogen when mixed. Therefore, care must be taken in the design to assure no contact between the two. LMFBR power plants will have a high thermal efficiency—about 40%, which is better than the 30% efficiency of LWRs.

Another type of fast breeder reactor design is the GAS-COOLED FAST REACTOR (GCFR), which uses high-pressure helium as the coolant. The GCFR is a combination of the LMFBR and HTGR systems. Helium is a very low density gas, and as a result it does not

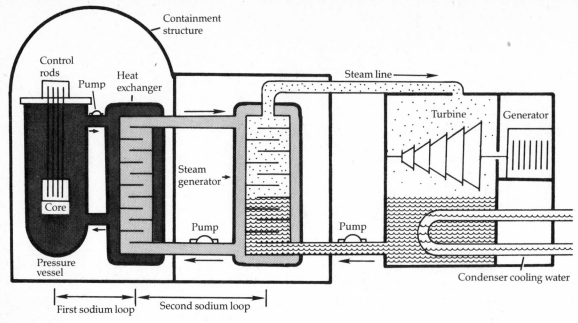

FIGURE 5. Schematic drawing of a liquid metal-cooled fast breeder reactor and turbine generator units. Because the sodium in the reactor core becomes radioactive, it is isolated from the steam system by an intermediate sodium heat transfer loop.

slow down neutrons as much as sodium does. This leads to a higher conversion ratio than that of the LMFBR. Because of its high core power density and use of helium as a coolant, a temporary loss of coolant is tolerable for only a few seconds before fuel melting occurs. Elaborate safeguards are required with the GCFR.

The first demonstration of a breeder reactor was at Los Alamos, New Mexico in 1946, to be followed by an experimental breeder reactor in 1951 that was the first to produce electricity. Many countries have taken up the development of breeder reactors, while the United States has fallen far behind. As recently as 1980, experts were estimating that the world would have 50 gigawatts of breeder reactor capacity by the year 2000. But that estimate is now changing. Many economists claim that breeders will not be competitive

with light water reactors for at least the next 50 years. It was argued until recently that the world would be short of uranium fuel in the near future and that breeder reactors were urgently needed. But, there now seems to be sufficient uranium to last through the first decade of the next century, so that the need for the breeder is diminishing rapidly.

France is the world leader, having built and operated the first prototype fast breeder, the 250-MWe Phenix reactor that has been generating electricity since 1973. However, in April 1982, Phenix sprang a leak in its cooling system, leading to a fire when the sodium coolant came in contact with the air. The accident put the reactor out of operation for several weeks. Until that time, this reactor had achieved the highest reliability of any power plant in the world. Since it was originally commissioned, Phenix has supplied more than 6 billion kilowatt-hours of electricity, with a load factor of more than 53%. The French, along with German and Italian partners, have designed and are constructing Super-Phenix, a 1200-MWe LMFBR, scheduled to become operational in 1985. In five years, the cost estimate for Super-Phenix has escalated threefold and will end up costing about twice as much as a conventional light water reactor of the same capacity.

The Germans are still trying to complete their own prototype, a 300-MWe sodium-cooled reactor that was started in 1973 at Kalkar, Germany. Belgium and the Netherlands are each contributing 15% of the cost, and Britain 2%. Originally planned for completion in 1978, it seems that it will not be operational until 1986 or later. Cost estimates have increased more than fivefold since 1973. Because of a number of important economic and safety questions, there may be further delays. Design of a commercial LMFBR of the same size as the Super-Phenix has been completed. Known as SNR-2, it is jointly financed by Germany, Italy, and France. However, construction may not begin until the 1990s, particularly considering current public attitudes toward breeder reactors.

The United Kingdom has been in the forefront of breeder development. There has been a 250-MWe fast breeder reactor in commercial operation at Dounreay in the north of Scotland since 1975. This reactor has had an erratic operating record, achieving only 6.8% capacity over its first eight years of operation. Britain is working on a design for a 1300-MWe commercial demonstration fast breeder reactor. No one knows how rapidly this breeder will be developed, but it is entirely possible that no more breeders will be started in Britain within the next 20 years.

In the Soviet Union, a fast breeder reactor of 350-MWe has been operating since 1972 and a 600-MWe LMFBR is probably just now in operation.

In the United States funding for the controversial Clinch River LMFBR has been stopped. This 350-MWe demonstration reactor is comparable to the French Phenix. The design of the power plant is nearly completed, a design recognized by experts around the world as one of the most advanced and one of the safest.

The public debate over plutonium for nuclear weapons is making it increasingly difficult to get public support for construction and operation of the breeder reactor, both in the United States and abroad.

Fusion reactors

All stars, including the sun, run on fusion energy. FUSION is the opposite of fission. Instead of breaking a heavy nucleus into two or more fragments of intermediate mass, as the fission process does, fusion brings together nuclei of light mass at high temperature and pressure, such that when they collide they fuse together into a heavier nucleus and release energy. The nuclei must collide at very high speed to overcome their mutual electrostatic repulsion. The sun fuses hydrogen nuclei into helium at interior temperatures in excess of 20 million degrees centigrade. A system using the same energy production method as the stars is seen as the ultimate source of energy.

Here on Earth it is possible to fuse together the heavier isotopes of hydrogen —deuterium and tritium—into helium at a temperature of about 100 million degrees centigrade. The reaction is

$$\text{deuterium} + \text{tritium} \rightarrow \text{helium} + \text{neutron} + \text{energy}$$
$$^2_1\text{D} + ^3_1\text{T} \rightarrow ^4_2\text{He} + ^1_0\text{n} + 17.6\,\text{MeV} \tag{1}$$

where MeV stands for million electric volts and 1 MeV is equivalent to 1.60×10^{13} joules.

The energy released by the fusion reaction is carried off as kinetic energy of the reaction products, for example, helium (an alpha particle) and a neutron. When these fast-moving atomic projectiles collide with other matter, their kinetic energy is transferred as heat, which in turn can be used to generate steam. The neutron carries off about 14.1 MeV of the 17.6 MeV released, with 3.5 MeV going with the alpha particle. The prolific release of neutrons in the deuterium–tritium reaction will induce radioactivity in structured materials of the reactor. These radioactive products, including tritium, pose a hazard to the public and must be handled properly. Some routine release of tritium to the atmosphere will occur.

Deuterium is found in water as 1 part in 6500 parts and can be easily separated from hydrogen atoms. Tritium, a weakly radioactive isotope, is almost nonexistent in nature and must be regenerated in the fusion fuel cycle. When the lithium isotope of atomic mass 6 is bombarded with neutrons, tritium is generated by the reaction

$$\mathrm{^6_3Li} + \mathrm{^1_0n} + \mathrm{^3_1T} + \mathrm{^4_2He} + 4.8\,\mathrm{MeV}$$

When a lithium blanket surrounds the core of a breeder reactor, the neutrons released generate additional tritium. Then, when the tritium is mixed with deuterium, a fusion reaction becomes possible. Thus, a fusion reactor is a breeder reactor fed by deuterium and lithium.

Lithium is found in seawater in about 1 part in 10 million parts and in the crustal rocks in about 20 to 32 parts per million. However, the lithium 6 isotope constitutes only about 7.42% of natural lithium. Hence, lithium 6 is present in a concentration of about 7 parts per billion of seawater and about 2 parts per million in surface rocks. Lithium is found concentrated in higher amounts in some minerals and is found, for example, in the water of the Great Salt Lake, Utah at 0.006%. Lithium is currently being mined and processed, and there would appear to be a supply on Earth of the same order of magnitude, in terms of available energy, as the world's initial supply of fossil fuel. When the world runs out of lithium, then the deuterium–deuterium reaction must be used as an ultimate source of fusion energy. This reaction is

$$\mathrm{^2_1D} + \mathrm{^2_1D} \rightarrow \mathrm{^3_2He} + \mathrm{^1_0n} + 3.2\,\mathrm{MeV}$$

or

$$\mathrm{^2_1D} + \mathrm{^2_1D} \rightarrow \mathrm{^3_1T} + \mathrm{^1_1H} + 4.0\,\mathrm{MeV}$$

The tritium atom reacts with a deuterium atom (equation 1) and releases 17.6 MeV of energy. The net result of all three reactions is that 24.8 MeV is released, but because five deuterium atoms are involved, 4.96 MeV of energy is released per deuterium atom.

Fuel cycles other than the deuterium–tritium cycle have been proposed for a fusion reactor. These include deuterium alone, or deuterium with heavier elements such as lithium, beryllium, or boron. By these reactions more energy is released as kinetic energy of charged particles and less in the form of neutrons. The amount of radioactivity induced in the structural materials of the reactor might be very low.

To make the fusion reaction work, it is necessary to bring the plasma (ionized matter) containing the deuterium and tritium nuclei to a temperature of 100 million degrees centigrade or more—a truly staggering temperature. The deuterium and tritium nuclei must collide with one another millions of times before fusion reactions occur. The plasma must be contained at this extremely high temperature for a sufficient period of time if any appreciable fusion energy is to be released. One cannot allow the plasma at this incredible temperature to strike the walls of the vessel or it would melt them. Because a plasma is an ionized gas it can be manipulated by a magnetic field. When the magnetic field lines have the right shape, they are able to confine the high temperature plasma to the center of an evacuated chamber. The design of a suitable magnetic field, or MAGNETIC BOTTLE as it is called, is very difficult because the plasma has its own magnetic field that tends to push back on the field of the magnetic bottle and distort it, thereby allowing the plasma to escape (Duderstadt and Kikuchi, 1979). The most successful magnetic confinement experiment has been the Tokamak project in the Soviet Union. Several fusion projects are underway in the United States, including machines at Los Alamos, New Mexico; Livermore, California; Princeton, New Jersey; and the Massachusetts Institute of Technology, Cambridge, Massachusetts.

Fusion reactors, when they are achieved, will be extremely expensive and complicated devices. Certainly research on fusion power is spurred on by the notion that it is virtually an inexhaustible source of clean energy. Nuclear fusion reactors can use as fuel the unlimited quantities of deuterium in ocean water. Although a fusion reactor will induce radioactivity in component parts from neutron bombardment, the amount of radioactivity will be significantly less than that associated with fission reactors. Fusion reactors will not utilize materials or generate isotopes that may be used in nuclear weapons; therefore a proliferation of strategic nuclear materials will not occur. Unfortunately, the dream of producing fusion energy has been far more technologically difficult to achieve than anyone anticipated.

RADIOACTIVE WASTES

One of the greatest concerns of modern societies, in addition to those of population growth, food supply, and nuclear war, is the disposal of nuclear wastes emanating from nuclear reactors, both civilian and military. RADIOACTIVE WASTES are produced by burning fissile material, and to a lesser extent by neutron bombardment of otherwise neutral materials within the reactor, including reactor metals, coolant fluids, and air or other gases. It turns out that the mass of radioactive fission products produced in a reactor is very nearly equal to the mass of fuel consumed.

After the fuel rods in a reactor have been used down to a certain level of activity, they must be removed and taken to a fuel-processing plant, where the fission products are separated chemically from the unspent fuel and refabricated into new fuel elements. The fission products comprise a large number of different isotopes that, taken collectively, are highly radioactive.

In today's power reactor, about 25% of the fuel must be replaced each year to sustain useful rates of heat release. When discharged, over 96% of the nuclear material in the spent fuel is reuseable uranium or plutonium. About 3.5% of the discharged material in a reactor is waste (mostly radioactive fission products) and a very small proportion of unuseable irradiation products (the transuranic isotopes).

Figure 6 depicts graphically the toxicity of wastes from a light water reactor as a function of the time in years following their discharge from the reactor. Those isotopes of relatively short-term radioactivity are fission products of medium atomic weight formed by fission of uranium or plutonium. Strontium 90, cesium 137, and, to a lesser extent, krypton 85 are the main fission products; zirconium 93, tellurium 99, iodine 129, cesium 135, and many others are of lesser importance. Most of these have half-lives not greater than 30 years. In 600 years, less than one-ten-millionth of their initial radioactivity would remain and would no longer be of concern. Some of the products of fission and breeder reactors are listed in Table 1.

The other major components of radioactive wastes are the transuranic actinides (isotopes of actinium, thorium, uranium, neptunium, plutonium, and others) that are

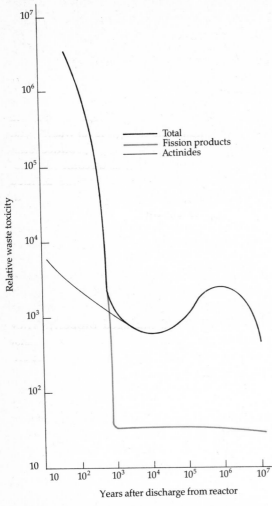

Relative waste toxicity

Years after discharge from reactor

- Total
- Fission products
- Actinides

FIGURE 6. Toxicity of wastes from a light water reactor for an equilibrium fuel cycle with 99.5% removal of uranium and plutonium. Each metric ton of fuel is assumed to deliver 33,000 megawatt-days during its operating lifetime. The hump, or increase between 10^5 and 10^6 years, arises from the growth of daughter products not present in the original material. (Redrawn from Kubo and Rose, 1973.)

Table 1. Fission products of fission and breeder reactors and product half-lives and activities

Isotope	Half-life (years)	Activity (curie/g)
Americium 241	433	3.2
Americium 243	7,900	0.19
Carbon 14	5,730	
Cesium 134	2.2	
Cesium 137	30	
Cobalt 58	71.3	
Cobalt 60	5.3	
Iodine 129	17×10^6	1.6×10^{-4}
Iodine 131	8 days	
Krypton 85	10.4	
Neptunium 237	2.13×10^6	7.1×10^{-4}
Plutonium 239	24,400	6.1×10^{-2}
Plutonium 240	6,600	0.22
Strontium 90	28	
Technetium 99	210,000	1.7×10^{-2}
Tritium 3	12.3	
Xenon 133	5.3 days	
Xenon 135	0.4 days	

formed, not by fission, but by neutron absorption into the original uranium or thorium fuel. All are extremely toxic and have long half-lives. For example, plutonium 239 has a half-life of about 24,400 years. It is the most abundant transuranic actinide formed in light water reactors and also in liquid metal fast breeder reactors. The actinides emit very low energy beta rays that are much less penetrating than the gamma ray radiation from fission products. Figure 6 shows that among all the nuclear radioactive wastes, the transuranic actinides dominate after 500 years because of their much longer half-lives. Daughter decay products of these actinides cause an increase in toxicity after about 100,000 years. Clearly the wastes with very long half-lives must be disposed of in a manner different from that used for wastes with relatively short half-lives.

When spent fuel is discharged from a reactor, it is stored in water-filled basins at the reactor site until the shorter-lived radioactive material decays. A year or so later, the total radioactivity level of the waste is only about 12% of the level at the time of discharge. After five years, it is down to about 3% of the level at discharge. The total activity of the waste continues to decline with time.

Closing the nuclear fuel cycle by disposing of radioactive waste safely and permanently is absolutely essential if humanity is to continue its use of nuclear reactors. Some people believe it is practical to do so, whereas others doubt the ability of humanity to safeguard its nuclear waste for perpetuity. (For further discussion of radioactive waste, see Murray, 1982.)

Quantities of radioactive wastes

A 1000-MWe light water reactor power plant produces about 900 kg of radioactive waste material each year; when solidified into glass, this waste occupies about 2.7 m^3. It is estimated that by the year 2000 the U.S. nuclear power industry will have generated about 14,000 m^3 of radioactive waste, an amount that would fit into a volume 24.1 m on a side; the high-level waste in this would fill a volume about 8 m on a side. The volume of radioactive waste generated by nuclear power production is actually quite small when compared to the volumes generated by coal-burning power plants. The problem, of course, is the nature of the waste from the nuclear power industry. Radioactive waste must be disposed of in such a manner as to assure its isolation from the biosphere for a sufficiently long time to allow decay to harmless levels.

Each U.S. citizen consumes about 7000 kWh of electricity per year. This amount of energy is derivable from 1 gram of nuclear fuel. The burning of this nuclear fuel leads to a radioactive waste of about 1 gram of fission products per person per year.

Approximately 7000 to 8000 metric tons of spent fuel are now being stored at reactor sites across the country. This has increased at about 1000 tons per year. It is estimated that on-site storage facilities will be saturated by the mid to late 1980s (see Krugman and Von Hippel, 1977 for more information).

Levels of wastes

Radioactive wastes range from dilute and short-lived materials to high-level wastes that are highly radioactive. These produce substantial quantities of heat as a result of radioactive decay. Radioactive wastes are divided into four categories and disposed of accordingly: high-level, transuranic, mill tailings, and low-level. Each of these types of wastes has characteristic properties and is managed differently.

HIGH-LEVEL RADIOACTIVE WASTES have a high specific activity, are thermally hot, are long lived, and pose serious health risks if not controlled and managed properly. They are generated by nuclear reactors and consist solely of waste from reprocessed fuel or spent reactor fuel.

TRANSURANIC WASTES are long-lived wastes consisting of isotopes with atomic numbers greater than 92, for example, plutonium. (Transuranic means those elements that are beyond uranium in the periodic table.) They are not thermally hot. They are produced in the nuclear weapons program and by nuclear power generation and generally have much lower levels of radioactivity than do high-level wastes.

MILL TAILINGS are the debris remaining at the mill after the uranium ore has been extracted from the parent rock. Mill tailings are generally of low specific activity and long lived, containing naturally occurring radioisotopes of uranium, thorium, and disintegration products called daughters. They are treated differently than other radioactive wastes because of the tremendous volume generated (around 10 to 15 million metric tons annually) and because they contain only natural isotopes.

LOW-LEVEL WASTES are defined as anything not high-level, transuranic, or mill tailings. Low-level wastes are subdivided into two broad categories. The first includes those low-level wastes with very short half-lives or those that are in minute concentrations and present minimal health risks. The second category of low-level wastes present longer term or more serious health effects and must be contained in relatively permanent repositories, usually shallow land-burial sites or ocean sites. Low-level radioactive wastes are generated by the nuclear power industry and by the use of radioisotopes in hospitals, medical research facilities, and certain industries. Materials include refuse from areas where radioactive isotopes are used, including paper bags, rubber gloves, floor sweepings, wood, glassware, carcasses, excreta of animals, protective clothing, and contaminated equipment. Low-level wastes are far less hazardous than high-level wastes, but they are bulky and plentiful. They must be handled and disposed of carefully. There has been public concern with the operation of low-level waste disposal sites, with the result that all but a couple of them in the United States have been closed.

Radioactive wastes also may be classified according to the levels of radioactivity. When the radioactivity is a hundred to thousands of curies per cubic meter, the material is a high-level waste; when it averages a few microcuries per cubic meter, it is a low-level waste. The radioactivity of intermediate-level waste falls between those of high-level and low-level wastes.

Low-level waste disposal

In 1980, some 100,000 m^3/yr of low-level wastes were generated in the United States. This amount is expected to increase to nearly 230,000 m^3/yr by the end of the century. Until a few years ago, the U.S. nuclear industry had six commercial disposal sites for low-level wastes, but three of these have now been permanently closed. The three closed sites are at Sheffield, Illinois; West Valley, New York; and Maxey Flats, Kentucky. The three operating sites as of 1984 are Barnwell, South Carolina; Hanford, Washington; and Beatty, Nevada. These are land-fill surface sites where the waste is deposited in trenches.

The Conservation Foundation Report (1981a) concerning a national policy for managing low-level radioactive waste makes the following statement about the low-level disposal sites that have been shut down: "If costs prove to be large, do the sites present sufficient risks to justify spending huge amounts of money to clean them up? While closed sites may present only minimal health risks, the outrage and frustration expressed by citizens and elected officials over host states being left with these burdens have created serious political problems and raised doubts about the entire low-level waste management system. Whether valid or not, some view these sites as ticking time bombs, while others worry that their communities may risk similar fates in the future."

The Conservation Foundation Report (1981a) recommends separating low-level wastes into four categories according to the half-life of various materials:

1. *De Minimis* are those having ultra-low activity levels below natural background. These should be disposed of in the same way the nonradiological hazardous materials they contain are disposed of.
2. Short-term wastes are those that have half-lives less than 14 days and that present no health risk after a reasonable holding period. They should be disposed of according to the nonradiological hazardous material they contain.
3. Medium-term wastes are those that re-

quire holding for about 20 years before they decay to harmless levels. After decay, they should be handled as any other toxic waste—according to their content.

4. Long-term, low-level wastes are those that have long half-lives or are otherwise inappropriate for decay storage. For these wastes, a shallow land-burial facility similar to those in which most low-level wastes now go is the appropriate disposal method. They should be buried in a manner that prevents leaching beyond the site.

High-level and transuranic waste disposal

By definition, high-level radioactive waste is highly dangerous to life because of its high specific activity, long half-life, and heat-generating capacity. Because the material is so highly radioactive, it is not possible to dispose of it by pure dilution, such as by dispersal in the ocean. High-level waste must be disposed of by isolation in some remote location, where there is no way it can migrate into the biosphere. A logical solution is burial of the waste in a geologic formation or other structure of absolute security. The only viable alternative to this would be 100% recycling of all the wastes, but this does not appear to be technologically feasible at the present time.

High-level waste contains essentially all the nonvolatile fission products (about 0.1 to 0.9% by weight of the uranium and plutonium concentrations in the spent fuels) and all the other actinides formed by transmutation of the uranium and plutonium in the reactors. Plutonium and uranium are segregated to recover as fissile fuel for reuse in nuclear reactors. However, not all of the plutonium and uranium can be recovered and some remains in the waste material.

The disposal methods and site for the long-lived, high-level and transuranic radioactive wastes must have very special characteristics. The repository site not only must be remote from habitation, but it must have predictable thermal behavior and chemical characteristics, long-term geologic stability, low water content, and very low water flow rates. The site also must be accessible so that the wastes are retrievable (DeMarsily et al., 1977.)

Many questions need to be addressed, according to Angino (1977), with respect to the geologic disposal of high-level and transuranic wastes:

> (i) Can high-level radioactive wastes be stored (isolated) safely underground for long periods of time? (ii) What geologic medium is the safest for long-term burial? (iii) Can the waste be properly stored so that it does not contaminate man's environment? (iv) In what form should the wastes be stored—liquid or solid? (v) Can the wastes be stored in such a manner that they will be retrievable at some future date, if the need arises?

The management of high-level radioactive waste is discussed in a report issued by the Conservation Foundation (1981b).

Five primary methods of storage are suggested for high-level and transuranic wastes. These methods are (1) mausoleum, (2) geologic, (3) ice sheet, (4) seabed, and (5) unsaturated zone storage. Angino (1977) discusses 2, 3, and 4 in detail. Winograd (1981) describes the unsaturated zone type of burial, and Starr and Hammond (1972) recommend consideration of the mausoleum or great pyramid type of storage. Both high-level and transuranic wastes contain isotopes with long half-lives, but heat is generated by the high-level wastes and not by the transuranic ones. The need for a cooling mechanism is different in the two cases.

It is most desirable to store high-level and transuranic wastes in solid form, rather than as a liquid, because in liquid form they are more mobile and can contaminate the environment. Methods for converting liquid wastes to a solid borosilicate glasslike structure have been thoroughly researched. The reprocessing of one metric ton of spent nuclear fuel generates approximately 500 liters of liquid waste, and finally 90 liters of glass. The glass is enclosed in cylindrical, stainless steel containers measuring 1 m in height and 0.5 m in diameter, with a capacity of 150 liters. The production of 1 gigawatt-year of electric energy produces about 2 m^3 of glass waste. These canisters would then be placed in some type of subsurface cavity. It may be a mine, vault, or drill hole in a deep stable formation of carefully chosen composition, such as salt, dolomite, or granite.

Mausoleum storage. Starr and Hammond claim that a stone object the size of the great pyramid of Cheops, which is about 240 m^2 at the base, could be arranged to contain small vaults to house all the nuclear wastes generated by the U.S. electric power production for over 5000 years, if the current rate of generation remained constant. Additional pyramids would be needed as electrical loads increased. Some of the spent waste could be eventually removed and buried elsewhere, to make room for new wastes. Such a mausoleum would need to have a cooling system to carry away the heat generated by the radioactivity. This would require an enlargement of the building. Such structures could be located in one of the deserts where the material could be safeguarded and recovered, if necessary. If a better storage system were established later, the radioactive waste could be retrieved and placed elsewhere.

Geologic storage. Emplacement of the steel canisters containing the borosilicate glass waste in deep rock formations requires that the following criteria be met: the rock unit must have high structural strength in terms of compressibility, good thermal conductivity, and high heat capacity, as well as the absence of subsurface water flow. It must have low permeability, low porosity, and high sorption capacity. Only the properties of salt have been fully explored with respect to suitable characteristics for the storage of high-level wastes, and very little is known of the effects of the wastes on granites, basalts, clay minerals, limestones, dolomites, or anhydrites. Structural stability suggests that seismic shocks, rock slippage, or faulting must not be allowed to break the canisters. It has been concluded by a National Academy of Sciences study that salt is the most desirable rock type in which to bury high-level radioactive wastes for extended time periods. Salt is approximately equal to concrete in its ability to shield harmful radiation. Burial would be at a minimum depth of 300 m below the surface and as deep as 3000 m.

Figure 7 shows a conception of a deep-mined, high-level radioactive waste disposal site with a surface receiving and handling facility, a subsurface handling facility, a means to transport the wastes underground, and finally placement of the canisters in holes drilled in the floor of the tunnel. There are shafts for access to and ventilation of the subsurface depots. The subsurface mine area will occupy about 800 hectares. Main access corridors will run from the receiving area to waste disposal rooms. The rooms will be separated from one another by solid walls about 22 m thick, depending on the type of rock. The waste emplacement holes will be distributed in each room so that the heat generated does not damage the rock. The exploratory boreholes and the repository access shafts will later be sealed to prevent a movement of water.

There is always the possibility that the glass matrix or ceramic container confining the radioactive isotopes will leak at some time during the ensuing half million years. Highly absorbed elements, such as plutonium, can be retained in the geologic environment if the ion exchange with the rock minerals is sufficient. A geologic formation should not be considered as a confining barrier for radionuclides with very long half-lives for which it has no exchange capacity, if ground water can reach the waste or if a hydraulic gradient exists. Deep geologic disposal is a method for putting high-level radioactive waste out of reach and of delaying the release of elements that may escape from the glass or the ceramic canisters. There are also physical means to create another barrier around the waste canister so as not to rely entirely on the properties of the geologic formation.

The geologic storage of high-level radioactive waste in a granitic rock mass of an iron mine in Sweden has been explored. Field tests indicate the importance of hydrology and the difficulties with predicting the thermomechanical behavior of fractured granitic

rocks. (Witherspoon et al., 1981 describe the many uncertainties associated with this type of storage for high-level and transuranic radioactive wastes.)

Salt deposit storage. A secure storage bed for high-level and transuranic radioactive waste must prevent the wastes from wandering away from where they are placed. Beds of salt seem most promising because they have a number of desirable characteristics that could help contain any waste placed into them. One is that salt has no cracks in it to allow the flow of water that might carry any radioactive wastes leaking from the canister. Salt flows under pressure, so that fractures deep in the Earth tend to seal themselves. Salt has been shown in the laboratory to be among the most impermeable rocks to be

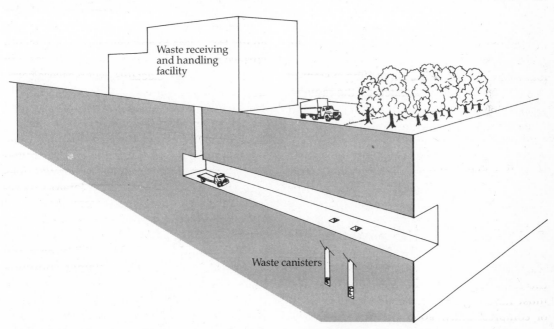

FIGURE 7. A deep mine high-level radioactivity storage site. Wastes are trucked to the surface receiving and handling facility, passed to a subsurface handling facility, and placed in canisters in the floor of the mine.

found. The persistence of salt deposits for 200 million years or more demonstrates their isolation from circulating groundwater. Unfortunately, salt has a high solubility in water, a property that does allow groundwater to penetrate and dissolve salt. Pockets of brine are sometimes left behind after the evaporation of seawater has formed the salt beds in the first place and this brine could transport leaking wastes. Pilot test drillings of the salt deposits in the Delaware basin in southeastern New Mexico have indicated some dissolution of the salt and some potential water movement. Also there are many boreholes that have been drilled in salt beds since the early part of the century in search of mineral deposits. These could lead to the infiltration of water; however, by allowing a one-mile buffer zone around each hole, infiltration by water to the burial site can be avoided.

Several other problems seem to prevail with regard to salt repositories. In the Delaware basin salt beds, there are collapse chimneys. There are columns of rubble 100 to 500 m in diameter that extend vertically thousands of meters through the salt beds. There are also brine inclusions in the salt bed that may be heated by the radioactivity of the waste, thereby bursting the inclusions and causing the brine to migrate toward the waste-containing canisters. If brine migration could actually affect containment of wastes, through corrosion of the waste container or by a weakening of the salt formation, this could be mitigated by letting the waste cool more before emplacement, by putting less waste in each canister, or by spacing them farther apart. The canisters, consisting of glassy cylinders of waste sealed in a stainless steel jacket, could also be given greater protection by means of a protective shield of an alloy of 99% titanium.

A great deal needs to be learned yet concerning the geologic properties of salt beds and domes. Many people believe that salt may still be the best repository for the long-term storage of high-level and transuranic wastes.

Ice sheet or Antarctic rock storage. Suggestions have been made for the storage of high-level radioactive wastes in the rocks beneath the ice sheet of Antarctica or in the ice sheet itself. There are proposals to let the heat from the radioactive waste melt them into the Antarctica or Greenland ice caps. The volume of ice that could be melted by all the wastes generated between now and the year 2000 would be less than 0.04 km^3. The bottom of the Greenland ice sheet is bowl shaped and below sea level. Nuclear wastes having a half-life of 700 years or less should be relatively safe there, when buried in stainless steel canisters. An international treaty prevents storage of radioactive waste in Antarctica by all nations. Ice cap disposal has several drawbacks. Any waste containing the actinides requires a period of storage of 100,000 years or longer. If the climate warms and melting of the ice caps accelerates, then clearly storage in them is not satisfactory.

Transportation and working conditions in Antarctica or Greenland are difficult and hazardous. Because of the political problems and the other concerns, these storage schemes have been dropped.

One proposed plan was to bury high-level waste at a modest depth in Antarctic rocks. All groundwater is frozen in Antarctica to a depth of 1 km. Insertion of wastes might then produce only warm inclusions in the totally frozen surrounding material.

Another proposed scheme was to remove the actinides before placing the wastes in deep holes in the long-lasting ice sheets of Greenland or Antarctica. Even if the actinides were removed, heat generated by the wastes might require dispersal of the wastes over a large area (Kubo and Rose, 1973). And, of course, the actinides would need to be fully reprocessed and recycled.

Seabed storage. The bottom of the ocean is seen by some experts as a suitable long-term repository for high-level and transuranic wastes that is less complicated and more predictable than other disposal sites. The seabed itself may not be suitable, but, if the wastes can be inserted beneath the seabed, they may avoid the small currents and mobile sediments of the ocean bottom. Also, oceanographers have discovered that animals on and near the deep sea bottom move about a great deal and that bottom animals may be linked to the surface through overlapping habitats of predator and prey. It is for these reasons that storage beneath the seabed is considered more reliable.

Red clay deposits have been found in the center of large ocean basins. These clay sediments have been relatively undisturbed for millions of years. One particularly long core sample from the center of the clay deposit showed it to be undisturbed throughout the past 70 million years. The red clay has highly favorable properties for the long-term storage of high-level nuclear wastes. Like salt, it flows, under pressure, although much more readily. Cracks should not occur because the plasticity of the clay tends to seal them, as well as to seal waste canisters after emplacement. One means of emplacement being considered is to drop with high precision a long, bullet-shaped waste cannister that would penetrate about 30 m into the bottom. The clay would flow behind the canister as it penetrated the bottom and immediately close up the hole. The permeability of the clay has been found to be so low that it would take over 100,000 years for any waste material escaping from the canister to move by diffusion through 30 m of sediment to the surface. The high absorptivity of red clay also makes the movement of any material extremely slow. As a result, it would require millions of years for strontium 90, for example, to move to the surface. The effects of heat and radiation on the fluid properties of red clay are essentially unknown. The red clay could be turned into a convecting viscous fluid by the heat from the radioactivity. However, calculations indicate that this is not likely to be a serious problem. Many nuclear engineers are quite optimistic that nuclear wastes might be stored successfully beneath the deep seabed. This method of disposal is much favored by Japan and Great Britain. These countries have relatively few potential storage sites on land.

Unsaturated zone storage. There are thick (as much as 600 m), water-unsaturated rock strata in the southwestern United States that are potentially suitable environments for the disposal of solidified high-level and transuranic nuclear wastes. Such arid zones occur in some of the mesas of the Great Basin in Nevada and Utah. According to Winograd (1981), the favorable points of waste burial at relatively shallow depths (hundreds of meters above the water table rather than in deep mines hundreds of meters below it) are (1) extremely low amounts of water in the unsaturated zone under present arid and semiarid climatic conditions; (2) a highly sorptive capacity for radionuclides in the sediments of the unsaturated zone; (3) an ability to bury the wastes in a human-made repository within the rock formation; (4) absence of water-related problems during construction and after sealing of a shallow repository; and (5) accessibility of the wastes for monitoring or for removal.

Shallow burial in unsaturated rock strata is particularly advantageous for retrieval if any one of several events should occur prior to decay of the actinides to acceptable levels of radioactivity. Among those events are erosion, earthquake activity, meteor impact, and possible flooding or flushing of the site by a major rise of the water table or by increased deep percolation of precipitation during any time in the future when the climate may be wetter.

In 1962, a nuclear detonation of 100 kilotons at Yucca Flat in Nevada produced Sedan Crater, which has a diameter of 370 m, a depth of 98 m, and a volume of 5.0×10^6 m³. The detonation occurred at a depth of about 190 m in Quarternary-Tertiary valley-fill deposits that underlie the site to a depth of 250 m. The water table is at a depth of about 580 m. The climate at the site at the present time is arid. The volume of the Sedan Crater is about 25 times the total volume of high-level and transuranic wastes expected to be generated by both commercial and military reactors in the United States between 1980 and 2000. This crater would seem to be ideal for the burial of high-level and transuranic nuclear wastes.

Storage of spent fuel

When spent fuel rods are removed from a reactor, they are highly radioactive. The radiation dosage to a person in direct contact with the surface of the fuel assembly could be millions of rems per hour. (The definition of a rem is given in this chapter in the section Radioactivity and Exposure Units.) Because an exposure of 400 rems is considered a lethal dose, people must be isolated from the fuel assemblies.

When spent fuel assemblies are removed from a reactor, they are stored under water in a deep pool at the reactor site. The water keeps the fuel assemblies cool and shields people nearby from the ionizing gamma radiation. The assemblies are kept separated in the pool by metal racks to avoid accidental criticality. Boron in the metal grid absorbs neutrons and prevents their multiplication.

Storage in on-site pools is only temporary because additional assemblies must be discarded each year and space is limited.

Reprocessing

About 25% of the fuel assemblies in a nuclear reactor must be removed each year. For a large burner reactor, 50 to 60 fuel assemblies are replaced. This material may be stored indefinitely or it may be reprocessed. At a reprocessing plant, spent fuel assemblies are cut into pieces and the contents of the fuel rods are dissolved with nitric acid. A solvent extraction process is used to separate the three main groups of materials: uranium, plutonium, and fission products. The uranium and plutonium can then be recovered for fuel. (Bebbington, 1976 describes these extraction processes in considerable detail.)

When the nuclear power industry began to build reactors, they had every expectation that uranium and plutonium would be recovered from spent fuel rods by reprocessing plants. It was assumed in the 1960s that a rapid buildup in the number of light water reactors would cause the price of uranium fuel to increase prohibitively. Reprocessing was thought to be a means, not only to salvage spent fuel, but also to reduce the quantities of high-level radioactive waste. In addition, it was assumed that the plutonium recovered would be needed to fuel the breeder reactor. Meanwhile, until breeder reactors were fully operational, the plutonium could be used in existing light water reactors by being mixed with low-enriched uranium fuel in what is called a THERMAL-RECYCLE.

Only one private plant in the United States was ever licensed to operate, and it was shut down in 1972. Then construction of a reprocessing plant was begun at Barnwell County, South Carolina to receive spent fuel. But in 1977, President Carter issued a national policy to ban reprocessing because the plutonium recovered could be used by terrorists to make nuclear weapons. As a result, spent fuel from this nation's 85 operating fission power reactors has been piling up at storage sites and the current inventory is over 8000 metric tons. With no secure long-term storage facilities for high-level radioactive waste and no reprocessing plant in operation, the United States is going to find its on-site storage facilities saturated in the near future.

Of five reprocessing plants operating in Europe and Japan, four have been shut down after only a few years of operation. The fifth, at La Hague in France, has been operating at about 10% capacity since its completion in 1976. In the United States, the future of the Barnwell reprocessing plant is in part tied to the future of the Clinch River breeder reactor.

IONIZING RADIATION

Organisms on Earth are subjected continuously to a low level of damaging radiation from radioactivity inside their tissues, from their immediate environment, and from cosmic rays. Organisms living at high mountain altitudes or in certain geographic areas are subjected to much higher levels of ionizing radiation than those living elsewhere. Humans, plants, and animals have evolved physiological mechanisms to repair tissue damage from background radiation at about the same rate as the damage occurs. However, when the ionizing radiation rate is significantly increased, such repair is no longer possible and permanent injury results. Depending on the nature of the exposure, radiation damage to organisms may take many forms, ranging from small and long-delayed effects to short-term lethal effects. A subtle and serious consequence of some radiation damage is the genetic transmission of induced physiological defects. Radioactive wastes differ from all other chemical wastes in that there is no way of defusing their radioactivity and potential biological harmfulness. Radioactivity cannot be changed by any process less drastic than bombarding nuclei with energetic particles that are accelerated inside of a nuclear accelerator. Each radioactive isotope decays at a fixed negative exponential rate and at a rate unique to itself.

Molecular and cellular damage

We know, of course, that ionizing radiation is damaging to organisms above a certain threshold level of exposure. However, some scientists believe that even very low levels produce some effect, if only in a few individuals. Our knowledge concerning the response of plants and animals (wildlife) to ionizing radiation is very inadequate. (See Upton, 1982 for a good, straightforward description of the biological effects of ionizing radiation.)

What is it that ionizing radiation actually does as it passes through biological tissue? The ionizing radiation gives up its energy through a series of random collisions with the atoms and molecules of biological tissue. The collisions produce ions and reactive chemicals that in turn break chemical bonds. The breakdown of chemical bonds causes a denaturation of tissue or its transformation to substances foreign to healthy, normal tissue. An injury to a cell results from the bond-breaking events within molecules. When the DNA molecule is damaged, serious genetic consequences may occur. When X rays or other ionizing radiations bombard the whole body of a person at a lethal level, hundreds of breaks in DNA molecules in every cell of the body take place. Breakage and rearrangement of chromosomes also can occur.

Genetic effects

Ionizing radiation has a greater effect on reproductive cells than on asexual or somatic cells. Because the genetic material of cells persists throughout the reproductive life of an organism, the incidence of radiation-induced events that result in mutations and birth defects might be expected to show up in the population. The U.S. human population is the most X rayed population in the world, yet there is no overt evidence of any increase in mutations, birth defects, or shorter life spans attributable to this cause. There is no supporting evidence of low levels of ionizing radiation affecting wildlife (plants or animals) in this manner.

The majority of scientists reviewing the long-term genetic damage to species other than humans have discounted the risks. They simply do not find serious effects. Deleterious mutations, unacceptable in human populations, are eliminated by natural selection in other species and have relatively little effect on population viability and fitness. Moreover, natural selection has been found to maintain high levels of genetic variation.

MUTAGENIC EFFECTS in animals can be produced by ionizing radiation. Although no heritable effects of irradiation of humans have yet been demonstrated, many such effects are thoroughly documented using experimental animals. In considering possible mutagenic effects produced by ionizing radiation on humans, one is reminded of the atomic bombings at Hiroshima and Nagasaki, Japan. Absolutely no genetic effects have shown up among the descendants of the survivors of these bombings. This issue is of such enormous importance that a quote from the report by Schull et al. (1981) on this subject is in order. "In no instance is there a statistically significant effect of parental exposure; but for all indicators the observed effect is in the direction suggested by the hypothesis that genetic damage resulted from the exposure." Indicators of possible genetic effects used in this study included sex, birth weight, viability at birth, presence of growth malformation, occurrence of death during the neonatal period, and physical development at age 8 to 10 months. They set the dose threshold—or what they term the genetic doubling dose for radiation—at 156 rems. This is four times higher than the results from experimental studies with mice using comparable sources of radiation. In fact, it is the studies using mice as the experimental animal that have been the principal guide to the inferred human sensitivity. Schull et al. (1981) conclude that if mice and people have similar responses, the doubling dose for longterm, low-level exposure of humans suggested by the atomic bomb data is 468 rems, in contrast to an estimate of 100 rems for low-dose exposure adopted by a committee of the International Commission on Radiological Protection. The atomic bomb exposures by the people of Hiroshima and Nagasaki were short-term or acute exposures. People working around nuclear plants or with nuclear waste are subjected to long-term or chronic exposures. Hence the difference in the estimated doubling dose levels given (156 and 468 rems) for short-term and long-term exposures, respectively.

Cancer induction

In contrast to the apparent lack of genetic damage generated by the atomic bombs, there is substantial evidence for the occurrence of increased numbers of cancers in the population. The cancer mortality data from Hiroshima are the most important anywhere. Unlike the data from Nagasaki, the Hiroshima data reveal a clear relationship between doses of radiation received and illness. The data indicate that any increase in radiation, no matter how small, above the natural background level, directly increases the risk of getting cancer. Marshall (1981) discusses a reevaluation of the Hiroshima and Nagasaki exposure data. (Some of the conclusions drawn by Marshall are challenged in a note by Radford, 1981.) Earlier research seemed to indicate that leukemia mortality was higher for a given dose received from the atomic bomb at Hiroshima than for that at Nagasaki. The reevaluation discussed by Marshall (1981) and reported from the American Institute of Physics (1981) show the dose response curves to be identical for the two cities.

The basic problem with estimating cancer induction from exposure to ionizing radiation is that the data base is much better for high dose rates than for low dose rates. A report issued by the National Research Council (1980) on the effects on human and animal populations of exposure to low levels of ionizing radiation has attempted to resolve some of the controversy surrounding this topic. The scientific basis for estimating the carcinogenic risk of low-dose, whole-body radiation is grossly inadequate.

Half-lives

Health physicists and physiologists have determined standards for the maximum concentration of radioactivity from different materials that are considered safe for human health or other biological response. As a general rule they consider that high-level radioactive waste must be isolated for 20 half-lives before it can be considered as safe. A HALF-LIFE is the time for a radioactive element to disintegrate or decay to one half of its original mass. If an element has a half-life of 20 years, only half of its initial mass will remain after the first 20 years, the other half of its initial mass would have been ejected as particles or nuclei. After another 20 years (therefore, at the end of 40 years), only one-fourth of the original mass will remain, and after another 20 years only one-eighth of the initial mass will be present. Strontium 90 and cesium 137 have half-lives of 28 and 30 years. To be considered safe, they must decay to one part in a billion initial parts, a time requiring isolation for at least 600 years before they can be considered as biologically harmless. The half-life

of radium is 1622 years. Plutonium 239 has a half-life of 24,360 years; iodine 131, 8.05 days; iodine 129, 17.2 million years; krypton 85, 10.4 years; and krypton 88, 2.8 hours. Different radioactive isotopes have different half-lives. Some of the products of fission and breeder reactors are listed in Table 1.

Fission products

In nuclear reactors there will be fission products from uranium 235, uranium 233, and plutonium 239. Uranium 233 and plutonium 239 are produced from the nonfissionable radioactive isotopes, uranium 238 and thorium 232. The fission products of plutonium 239 will represent significant quantities of radioactive waste only if breeder reactors become commonplace. Global supplies of the fissionable uranium isotope uranium 235 are limited and within a comparatively short time will be exhausted. Breeder reactors will then be needed to convert uranium 238 and thorium 232 to fissionable plutonium 239 and plutonium 233, respectively. Among the radioactive fission products of this conversion are iodine 131, cesium 137, and strontium 90. These products are significant because they tend to be accumulated by organisms. Iodine 131, with a half-life of 8.05 days, is readily accumulated in the thyroid gland of humans and other vertebrates. Cesium 137 and strontium 90 are accumulated by organisms in the same way as potassium and calcium because of their chemical similarities, and therefore, they are concentrated in the bones and teeth of mammals, including humans.

Nuclear reactors release several radioactive gases into the atmosphere, thereby contributing to the ambient radioactivity. These gases are tritium, with a half-life of 12.36 years, and krypton 85, xenon 133, and xenon 135, inert gases that have no direct biological or chemical activity. Krypton 85 is produced in high yields from the fission of uranium

235, but it remains well locked inside the fuel cladding until the cladding has to be removed during the reprocessing of the fuel. Krypton is then released to the atmosphere. It is suggested by Boeck (1976) that the release of Krypton 85 on a continuing basis from reprocessing plants will cause the electrical resistance between the Earth and the ionosphere to decrease significantly within the next 50 years, and global weather patterns may be modified as a result.

Fusion was once thought to provide an energy source completely free of environmental contamination. However, the neutron flux in such reactors will make all structural materials intensely radioactive. Tritium will be produced in large quantities, and much will escape to the atmosphere, thereby increasing ambient levels of radioactivity.

Radioactivity and exposure units

The amount of radioactivity in any quantity of material is measured in CURIES. There is one curie of radioactivity present if there are 37×10^9 disintegrations per second taking place. The quantity of material may be a gram, a pound, or a tubfull. One curie was originally defined as the number of disintegrations per second occurring within one gram of radium.

Because extremely small quantities of radioactivity are involved in power plant discharges, the quantities released are often expressed as millicuries, microcuries, nanocuries, or even picocuries. These quantities are thousandths, millionths, billionths, and even trillionths of a curie, respectively. Radioactive wastes, for example, must be classified not only in terms of the half-lives of the isotopes they contain, but also by the amount of radioactivity in them.

IONIZING RADIATION is by definition that which imparts sufficient energy to an absorbing material to ionize it. By ionization we mean removal of electrons from or addition of electrons to the material. Various radioactive emissions produce different amounts of ionization of the matter through which they are passing. Some ionizing radiations, such as gamma rays or X rays, are electromagnetic radiation of short wavelength and high frequency. Other ionizing radiation is made up of subatomic particles such as electrons, protons, neutrons, and alpha particles. To measure the effectiveness of ionizing radiation, various quantities and doses have been defined.

The oldest unit by which one measures the amount of ionization resulting from X rays is the ROENTGEN (R). A roentgen of exposure measures the number of ionizations caused by radiation in air. One roentgen produces 1.6×10^{12} ionizations in one cubic centimeter of air.

The quantity of incident ionizing radiation absorbed by a material, including living tissue, is the RAD. With one rad of absorbed dose, one gram of material absorbs 10^{-5} J of energy, or 100 ergs of energy, per gram. It is conventional to speak of the number of rads of energy absorbed per roentgen emitted. Another similar unit for a measure of an absorbed dose is the GRAY. One gray equals one joule of absorbed energy per kilogram of tissue, or 100 rads.

Radioactive isotopes emit a variety of different radiations, such as gamma rays, protons, neutrons, electrons, and alpha particles; and each of these causes somewhat different effects on organisms. For a given dose, particles generally cause more radiation damage than do X rays or gamma rays. A unit called the ROENTGEN-EQUIVALENT-MAN, or REM, represents the amount of radiation that, if absorbed by a human body, will produce a biological effect equivalent to an agreed-upon standard (the same theoretical effect as one roentgen of gamma rays or X rays). For gamma rays and beta rays (electrons), 1 rem

= 1R. For alpha particles (helium nuclei), 1 rem is greater than 1 R. However, as a general rule and for the accuracy with which we are concerned here, 1 rem = 1 R. One R produces with each human exposure approximately 1 rad of absorbed radiation dose. At 0 to 25 rem, there are no detectable human biological effects; at 25 to 50 rem, there are slight temporary blood changes; at 100 to 200 rem, nausea and fatigue develop; at 500 rem, half of the people exposed die; and at 10,000 rem, tissue is destroyed.

A larger unit than the rem is the SIEVERT. One sievert is the amount of any radiation that is equivalent in biological effect to one gray of gamma rays. One sievert equals 100 rem. A typical dental X ray delivers about one millisievert to the center of the cheek. Units employed for expressing collective doses are the MAN-REM and the MAN-SIEVERT. Each is calculated by the product of the average dose per person times the number of people exposed. For example, one rem exposure by each of 100 people would equal 100 man-rem or 1 man-sievert of collective dose.

Exposure standards

Several different standards for levels of exposure by humans to ionizing radiation are used in the United States. Three distinct, but interrelated numerical values are used.

The first limit applies to individuals. The limit for any one person's whole-body exposure is 500 mrem/yr. (An mrem is 1/1000 rem.)

The second limit is used when a group of individuals is at risk. The whole-body exposure for the average of the population exposed must not exceed 170 mrem/yr. It is assumed that the most exposed individual in the population is to receive not more than three times the average for a total of 510 mrem/yr. This last number agrees with the first limit.

The third limit is based on population genetics. The per capita exposure limit for the human reproductive organs is set at 5 rem in 30 years. This number averages out to 167 mrem/yr, a number close to the whole-body exposure limit for a population.

If the 170 mrem/yr limit for a population sample and the 500 mrem/yr level for an individual are adhered to, there is almost no possibility that any individual could receive the full per capita dose to the reproductive organs.

Ionizing radiation background

Across the United States the natural background of ionizing radiation averages about 106 ± 20 mrem/yr. About 43 mrem/yr of this comes from cosmic radiation and about 63 mrem/yr comes from terrestrial radiation sources of natural origin. Coastal areas of the southeastern United States have an average background radiation of 60 to 70 mrem/yr, whereas Colorado has 250 mrem/yr. In China, two populations of people have lived for 30 generations with an ionizing radiation background of 300 mrem/yr without any discernible deleterious genetic or physiological effects.

An additional dosage of ionizing radiation is received from human sources. About 64 mrem/yr is received from medical exposures, air transport, television, weapons fallout, and the nuclear fuel cycle altogether. A person crossing the Atlantic ocean in a jet plane will receive an extra 2.6 mrem. The entire nuclear fuel cycle contributes on the average less than 0.1 mrem/yr to our total dosage.

Dosages from nuclear power

Neyman (1977) discusses the public health hazards of releases from nuclear electric power-generating plants. One of his examples involves the release of radioactive noble gases from the Big Rock nuclear power plant near Charlevoix, Michigan. The annual eight-year average whole-body exposures have been between 0.18 to 1.21 mrem/yr, depending on the distance from the plant. Within 5 km of the plant, generally downwind, there were dose rates during elevated releases as high as 12 mrem/yr, but beyond that the dose rate was very low.

In 1977, approximately 72,000 workers at nuclear power plants received a collective dose of 32,700 man-rem, or an average per person dose of 454 mrem/yr. The U.S. population is exposed each year to ionizing radiation of 34 million man-rems (200 million people × 170 mrem). Approximately 44,000 nuclear power plant workers received more than 1000 mrem in 1977, with an average dose of 740 mrem. In that year, 270 workers each received between 5 and 10 rems. The International Commission on Radiological Protection (ICRP) estimates the cancer-dose risk coefficient involved with exposure to ionizing radiation. According to the ICRP numbers, the 44,000 workers would have three additional radiation-induced fatalities with the total annual dose of 32,560 man-rem. This is equivalent to about seven radiation-induced hypothetical fatalities per 100,000 people. The National Safety Council reports that in other occupations the fatalities per 100,000 workers per year are manufacturing, 9; government, 11; transportation and public utilities, 33; agriculture, 53; construction, 60; and mining and quarrying, 63. In the home, the risk is about 12 fatalities per 100,000 persons per year.

Iodine isotope release

Radioactive iodine isotopes can be released as a gas from nuclear reactor accidents and nuclear bomb testing. Two iodine isotopes are of greatest concern. Iodine 131 is a highly radio-active, short-lived form of iodine, having a half-life of only eight days. On the other hand, iodine 129 emits very low energy electrons and gamma rays but has a half-life of 17 million years.

Radioactive iodine is an important concern to humans when there is a severe reactor accident because it is volatile enough to escape a failed reactor containment and, unlike the noble gases such as krypton and xenon, it enters the life cycle through vegetation, is ingested by milk-producing animals that eat grass, and is passed to humans drinking the milk, where it concentrates in the thyroid gland. If the iodine is radioactive, it can induce cancers in the thyroid.

As our nuclear power plants become older, there is an increasing probability of a nuclear reactor accident; albeit still a low probability. If a serious reactor accident did occur, it could result in the release of considerable quantities of iodine 131 and iodine 129 into the atmosphere. To protect people living downwind from such a potential accident, it has been suggested that they be provided with capsules of potassium iodide (KI), which may be ingested at the time of the accident. This salt will saturate the thyroid gland with iodine so that radioactive iodine released from the reactor accident will not be lodged in human thyroids. The use of KI for thyroid blocking appears to be one of the few practical strategies for mitigating the consequences of a reactor accident. This is a highly controversial subject, and as yet no federal agency has been willing to undertake the task of making potassium iodide available to people downwind of nuclear power plants, or at least to have distribution plans in a state of readiness. (Further information on this subject may be found in the article by von Hippel, 1982.)

The accident at Unit 2 of the Three Mile Island nuclear power plant on 28 March 1979 resulted in the release of 1.4 million to 13 million curies of the radioactive noble gases krypton and xenon and of only about 13 to 17 curies of radioactive iodine. This is an insignificant exposure to the population. The very small release of iodine from the Three Mile Island accident demonstrated that the containment vessel remained sealed, except for the release of small quantities of liquids and gases through the auxiliary building. The fission products were released from the fuel at very high temperatures, perhaps 2000°C, but when they moved to cooler areas they came in contact with coolant water and nearly all the iodine dissolved in the water. This greatly reduced the amount of iodine that remained airborne and leaked to the environment.

The chemistry of iodine in the reactor core and containment building is quite complex and not thoroughly understood. Volatile molecular species of iodine, such as CH_3I, I_2, and HOI are produced. On 2 April 1979 the iodine compounds that collected on the Three Mile Island auxiliary building exhaust fan filters were 28% CH_3I, 19% HOI, 26% I_2, and 27% particulates. According to Mynatt (1982), the largest iodine release resulted from small leaks (a few cubic meters per hour) from various seals in the containment system. This experience demonstrates that experts have overestimated the quantities of iodine that will be released from a reactor accident. On the other hand, a plutonium production reactor at the Savannah River plant, Department of Energy, released 153 curies to the atmosphere in 1961 during a brief malfunction incident.

NUCLEAR REACTOR INCIDENTS

Accidents do occur to nuclear reactors and will continue to occur as the world accumulates more reactors and as they age. This chapter would not be complete without some discussion of the history of reactor accidents. I will not attempt to include all accidents, large and small, but only some of the more notable.

There is an enormous amount of success in the nuclear power industry. However, it is the reactor accidents that attract public concern. In the United States, as of 27 September 1984, there were 85 nuclear power reactors in commercial operation and 46 were under construction or on order. A total of 22 countries have operable nuclear power plants and 14 more countries have reactors under construction. Worldwide, approximately 242 nuclear plants were in operation at that time and another 291 were under construction or on order. The very few nuclear reactor accidents that occurred out of this total number of reactors is impressive and the fact that no human lives have been lost from a single reactor is strong evidence of the inherent care taken with their operation.

Enrico Fermi I

Under the leadership of the Detroit Edison Company and with the financial support of the electric utility industry, the first U.S. experimental fast breeder reactor, a LMFBR, using liquid sodium as a coolant went CRITICAL on 23 August 1963 near Detroit, Michigan. Testing was underway and was to continue for six months before bringing the unit up to full power, about 200 MW (thermal). Three years of tests were undertaken and a variety of operational problems resolved. The reactor had operated at one-half its rated core power level for over 55 hours from August 5 to 7, 1966, and it had produced 1 million kilowatt-hours of electric power. On 5 October 1966 two small pieces of sheet metal broke from their fasteners as a result of the constant vibration from the sodium stream, and one of them obstructed the flow of the coolant to several of the fuel elements. The temperature in this area rose abruptly, and there was partial melting of several of the fuel elements. Although significant reactor damage took place, no radioactivity was released from the containment vessel, and the danger to the public was minimal. Several changes in government policy occurred soon thereafter. The U.S. government decided to support breeder development within the federal laboratories and not with private industry. The Enrico Fermi plant then remained shut down and was subsequently decomissioned.

Three Mile Island

The most serious nuclear reactor accident to date in the United States was that of the Three Mile Island Unit 2 on 28 March 1979. A relatively small operational problem was compounded through human error and misjudgment into a near catastrophe. The result was a partial meltdown of the core and the release of some radioactivity into the environment. The whole-body dose of α-radiation downwind from the plant was 0.004% of the annual dose from natural sources. (See the section in this chapter, Iodine Isotope Release.) The impact on public attitudes toward nuclear power has been enormous, with the likelihood that public acceptance of nuclear power may be delayed at least two decades. The cost to the electric power industry was enormous because of the elaboration of safety requirements, new design criteria, and extensive constructive delays for new nuclear reactors. Many electric utilities have either canceled plans or put construction on hold for an indefinite period. It is entirely possible that, as a result of the great slowdown for nuclear power plant construction, the electric utilities of the United States will not be able to supply sufficient electric power to meet demand in 10 to 20 years.

The Three Mile Island accident started when a pump in the secondary steam loop stopped operating. This pump pumps water condensate back as coolant water for the core. By a series of other inadvertent events, a small primary release valve in the primary coolant system stuck open, causing a loss of coolant and slow depressurization of the reactor. Then, because of insufficient coolant, temperatures began to rise. Once core damage began, it progressed and was arrested only when coolant was restored to the system. Misoperation and lack of proper diagnosis and corrective action over a period of hours led to continued loss of coolant and the subsequent severe core damage (Mynatt, 1982). The reactor was operating at 98% of its rated power when the accident began. (See Duderstadt and Kikuchi, 1979 for a more detailed description.)

Robert E. Ginna plant

On the morning of 25 January 1982, there was a radioactive steam leak from the Rochester Gas and Electric Company Ginna nuclear plant in New York State. One of the thousands of small pipes carrying hot, radioactive water through the steam generator burst. Small pipes of this kind are subject to corrosion, denting, and pitting. After the accident, the Nuclear Regulatory Commission inspection team found metal debris in the steam generator system. It appeared that a workman several years earlier had allowed a small metal plate to fall into the bottom of one of the steam generators. The metal plate rattling around in the boiling water may have damaged the pipe. In any event, it seems there had been repeated problems with debris in the small steam pipes of several nuclear plants. Nevertheless, despite several small operational problems, the operators of the Ginna nuclear power plant responded quickly and brought the plant under control within four hours of the first sign of a leak.

According to Mynatt (1982), there have been several other small incidents involving anomalous operations at about half a dozen other nuclear plants in the United States since the Three Mile Island accident. There have been no serious releases of radioactivity, and no one has been endangered.

Tsuruga, Japan spill

The Japanese have 22 operating reactors and plans for 35 more to be built by 1990. Japan must import all of its energy in the form of fossil fuel, and therefore nuclear power is of great importance to them. On 7 March 1981 a worker in the waste treatment part of the Tsuruga nuclear plant was flushing some pipes and neglected to shut off the intake valve. A holding tank for radioactive water overflowed, covering the floor of one room and flowing down a corridor. A small portion of the radioactive water leaked into an adjacent laundry room, seeped through cracks in the floor, then into a storm sewer line from where it flowed into Urazoko Bay. It is estimated that about 15 tons of water spilled, having a radioactivity of some tens of milli-curies per ton, of which a very small portion got into the bay.

The problem with this accident was not only that it was not discovered by the utility until the next morning, but the accident was deliberately kept secret from the Japanese public for a couple of weeks. Then it turned out there had been four other spills at the plant during the previous five years and none of them had been revealed to the public. This was a classic case of human error, poor judgment, and bad administration. The Japanese people have been assured by the government that this sort of thing will never happen again. However, never is a long time and we all know that human frailties are such that it will almost certainly happen again, somewhere, sometime.

Nuclear reactor accidents are going to occur. The important thing is that all of the elements designed to protect the public, both physical ones such as containment and sociological ones such as regulation and management, be as secure as possible. There is absolutely no doubt that nuclear power plants must be, not only carefully designed, but also thoroughly monitored and regulated.

14

Alternative Energy

A Darrieus type of wind generator. This type of machine has a high power output and is competitive with the propeller type of machine. Courtesy of Sandia National Laboratories.

INTRODUCTION

The people of the world will need to use every conceivable source of energy to sustain a burgeoning population living on a planet of limited space and resources. Fossil fuels are clearly limited resources, and serious environmental and ecological effects are associated with their use. Nuclear energy, thought by some people to be the salvation of the world's energy needs, is inhibited in development by public apprehension concerning its safety. However, there are several renewable energy sources in addition to direct solar and biomass that may, in the future, contribute a very important mix to the world's energy supply. These alternative energy sources include ocean temperature differences, ocean currents, tides and waves, wind, geothermal, and hydro. These energy sources are generally dispersed, with the exception of geothermal, and have fewer associated environmental, ecological, health, and safety problems than do fossil or nuclear fuels. They are best utilized for generating electric power or for manufacturing energy-intensive products such as hydrogen gas or ammonia. Hydrogen gas may be conveniently transmitted through pipelines and burned without any carbon dioxide, carbon monoxide, or sulfur dioxide emissions. Hydrogen may also be used to energize fuel cells. Ammonia is in great demand for fertilizer and its production by alternative means to use as natural gas could save enormous quantities of petroleum.

Ocean thermal, ocean currents, tides and waves, wind, hydro, and geothermal power are self-renewing energy resources that may be tapped to various degrees for the extraction of power. In each case, nature provides a higher concentration of these resources at certain sites than at others. It is at these sites that extraction of some of the energy is most practical. Only certain estuaries possess the highest tides; certain coasts, islands, or mountain passes the highest winds; steep mountain valleys the greatest hydro potential; and hot magma areas near the surface the greatest geothermal possibilities. The equatorial and subequatorial oceans have the greatest temperature difference between surface and deep water. Great ocean currents sweep the coastlines of the world. Even the swiftest of these—those currents in equatorial regions along the western boundaries of the oceans—are of very low energy densities. It is unlikely that ocean currents will be used to generate much power. Tidal power is the most widely distributed of these, but it is not generally located at useful sites.

The total global hydrologic runoff energy is about 9×10^{12} W, but that which can be usefully collected is only about 30% of this, or 2.7×10^{12} W. The total global tidal power is about 3×10^{12} W, and of that only about 2% can be usefully collected, or 6×10^{10} W. The total global geothermal flux is about 2.7×10^{13} W, but only about 0.5% can be tapped. This amounts to 1.3×10^{11} W. Wind power on a global basis amounts to 1.3×10^{15} W, but only about 10% can be extracted for use, or 1.3×10^{14} W. [The preceding numbers, taken from Gustavson (1979), are summarized in Table 1, where they are also converted into quads per year.] Ocean thermal gradients have a total global potential power of 5×10^{13} W, but when the Carnot cycle efficiency is taken into consideration, something less than 2×10^{12} W can be extracted (Isaacs and Schmitt, 1980). Ocean currents have a total global potential of 10^{11} W, but only 5×10^{10} W are extractable. Ocean waves have a total global power of 7×10^{13} W, of which only 2.7×10^{12} W is potentially usable. Clearly, wind has the greatest power potential of all of these sources.

According to Isaacs and Schmitt (1980), any of these alternative forms of energy may be used for purposes other than generating

TABLE 1. Global power potentials of renewable resources

Resource	Total watts	Useful power watts	Potential quads/yr
Wind	1.3×10^{15}	1.3×10^{14}	3900
Hydro	9×10^{12}	2.9×10^{12}	86
Waves	7×10^{13}	2.7×10^{12}	80
Ocean thermal	5×10^{13}	2×10^{12}	59
Geothermal	2.7×10^{13}	1.3×10^{11}	4
Tidal	3×10^{12}	6×10^{10}	1.9
Currents	1×10^{11}	5×10^{10}	1.6

electricity or for manufacturing energy-intensive products. For example, thermal-gradient power can produce fresh water or air conditioning as a by-product; tidal flow can dredge harbors; waves and wind can be used for propulsion; and salinity gradients can desalinate brackish water. The salinity gradient is a possible power source only recently recognized, but its utilization is extremely difficult and may not become practical for many decades.

If large icebergs are transported from the Antarctic Ocean to the arid subequatorial regions to supply fresh water, they might be used to operate an ocean thermal energy conversion (OTEC) plant or be used as a heat sink for giant fossil fuel or nuclear electric generating plants; the heat would melt the ice and provide the required fresh water.

OCEAN THERMAL ENERGY CONVERSION

The tropical oceans are continually heated by the sun and have stored in their surface layers an enormous quantity of energy. However, at a depth of 1000 m or more, the tropical seawater is very cold. It might be possible to utilize this source of energy if a heat engine could be operated across this temperature differential. It might be possible

to do so if the proper working fluid can be found for the closed-cycle engine required to do the job. To understand how this may be accomplished, we will review briefly how a steam electric-generating plant works.

In a conventional electric power-generating plant, energy released from burning coal, oil, or gas converts water to high-pressure steam. The steam then expands against the blades of a turbine, exerting a pressure on them and causing the turbine to turn. The turbine mechanically rotates an electric generator. The expanding steam is converted to water at low temperature in the condenser of the power plant. The temperature difference between the steam and the condensed water may be 600°C or more. Water is known as the working fluid in a steam power plant. When a working fluid that has a low boiling point is used, temperature differences like those found in the tropical oceans can be used to expand and condense the fluid. For example, when ammonia is used, the temperature difference may be as low as 20°C, rather than hundreds of degrees.

The temperature of the warm, solar-heated surface waters of the tropical oceans ranges from 24°C to 30°C and the temperature of the cold, deep waters ranges from 4°C to 10°C. Temperature differences of 24°C are easily found in some tropical oceans, and differences of 20 to 22°C are commonplace (Figure 1). With such temperature differences, an ideal heat engine would have a thermodynamic efficiency of 7.9%, but in practice a heat engine would achieve much less, perhaps about 2.5%. Such a meager performance rating becomes useful when massive amounts of seawater are pumped through such an engine. The process of generating electric power from this temperature difference is called OCEAN THERMAL ENERGY CONVERSION or OTEC.

The concept

A simplified drawing of an OTEC plant is shown in Figure 2. Warm seawater near the surface is drawn through a heat exchanger containing liquid ammonia and the heat causes the ammonia to boil. The ammonia vapor turns a turbine, and the turbine drives an electric generator. The ammonia vapor is condensed in a second heat exchanger by cold seawater pumped from a depth of about 1000 m. The liquid ammonia is then pumped back to the evaporator, thereby completing the cycle. About one-third of the energy generated will be used for operating the pumps and auxiliary equipment and two-thirds will be available as net power output.

An OTEC system can deliver the electrical energy produced by cable to land areas close by; or the electrical energy can be converted to an energy-intensive product produced within the OTEC plant and shipped elsewhere. One such product is ammonia (NH_3), which combines nitrogen from the air with hydrogen produced by the electrolysis of seawater. It is a coincidence that ammonia is also the working fluid used in the OTEC plant. Ammonia is an excellent choice for an energy-intensive material because it can replace the ammonia now made from natural gas for fertilizer, serve directly as a synthetic fuel, and be used in fuel cells. Ammonia can be readily transported and stored at room temperature in liquid form. It can be decomposed into nitrogen and hydrogen and used in a fuel cell at high system efficiency to generate electric power at any time and place as needed; 635 billion ft^3 (18×10^9 m^3) of natural gas per year is now used for ammonia manufacturing in the United States alone.

OTEC plants are attractive because they use the sun as an energy source and the ocean as a heat sink. Because the ocean stores heat from the sun in its surface waters, an OTEC plant can operate 24 hours a day and provide baseload capability for electric utilities—something most solar power electrical generating units cannot do. A fossil fuel

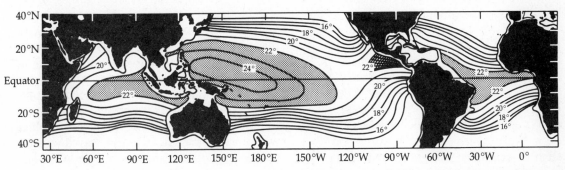

FIGURE 1. Mean annual temperature differences (°C) between the surface water and water at a depth of 1000 m in the ocean.

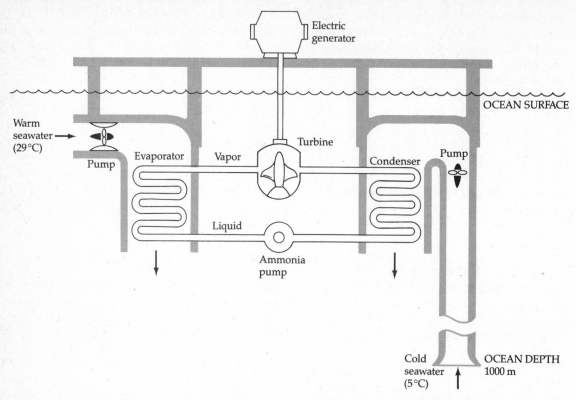

FIGURE 2. Schematic drawing of an Ocean Thermal Energy Conversion (OTEC) system that uses ammonia as the working fluid. (Courtesy of Solar Energy Research Institute, Golden, Colorado.)

or nuclear energy plant must be built and the fuel mined, refined, and shipped to the site to be burned. When all these expenditures are added up, a fossil fuel plant burns 2632 W of power for each 1000 W of electrical energy produced, and a nuclear plant burns 3000 W. For an OTEC plant it is estimated that 1700 W of power must be expended for each 1000 W of power output. In addition, the OTEC plant has a perpetual source of energy—the sun.

When an OTEC plant converts its electrical power into an energy-intensive product such as ammonia, it need not be rigidly anchored because it will not have a power cable to shore. In this case it can operate as a grazer, moving slowly through the tropical ocean, always seeking the warmest surface

water. Figure 3 shows an artist's drawing of a major OTEC plant. This particular plant has a vertical, semisubmerged structure comparable in size to offshore oil platforms. A plant of this particular design would be held in place by a single mooring line and a heavy anchor. Its electrical power would be delivered to shore by cable. Living quarters for two crews of 30 people each would occupy the uppermost part of the structure.

The fuel for the OTEC plant is supplied without cost. So, despite the very low thermal efficiency at which it operates it is possible to generate electric power productively by having a massive machine that moves great quantities of water. The cold water pipe would have a diameter of 30 m and go straight down 1000 m into the ocean. An OTEC plant producing 250 MWe of power would require a sustained water flow through the pipes of about 1000 m³ per second, or 264,118 gallons per second. This is comparable to two-thirds of the average flow of the Missouri River at Omaha, Nebraska. This is an enormous amount of water, but it can be handled by modern technology. A lesser amount of power production will require

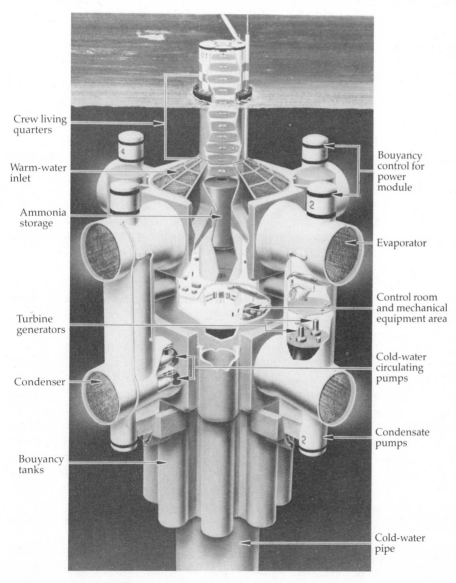

Crew living quarters

Warm-water inlet

Ammonia storage

Turbine generators

Condenser

Bouyancy tanks

Bouyancy control for power module

Evaporator

Control room and mechanical equipment area

Cold-water circulating pumps

Condensate pumps

Cold-water pipe

FIGURE 3. An artist's conception of a vertical semisubmerged OTEC plant. Living quarters for two crews of 30 people would occupy the upper-most part of the structure. (Courtesy of Lockheed Missiles and Space Co., Sunnyvale, California.)

comparably less water. In one plan, it would require 20 pumps, as large as anything available today, to draw in the cold water and another 20 pumps for the warm water. Other designs call for larger but fewer pumps.

Construction and operational problems

The pipes and heat exchangers are the major cost items. The heat exchangers alone would have surface areas of 700,000 m^2. If they are built of inexpensive aluminum, corrosion could be a problem. If built of noncorrosive titanium, they not only would be costly, but would consume the entire national production of titanium at the time of construction. The heat exchangers are so large that they would need to be manufactured at the shipping yard where the floating platform was assembled. The construction period, once the final design is finalized, is estimated at two years, comparable with the construction schedule for an offshore oil rig. The main platform of an OTEC plant should be good for a 40-year lifetime. Aluminum heat exchangers would probably require new tubing in 12 to 15 years, but titanium heat exchangers would be much more durable.

There is a tendency for all marine structures to accumulate slime on the metal surfaces. If slime, made up of algae and other organisms, accumulates to a thickness of ¼ mm, the performance of the power plant would be reduced by 60%. The slime would need to be cleaned off by using brushes or water jets. However, the surface areas involved are so enormous that it is hard to conceive of keeping all these surfaces clean. The necessary large-scale experiments are yet to be done; however, Mini-OTEC, a pilot plant of 50 kW tested off the coast of Hawaii in 1979, had no accumulation of scum on the metal surfaces of the heat exchangers. This was accomplished by adding a small amount of chlorine to the seawater flowing through

the system. The brief operation of Mini-OTEC was no real test of the potential biofouling that might occur within a fullscale OTEC plant. Encouraging, however, is the fact that tests near Panama City, Florida indicate that extraordinarily small amounts of chlorine are needed (only 0.5 ppm per 15 minutes per day) to keep the heat exchangers clean.

The very long (1000 m) cold water pipe must be able to withstand enormous stresses in the ocean. It must sustain the stresses produced by ocean currents and by hurricanes. Lightweight concrete and fiberglass are among the materials being considered for the cold water pipe, as is reinforced plastic.

History of OTEC development

The idea of harvesting energy from the warm surface waters of the oceans was first suggested in 1881 by Frenchman Jacques Arsène d'Arsonval. His idea was to operate a heat engine off of the temperature difference between the Seine River and a mountain glacier. His student, Georges Claude, actually operated a prototype ocean power plant off the coast of Cuba in 1950. He used a primitive power plant in which seawater itself was the working fluid for the turbines, and with this open-cycle system, his engine generated 22 kW of power. His cold water pipe was destroyed by a hurricane and his financial resources were exhausted, so he was unable to continue this development. His major technical problem was that seawater has too low a vapor pressure at ambient temperatures to generate sufficient pressure at the turbine to generate much power.

OTEC was revived as a viable concept in the 1970s following the 1973 Middle East oil boycott. A Mini-OTEC plant mounted on a barge was designed and operated off the coast of Hawaii in 1979 as a joint effort funded by Lockheed Missiles and Space

Company, the Dillingham Corporation, Alfa-Laval Inc. of Sweden, and the state of Hawaii—with no federal funding. Mini-OTEC generated about 50 kW, 40 kW of which were used to power the plant. The plant ran as well or better than its designers expected. None of the problems raised by OTEC skeptics actually occurred. This was a miniature version of what is envisioned as a full-scale OTEC plant. Warm surface water evaporated ammonia in a closed-cycle system. The vapor expanded through a turbine and was condensed by cold seawater pumped through a pipe from a depth of 700 m. Although Mini-OTEC was built hurriedly with off-the-shelf components, it did generate net power, although inefficiently. A carefully designed power plant built of specially designed components should do much better, leaving 65 to 70% of the power generated for utility use.

Figure 4 shows a schematic drawing of an OTEC plant ship that uses the electricity gen-

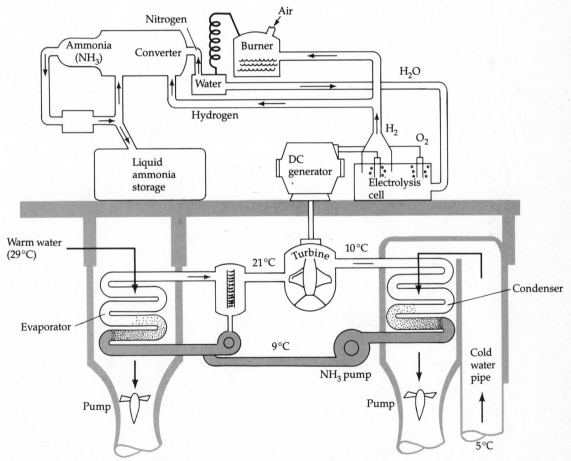

FIGURE 4. Schematic drawing of an OTEC plant ship that uses the electricity it generates to produce ammonia from nitrogen and water. (Courtesy of the Johns Hopkins University Applied Physics Laboratory.)

erated to manufacture ammonia from nitrogen and seawater. The water is broken into H_2 and O_2 through electrolysis, and the H_2 is combined with N_2 from the atmosphere to produce NH_4. The ammonia is then bottled and shipped to shore by tankers. An artist's drawing of an OTEC plant ship and tanker is shown in Figure 5.

OTEC systems are likely to be slow to develop because of their enormous cost. Nevertheless, in a world faced with critical fossil fuel energy shortages within the next century, the pressure on society to develop OTEC power plants will be ever increasing.

McNichols et al. (1979) have proposed using the temperature differences existing in hydroelectric reservoir thermoclines to generate electricity. They point out that the energy stored in reservoir thermoclines is enormous and, in fact, far exceeds the gravitational hydroenergy of the surface water that is now used for hydroelectric power production. (Further information concerning OTEC plants may be found in Avery and Dugger, 1980; Hartline, 1980; Marland, 1978; and Metz, 1977b.)

Ecological and environmental impacts

OTEC power plants will entrain and discharge enormous quantities of seawater. The plant will displace about 4 m^3 of water per second per MWe output both from the surface layer and from the deep ocean and discharge them at some intermediate depth between 100 and 200 m. This massive flow may disturb the thermal structure of the ocean near the plant, change salinity gradients, and change the amounts of dissolved gases, nutrients, carbonates, and turbidity. These changes could have beneficial or adverse effects on organisms in the local ecosystem.

The enrichment of the near-surface waters with the nutrient-rich cold water brought up from a depth of 1000 m is of particular significance. Natural upwellings of cold water from great depths in the ocean produce sites that are enormously rich in marine life. One of the well-known natural upwelling sites is where the Humboldt current off Peru enriches the surface waters. The productivity there is so high that almost one-fifth of the world's fish harvest comes from this region. It would be possible to use the cold water effluent from an OTEC plant for the cultivation of algae, crustaceans, and shellfish. For example, at an experimental station on the north coast of St. Croix, U.S. Virgin Islands, researchers have successfully cultivated shellfish with water pumped from a depth of 870 m to the surface. In the nutrient-rich water, unicellular algae grow to a density 27 times greater than the density in surface water and are in turn consumed by filter-feeding shellfish such as clams, oysters, and scallops.

Algal blooms may occur naturally in these enriched surface waters. However, if the algal bloom includes certain dinoflagellates, there may be problems. Shellfish consume the dinoflagellates and sometimes when humans eat the shellfish, they become very ill. This is known as a red tide condition, a situation often associated with upwelling off the coast of Florida. OTEC advocates hope that, by designing the OTEC plant to discharge its water below the photic zone (the region in the surface waters where photosynthesizing organisms live), the surface waters will not be enriched. Furthermore, the fishes living below the photic zone do not feed on these nutrients. If nutrient-rich water is discharged anywhere near the surface water intake valves, it could cause biofouling to develop inside the pipes.

Marine biota may be impinged on the screens covering the warm and cold water intakes of an OTEC plant. Small fishes and crustaceans may be entrained through the system, where they will experience rapid changes of

FIGURE 5. An artist's conception of an OTEC plant ship for producing ammonia and a tanker for conveying the ammonia to shore. (Courtesy of the Johns Hopkins University Applied Physics Laboratory.)

temperature, salinity, pressure, levels of turbidity, and dissolved oxygen. A major change occurring in the cold water pipe is the depressurization of up to 10^7 pascals in water coming from a depth of 1000 m to the surface.

Sea surface temperatures in the vicinity of an OTEC plant could be lowered by the discharge of effluent from the cold water pipe. This will have minimal effects on organisms and negligible effects on climate. The pumping of large volumes of cool, deep ocean water to the surface will release dissolved gases such as carbon dioxide, oxygen, and nitrogen to the atmosphere.

Biocides, such as chlorine, used to prevent biofouling of the pipes and heat exchanger surfaces may be irritating or toxic to organisms. If ammonia is the working fluid and it leaks out, there could be serious consequences to the ocean ecosystem nearby.

On balance, it is likely that large OTEC power plant systems will have less serious environmental and ecological impacts than any other large-scale energy conversion system, including nuclear power.

TIDAL AND WAVE POWER

Tidal power

All flowing water has a potential for turning a turbine and generating electricity, whether the water is in the open ocean, an estuary, or a river. Water flowing out of a reservoir is equivalent to water in a river coming out of a watershed. Tidal power utilizes the kinetic energy of tides generated by the gravitational forces of the sun and the moon on the oceans. The tides along most coastlines run about a meter high, but in constricted areas where they are amplified by a funneling action, they may run 10 meters or more. It is in these constricted areas that the most effective tidal power plants may be located. A dam or sluice gate is placed across an ocean bay or estuary. An incoming tide fills up the enclosed basin while passing through a row of hydraulic turbines. After the basin has filled with water, the gates are closed and the turbines are shut down. Then the turbine blades are reversed and the gates opened again to let the water surge out. The intermittent nature of the flow will cause variations in electric power output—a somewhat undesirable feature of this system.

The maximum potential for electric power generated by tidal action is estimated at about 60 billion watts or about 550 billion kilowatt-hours per year. Although these numbers seem impressive, they are somewhat misleading. The number of good sites in the world for the development of tidal power are extremely limited. They include the San Jose Gulf in Argentina; Severn estuary in England; several coastal sites in France; four locations in the Soviet Union; Cook Inlet in Alaska; Puget Sound, Washington; the Bay of Fundy in Nova Scotia and the adjacent Passamaquoddy Bay in Maine. As much as 29,000 megawatts could be generated from the massive tides of the Bay of Fundy; however, the three best sites there would generate a total of only about 9000 megawatts. Nevertheless, this could supply between 20 and 27% of the electricity consumed by New England and the Canadian Maritime Provinces.

A tidal power plant is in operation in the Rance River estuary at the head of Saint Malo Bay on the coast of Brittany in northwestern France. An 800-m dike and turbines generate 240 MWe out of an estimated 1500-MWe potential power output for that estuary. The average tidal range is 8.4 m. The dam encloses an area of 22 km^2. The power plant comprises 24 units, each of 10,000-kW capacity. The turbines have adjustable blades, thereby permitting operation during both filling and emptying of the basin.

Another tidal electric power project in operation is a Russian experimental station in the Kislaya Inlet on the coast of the Barents Sea northwest of Murmansk. Here there is a 400-kWe turbine of French manufacture in use.

Although tidal power may provide only a fraction of one percent of the world's power needs, it has the advantage that it produces no noxious wastes, produces a minimum of ecological disturbances, and uses no non-renewable energy resource.

Wave power

Rarely is the ocean without waves, within which a lot of energy is contained. The ocean surface is a vast collector of wind energy; it is a source steadier than the wind itself for waves convey wind energy over immense distances with only moderate dissipation. Waves do not contain as much energy as is sometimes attributed to them. Wave-energy fluxes in the open sea or against coasts may vary from a few watts to kilowatts per meter. They are smallest in summer and greatest in winter, mainly in the zones of the prevailing westerlies and the trade winds.

Wave motion consists of both vertical and horizontal movement of the fluid. Individual particles of water undergo an almost circular motion, moving up as the crest approaches, forward at the crest, down as it recedes, and backward in the trough. Generally, the scheme for extracting power from waves takes advantage of the vertical motion by using the movement of a float versus an anchor. The vertical displacement is used by the Japanese to energize navigational buoys; the Japanese are also constructing a larger power plant using this principle. In contrast to this, S.H. Salter of the United Kingdom proposes generating electric power from the horizontal component of wave motion. His mechanical system, known as Salter's Ducks, uses two sets of vertical paddles arranged so that one set receives the crests of the incoming waves while the other set is in the troughs. The to-and-fro motion of the two sets of paddles is converted to electricity by a turbine. A huge wave energy machine, the size of a super-tanker, might convert only about 50 MWe of power. A comparable-sized OTEC plant would deliver about four times this amount of power. Salter has tested a 1/10-scale field unit on Loch Ness. There is a serious problem in converting the rocking motion of the paddles into high-speed, unidirectional rotation of an electrical generator.

A very different idea for extracting energy from ocean waves is the Dam-Atoll device illustrated in Figure 6. This device does not respond to the up-and-down wave motion but utilizes wave refraction to alter the wave direction and concentrate the wave energy into a vortex that drives a turbine. The vortex in the central core of the device serves as a fluid flywheel. The flywheel turns a turbine generator. A key element in the Dam-Atoll device is the large dome. The dome must be about 100 m in diameter, comparable in size to the roof covering a football stadium. As a wave travels over the dome, the velocity of the wave slows as the water depth decreases. This causes the wave to bend, or refract, so that it is focused into the center of the dome. Vanes then help to guide the flowing water into the vertical cylinder and turbine. The Polynesian people apparently understood this principle well because they knew there was no safe lee side on a volcanic atoll. An ocean wave breaking over an atoll is always refracted around to the lee side, where it breaks on the shore, as well as breaking on the windward side. The Lockheed Missiles and Space Company has built and tested a 1/100-scale model to demonstrate proof of the concept. A full-scale unit, 80 m in diameter, is expected to generate about 1 MWe of power output. Much developmental research is still required for this device. Ecological impact should be minimal for the Dam-Atoll ma-

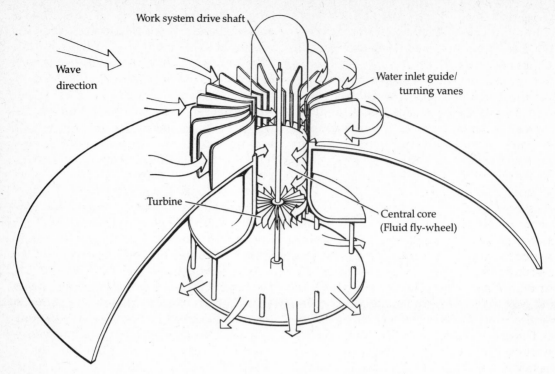

Work system drive shaft

Wave direction

Water inlet guide/ turning vanes

Turbine

Central core (Fluid fly-wheel)

FIGURE 6. A Dam-Atoll system for generating electric power. One 80-m diameter Dam-Atoll unit could produce 1 to 1.5 MWe power in waves with a 7- to 10-second average period. This amount of power would be enough to supply the electrical needs of about 480 homes. (Courtesy of Lockheed Missiles and Space Co., Inc., Sunnyvale, California.)

chine. Whereas ocean thermal energy conversion looks quite practical, and tidal power systems are in operation, wave power seems to be a long way from practical utilization.

WIND POWER

There is nothing new with using the wind as a source of power. Human utilization of wind power is as old as civilization. There is a picture of a windmill on a Chinese vase dating from the third millenium B.C. Early windmills in Persia and China were horizontal wheels carrying sails, rotating around a vertical shaft. The Persian use of windmills began sometime around the seventh century A.D. The first account of windmills in Europe was in the twelfth century, and it was about this time that a horizontal shaft was used. Wind powered the very earliest of sailing ships. The Egyptians used sailing ships in 2800 B.C. The mechanical use of wind for driving ships, pumping water, grinding grain, turning the machines of factories, and doing a large variety of other tasks advanced steadily over the centuries. Finally, with the invention of the electric generator in the nineteenth century, wind was used to generate electricity. The windmill is often credited with sparking the industrial revolution.

Using the wind

Wind is a truly ubiquitous, albeit a variable, source of power. Wind, of course, is a form of solar power. The Earth's atmosphere is a gigantic thermodynamic machine driven by the sun. This suggests that wind as a power source has the same inherent disadvantage of all other solar energy-derived sources; namely, it is a dispersed source of power. This dispersal of energy severely limits its applications and suggests that to harness any considerable amounts of power from this source massive wind machines must be used. Rather than trying to generate an enormous amount of power from a single site, it is more practical to use wind machines to pump water and irrigate fields. This was exactly what the pioneer settlers did in America. They used windmills to grind grain, pump water, and irrigate fields. Windmills were seen on every ranch, farm, or homestead. Approximately 6.5 million windmills were made in the United States between 1880 and 1930. Most of these windmills were used for pumping water and running sawmills, but some were turning small turbines to generate electricity. It was taken for granted that certain tasks were performed by windmills. Gradually, however, rural electrification spread across the land and windmills became a thing of the past—abandoned and broken, they littered the landscape. Suddenly in the early 1970s, there was an international energy crisis and the people of America looked around to find themselves without alternative energy sources. The windmills were gone. A new idea occurred: Why not generate power using wind machines?

Engineers had to reinvent the windmill and then carry the development of wind power technology far beyond anything previously devised. This is not the place to describe the technology of modern wind machines, except to indicate the levels of power production achieved. (See Inglis, 1978 for a thorough description of wind machines.)

Most wind turbines built during the 1960s and 1970s used a horizontal axis, and most had two blades, although some very large machines had three and even four blades. Most wind turbines produced alternating current and successfully synchronized its input to a commercial electrical grid. The development of wind turbines has taken place mainly in France, Denmark, Germany, Great Britain, and the United States. Many of these machines had blades of constant cross section and were not twisted. However, considerable improvement in efficiency can be achieved by twisting the blade so that it is edgewise to the wind near the hub, where the speed during rotation is low, and flat to the wind near the tip, where the speed during rotation is high.

Power in the wind

The winds of the world amount to an average power of 2.5 Wm^{-2} of Earth's surface. At many windy sites throughout the world, the flow of wind through a vertical square meter, at a height of about 25 m, may reach a yearly average of 500 Wm^{-2}. An efficient wind turbine might convert to electricity about 175 Wm^{-2} in the area swept by its propeller. Theoretical considerations show that the ideal windmill could extract only about 59% of the kinetic energy of the wind passing through the area of its blades. Good aerodynamic design will achieve less than 70% of the theoretical maximum. Enthusiasts for wind power visualize large arrays of wind machines strung out all over the landscape. They then calculate the total energy that might be extracted from the wind and get truly impressive numbers. Gustavson (1979) shows that if only 10% of the wind energy available can be extracted, 2×10^{12} W can be generated in the

United States alone. According to Gustavson, the total amount of global wind energy that might be extracted is 1.3×10^{14} W. Converting these energy amounts to quads—to compare them with the total current energy demand by the people of the world (250 quads) and by the United States (80 quads), we get 3900 quads (global) and 60 quads (U.S.) of wind energy potentially available. For many reasons I believe these numbers are gross overestimates of the amount of wind energy ever likely to be extracted. For example, in the United States alone, 2 million 1-MW wind machines would need to dot the landscape in order to achieve the necessary energy extraction, far beyond any likelihood. Gustavson also comes to the same conclusion.

There are a number of limitations on the extraction of energy from the wind. These limitations include the practical size of wind machines, their density, friction losses in the rotating machinery, and efficiencies of conversion from rotational energy to electrical energy. Other limitations include the competition for space with housing, industry, agriculture, transportation, and other human endeavors; aesthetics; and environmental and ecological impacts.

Wind turbine sizes

Wind machines for generating electricity may usefully be grouped into three categories (Sorenson, 1981). The first are miniconverters, which have an average output of about 1 kWe and can be used to generate electricity to run small irrigation units, lighthouses, and ranger stations. The second are machines with an average output of 50 kWe or more; these machines have applications for rural villages in developing countries, farms, rural industries, and isolated individual homes. The potential market for this second category is about 10% of all stationary energy used in

rural areas of North America and Scandinavia. This would include energy for space heating, hot water generation, and pumping of irrigation water.

Finally, wind machines in the third category have average outputs that range from about 500 kWe to many megawatts; they supply power to an electrical power grid. In those parts of the world where elaborate power grids exist (grids now fed by fossil fuel or nuclear power plants), wind power could deliver 20 to 25% of the electricity demand. Sorenson (1981) writes that this would correspond to about 2000 terrawatt-hours (TWh) annually, or about 15% of the present world use of electricity. Among the industrialized nations, a wind–hydro combination could produce an additional 1200 TWh per year from wind. This assumes a pumped-storage type of system at prime wind sites such as seashores, the Great Plains, and mountain passes or ridges. Whether or not these goals are ever achieved remains to be seen. However, the potential for electric power production by wind is very substantial. As fossil fuel becomes scarce and prohibitively expensive and if nuclear energy is not desired, wind power may generate a greater share of the world's electrical power needs.

History of wind turbines

Many wind turbines are in operation today throughout the world; most of them are small. The history of large wind turbines is a relatively recent one. In Denmark engineers built a number of wind dynamos. Between about 1900 and World War II, these generated collectively up to 100 kWe of electricity and fed this into an electric power grid. The Soviet Union built a 100-kWe machine in the Crimea in 1931. The Germans used wind turbines to generate electricity in place of scarce fuel during World War II. The most impressive of all the early wind turbines was a dynamo built and operated in 1941 on a mountain in Vermont known as Grandpa's Knob. This wind turbine stood 33.5 m high and had blades 53.4 m from tip to tip. It fed 1.25 MWe into the commercial power grid of the Central Vermont Public Service Company for one month before a propeller broke from metal fatigue. Because World War II was in progress, replacement materials could not be obtained and the project had to be abandoned.

Several wind dynamos of a variety of designs were built and tested in England in the 1950s. These machines are about 100 kWe in size. Many machines of larger power output were designed but never built. In Denmark a 200-kWe wind generator was built in 1956 at Gedser on the Baltic coast. This machine was highly successful. It had three blades, each nearly 12 m long mounted on a 23-m tower of prestressed concrete. In Denmark the greatest winds are in the winter when electrical heating demand is the greatest. A 100-kWe wind dynamo was built near Stuttgart, West Germany in 1959 and fed electricity into a power grid until 1969. The French built and operated an 800-kWe wind dynamo at Nogent-le-Roi in northern France about 1960. It was operated for 10 years. The low cost of petroleum and the installation of nuclear power plants throughout many of these countries forced the abandonment of the development of wind-powered electrical generators.

Future of wind power

A study by an advisory committee to the U.S. Department of Energy in 1980 showed that wind power was the most promising source of alternative energy for the near and intermediate term. There is no doubt that wind power has the potential to produce prodigious amounts of energy. One of the problems is to build wind machines large enough to make the electrical power output worthwhile. Another problem is to generate electricity reliably despite uncertain and highly variable winds. Advantages are that wind power is continually regenerated in the atmosphere under the influence of solar heating and that wind is a clean power source. There is no waste product to dispose of, except during the manufacturing of the materials going into the wind machine.

In 1972 Professor William E. Heronemus of the University of Massachusetts proposed a gigantic scheme of hundreds of huge windmills strung out across the great plains of mid-America. Heronemus planned to mount 20 wind turbines on a tower 115 m tall. Originally, Heronemus visualized some 300,000 towers strung out from Texas to the Canadian border. Now, to conserve land, his plan calls for towers 240 m tall and straddling the highways at half-mile intervals, with the 20 turbines slung from cables strung between the towers. He visualizes 2000 windmills pumping an annual 1.5 trillion kilowatt-hours per year into the national power grids. This is nearly as much as the yearly amount of electricity generated in the United States today. Whether this ever comes about remains to be seen. However, people are taking the idea more seriously as other energy sources become more expensive and more precarious. Within a decade or so, we may see the development of very large arrays of wind turbines to generate electricity. (McCaull, 1973 gives a good description of the Heronemus proposal.)

On a much less grand scale, real progress is being made with the design and testing of modern wind turbines. Several large U.S. corporations are competing to design and sell wind machines for the production of electricity, and several major power companies are purchasing machines to feed electricity into their power grids. These machines range in size from about 50 kWe to 200 kWe. A 100-kWe experimental wind turbine is being tested by researchers of the National Aeronautics and Space Administration at Plum Brook, Sandusky, Ohio. The two blades on this machine are joined at the hub and are 38 m from tip-to-tip. The tower is approximately 30 m tall. Three wind turbines, each rated at 200 kWe peak power output, are operating at Clayton, New Mexico; Culebra Island, Puerto Rico; and Block Island, Rhode Island; and a fourth is to be installed at Oahu, Hawaii. Another wind turbine, rated at 2 MWe, is undergoing initial testing at a site near Boone, North Carolina. Plans have been underway for installing three wind machines, each producing 2.5 MWe, at a site on the Oregon–Washington border, with the output to be fed into the grid of the Bonneville power administration. These machines are to be the largest ever built, with a wingspan of 91 m.

Wind sites and properties

Wind speed varies greatly over the Earth's surface. Some sites are persistently windy, whereas others are usually without much wind. In Chapter 2 we described the winds that flow over the world as a result of temperature differences and planetary rotation. We described there the polar easterlies, prevailing westerlies of intermediate latitudes, and the northeast and southeast trade winds of the subtropical latitudes. Also discussed were the winds that blow along coastlines and result from temperature differences between sea and land. In fact, the wind is often most persistent along coastlines.

The Earth's surface is rough. It is made up of obstacles such as grasses, shrubs, trees, rocks, hills, and mountains. All of these obstacles interfere with the flow of air across the surface. Air is a fluid with viscosity. Although its viscosity is low compared with water, oil, or molasses, air is nevertheless viscous. It sticks to surfaces, and these produce a drag on it. Hence, air near the ground is

moving slowest; so wind speed increases with height. Wind speed increases rapidly in the first few meters above the ground and then reaches a steady-state value many meters above the surface. Hence, wind machines must be mounted on towers in order to get above the boundary layer, the transitional layer near the surface. As far as a wind machine is concerned, the power content of the wind is proportional to the cube of the wind speed. When the wind speed doubles, the power available to a wind machine increases eight times. A wind generator producing 1 kW in a 10-mph wind will produce 8 kW in a 20-mph wind. Increased height of the wind machine above the surface can result in an enormous increase in the power output.

Coastlines, mountain tops, and grasslands are generally windy places. Remote islands located in the persistent trade wind belts are also good wind sites. These are of particular significance because the people living on remote islands usually must import all of their energy in the form of oil transported over large distances. The three most favorable sites for large wind power installations are (1) mountaintops and ridges; (2) the western Great Plains; and (3) offshore. A wind site, to be useful, must be accessible and must have consumers located not too far away. It must also be located such that the aesthetics of wind machines are acceptable to the people of the area.

The distribution of average available wind power over the coterminous states of the United States is shown in Figure 7. There are three areas where wind power is particularly intense. Two of them are offshore, one off the New England coast and the other off the coast of the Pacific Northwest. The third region is the western Great Plains in southwestern Kansas and western Oklahoma and also in southern Wyoming. Fortunately, all of these high wind areas are where the installation of wind machines would cause relatively little conflict with existing human activities. In the western Great Plains, much of the land is used for cattle grazing and grain growing. Some of these areas would be far from large population centers and would require long transmission lines to the consumers. However, from southern Wyoming, the transmission lines could deliver power to the rapidly growing population centers just east of the continental divide in Colorado.

The other highly favorable location for the installation of wind machines is about 170 km east of Cape Cod, where power densities average 700 Wm^{-2}. (Compare this with an average power density of 400 Wm^{-2} in southern Wyoming.) The location of windmills on George's Bank just north and east of Cape Cod would provide electric power to the high-density human population along the northeastern coast of the United States. Sev-

eral thousand offshore windmills might be located there. They could be mounted on floating platforms or on tower structures that are rigidly set on the sea bottom.

Examples of modern wind turbines are shown in Figure 8.

Storage

The wind does not blow all the time. Thus, the power output of wind turbines is highly variable and, sometimes, unpredictable. It is possible to feed an electrical grid with the power output of wind turbines without the use of special wind-storage facilities. However, storage would make for a smoother, more dependable electric power system.

When we think of electrical storage, we automatically think of batteries, but this is not necessarily the only form of storage for the power output of wind turbines. In addi-

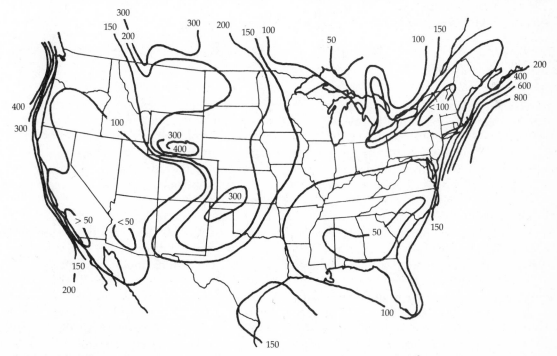

FIGURE 7. Annual average available wind power in the United States in W/m^2.

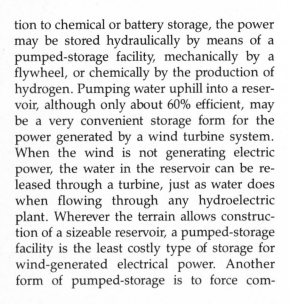

FIGURE 8. Two examples of modern wind machines for generating electric power.

tion to chemical or battery storage, the power may be stored hydraulically by means of a pumped-storage facility, mechanically by a flywheel, or chemically by the production of hydrogen. Pumping water uphill into a reservoir, although only about 60% efficient, may be a very convenient storage form for the power generated by a wind turbine system. When the wind is not generating electric power, the water in the reservoir can be released through a turbine, just as water does when flowing through any hydroelectric plant. Wherever the terrain allows construction of a sizeable reservoir, a pumped-storage facility is the least costly type of storage for wind-generated electrical power. Another form of pumped-storage is to force com-

pressed air into a gigantic underground reservoir and later release this air through a special kind of turbine. This technique is being used in Sweden. (See Chapter 12 for a discussion of the advantages and disadvantages of hydroelectric pumped-storage systems.)

A huge spinning flywheel may be used to store wind power. The rotational power of a wind generator turning at 120 rpm might be stepped up through a motor generator to turn a flywheel at 24,000 rpm. The rotor of such a flywheel would be made of heavy material, such as pressed wood or fiberglass, be about 5 m in diameter and weigh about 181,000 kg. Generally, flywheel storage is only useful for daily peak smoothing and not for the several-day storage needed when the wind is not blowing. However, some engineers believe that flywheel storage can be designed to provide power for as much as a week of windless days.

The Earth's most abundant element—hydrogen—can be produced by the electrolysis of water. The hydrogen gas generated can be stored underground and drawn upon as needed. The hydrogen gas can be burned directly and the hot vapor used to drive a turbine, or it can be used in a fuel cell. The main disadvantage of this cycle is its inefficiency; about half of the input energy is lost in the process. Hydrogen gas can also be put directly into existing gas distribution pipelines and burned as a high-grade fuel.

For small wind turbine systems, lead–acid batteries are the most dependable form of energy storage. They can go through thousands of cycles, charging and discharging, without serious deterioration. Batteries are expensive and may cost as much as the wind generator itself. However, research on improved battery design is progressing, and costs may be reduced. Batteries have been shown to store as much as 85% of the power delivered to them; but when operated over long periods of time without charging, there may be a loss of charge in the battery and efficiencies may fall to about 50%. This drop of efficiency will double the cost per kilowatt based on the annual energy produced by the wind generator.

Environmental impact

Important concerns about the environmental impacts of wind machines are safety, noise, and electrical interference, as well as visual and climatic impacts. Safety hazards are similar to those in the building industry, including the risk of falling from high buildings during construction and repairs or the failure of parts through fatigue or design. Turbine blades sometimes fail, but no serious accidents are known to have resulted from this. Large wind machines are generally located in sparsely populated areas and the risk of human accident is much reduced. More accidents are likely with the large number of small wind machines located within higher population densities.

The noise generated by a wind machine increases with the wind speed, and some of this noise is of infrasound at frequencies below the audible range. This infrasound may cause houses and other structures to vibrate. The infrasound is caused by the turbulence that is set up by the rotating blades and interacts with the tower structure. These low-frequency waves can be eliminated by careful design considerations. Measurements for a number of wind rotors show that audible noise diminishes rapidly with distance, and acceptable levels are reached less than ten rotor-diameters away from the turbines.

Wind turbines can create a generally broad signal that can interfere with television reception. The problem is caused by the rotor blades, which intercept the television beam, and by the tower structure, which scatters the beam. If television reception is required within the interference area of the wind generator, a cable connection should be provided to an antenna located beyond the interference area.

Rotating blades of wind machines will kill some birds, bats, and flying insects. Interference with the free movement of wind across the landscape may perturb precipitation or evaporation patterns. However, measurements indicate that no serious effects will occur, other than strictly localized ones that will happen within a couple of rotor-diameters of the machine. The presence of wind machines will change the aesthetics of the landscape. However, with careful attention to design, the towers and machines may be given an acceptable appearance.

GEOTHERMAL POWER

Geothermal energy is the natural heat of the Earth that is either conducted or convected to the surface through volcanoes and hot springs. Radioactive decay, as well as frictional forces from the movement of crustal plates and energy dissipation of solar and lunar tides, creates a hot interior of the Earth's molten center in excess of 1000°C. When we drill into the Earth, we find that the temperature of the rock strata increases 30°C/km and that, on the average worldwide, heat is flowing outward toward the surface at the rate of 0.063 Wm^{-2}. By comparison, the average annual amount of solar energy incident on the Earth's outer atmosphere is 340 Wm^{-2}. The average daily amount of solar energy at the Earth's surface is about 200 Wm^{-2} (averaged over a 24-hour period)—a value 3175 times larger than the geothermal flux. Hence, the average amount of geothermal energy flowing to the Earth's surface from inside is not very impressive when compared with the average solar flux incident from above.

Studies have been made to see if it is worthwhile passing water down a hole drilled into the warm interior of the Earth, there to be heated and returned to the surface where it would be used as steam to generate electricity. It turns out that it is not economically feasible to do so in an arbitrary area. However, there are special geologic situations where there are hot spots near the Earth's surface, revealed by volcanoes, geysers, or hot springs. The use of hot, dry rocks underground as an energy source to be extracted by injecting water and returning it as steam to the surface is discussed by Heiken et al. (1981). They also discuss lower temperature, dry geothermal sources.

The Earth's crust is made up of gigantic plates, of continental size, that move relative to one another and have gaps or fissures between them. The hottest material beneath the Earth's crust rises by convection in circulation patterns along the joints between these plates. Along uplifted areas (mountain chains), Earth's crust is thinner than elsewhere, and it is fractured. In these regions the heat flow is much greater than average. One such region is the San Andreas fault zone in California. Other regions are located in Italy, New Zealand, Iceland, Japan, and the Soviet Union, places where geothermal power has been produced.

When wells are drilled to tap a geothermal resource, one of several situations is encountered. In some cases, dry steam is released unaccompanied by liquid water. This was the situation at Larderello, Tuscany province, Italy, and is also the case for the geyser fields of northern California. In other cases, the geothermal wells encounter a reservoir of extremely hot water. When the superheated water comes up the well, it expands and partially changes to steam and emerges as a mixture of steam and water. These can be separated and the steam used to turn an electric generating turbine and the hot water used for heating. This is the situation in Iceland and New Zealand. In some locations there are hot rocks or magmas not in contact with underground water. At these locations holes can be drilled from the surface to convey water forced down the drill hole, where it contacts the hot rocks and becomes superheated. It then can be returned to the surface to use in a steam electric turbine.

Potential geothermal power

White (1965) has estimated that the amount of heat stored in the Earth within 3 km of the surface is 8×10^{22} J. Within 10 km of the surface it is 4×10^{22} J. To obtain an order of magnitude estimate for worldwide geothermal power, White assumes that about 1% of the heat stored can be converted into electrical energy. The available thermal energy within

10 km of the surface is 4×10^{20} J. For a 25% conversion factor, this would represent 10^{20} J of electrical energy, or about 3×10^6 MW-yrs. If this amount of energy were to be withdrawn during a period of 50 years, the average annual geothermal power would be $3 \times 10^6/50 = 60,000$ MW, or about 60 times the present installed geothermal capacity.

The quantity 60,000 MW is really not all that much power. The world today uses energy at a rate of 8.35×10^{12} W, or 8.35×10^6 MW. Therefore, geothermal power might only supply 1/139 of this, even when fully developed; by then, the world will, no doubt, be using more power. Although the potential global geothermal supply is comparatively small, it is particularly significant to isolated island communities like Iceland, where it represents a major share of their power demand.

Some estimates for the quantities of geothermal reservoirs available in the United States are considerable. The National Research Council (1980) gives the following total potentially producible thermal energy from accessible U.S. geothermal reservoirs: hot water, 6009 quads; natural steam, 45 quads; geopressurized reservoirs, 2421 quads; normal gradient, 12,530 quads; hot dry rock, 1635 quads; and molten magmas, 35 quads—or a total of 22,675 quads. Ultimately, whether or not this energy is used depends on the economics of the competitive fuels. The National Research Council report discusses the cost figures and points out that they mean very little by themselves without consideration of use efficiency. The highest grade geothermal resources, such as natural steam fields, allow us to obtain a generating efficiency of only about 20% as compared to 35% or more for conventional thermal power plants. Hence, to be competitive, geothermal heat must cost, at most, about half the cost of competing fuels on a Btu for Btu basis. Most geothermal sources, when used to produce electricity, will have a net generating efficiency of only about 10% because of the low temperature differentials involved. If the geothermal source can be used directly for heating dwellings, the use efficiency will be quite high and geothermal energy will be more than competitive with other fuels. However, the number of instances where this can be accomplished are relatively few because dwellings are rarely located near geothermal sources.

History of geothermal power

The first commercial utilization of geothermal power was at Larderello, Tuscany, Italy in 1904. The capacity of the power plants there was increased to 169 MW in 1969. Italy today has a total of 44 geothermal wells that deliver a total power of nearly 500 MW.

The second largest development of geothermal power is at Wairakei, New Zealand. The first power plant began operation about 1958 and now has a capacity of about 290 MW. The Wairakei plant is 10 km north of Lake Taupo in the volcanic zone of North Island. The lake's outlet river provides the cooling water for the power plant. At the wellheads, the geothermal fluid is a mixture of one part of steam to four parts of water. The steam is separated from the water by cyclone separators. The power plant is base-loaded, that is, it produces continuously at full power and has an annual load factor of 0.9. (Axtmann, 1975 gives a good description of this plant and its environmental impacts.)

Power has been produced by a consortium of three companies delivering steam to the Pacific Gas and Electric Company from the Geysers area of northern California since 1960. The steam is of relatively low pressure (689,000 pascals) and temperature (205°C) compared with the pressure (2,067,000 pascals) and temperature (550°C) of steam generated in modern fossil fuel power plants.

Because of the lower pressures and temperatures at which they operate, the turbines at the Geysers are about one-third less efficient than those of conventional power plants and, according to Hammond (1972), require about 450 MW of heat to produce 100 MW of electricity. The present capacity of the Geysers is 1400 MWe and increasing. This resource is estimated to have a total capacity of 2500 MW or more. This is a field of dry steam, a relatively rare situation. How long this steam supply will last is not known.

Southern California Edison Company has dropped its plan to build a 47-MWe geothermal plant about 10 km north of the Mexican border near the town of Calexico and near the Salton Sea. This plant would have supplied electrical power to about 50,000 people. However, the cost of this geothermal power is expected to be much too high at the present time.

In Japan, geothermal power production was begun at Matsukawa in 1966 and at Otake in 1967. The capacity was 20 and 13 MW, respectively, in 1969, but each was expanded to about 60 MW later. Japan has a total geothermal power production today of about 215 MW.

In Mexico, near Cerro Prieto, just south of the U.S. border, a 425-MW power plant is in operation. It is a wet steam plant; this means that they must separate the steam from the hot water. In addition, there are several small geothermal power plants in operation in Mexico.

Iceland has been well known for its use of geothermal power. Two-thirds of its dwellings are heated geothermally. The steam from one of its large geothermal reservoirs has been used for heating the entire town of Hveragerdi, and a 17-MWe power-generating plant was installed there in 1969. The island that is Iceland is an exposed segment of the mid-Atlantic ridge, which is a boundary between the Eurasian and American continental plates. Iceland is literally being pulled apart with the movement of these two gigantic tectonic plates in opposite directions. The result is hot basalt flowing to the surface through fissures underneath. For example, Surtsey, a volcanic island, was formed in 1963 in the ocean south of Iceland when 300 million m^3 of lava and 600 m^3 of ash poured out of the sea floor. Most of the houses of Reykjavik, Iceland are heated by hot water tapped from hot lava nearby.

In the Soviet Union geothermal power is produced at three small plants in Kamchatka, with a total capacity of less than 40 MW.

In the Puna district on the big island of Hawaii, a 3-MWe plant is generating electricity from geothermal power for delivery to about 800 homes. There are plans to build a 25-MWe plant there. However, the power is not needed on the island of Hawaii but on the heavily populated island of Oahu, where 80% of Hawaii's energy consumption occurs. However, transfer of the power would require a cable between the two islands, a cable covering a distance of 250 km and lying at a depth of 2100 m. Hawaii has an abundance of renewable energy in the forms of wind and biomass. On the windward side of Oahu, the world's largest wind turbine farm is under construction; it is expected to provide 80 MWe by 1985. Another alternative is to use the bagasse (sugarcane waste) to burn for fuel for electric power production and process heat. Hawaii has also been working with the Mini-OTEC plant off its coast. It seems that Hawaii has so many sources of renewable energy other than geothermal that it is not likely to develop this resource extensively.

Many towns in the United States heat some of their houses and commercial buildings with geothermal energy. Among these are San Bernardino, California; Pagosa Springs, Colorado; Boise, Idaho; and Klamath Falls, Oregon. Some of these were constructed as early as 1890.

The total installed geothermal power production in the world today is little more than 1000 MW, a rather small quantity when one considers that a single nuclear power plant may exceed this level of power output. Geothermal energy now contributes less than 1/40 of a quad annually in the United States.

Hot, dry rock energy extraction

Most of the heat underground is in rocks not associated with permeable fracture systems filled with water; rather it is associated with volcanic fields or other anomalies or it is an integral part of the Earth's crust. Hot, dry rock geothermal resources are of three general types: a high-grade resource that has temperatures above 150°C and is useful for generating electricity; an intermediate-grade resource that has temperatures between 90° and 150°C and could be used directly for heating dwellings or for industrial process heat; and a low-grade resource that has temperatures below 90°C and, in general, is not worth exploiting. To be useful, these resources must be at depths less than 6 km. Heiken et al. (1981) estimate that the amount of this high-grade hot, dry rock resource in the United States is about 102×10^{21} J—the equivalent of 191.9×10^{11} barrels of crude oil and a very substantial amount of energy.

Some hot, dry rock sources are clearly discernible from the surface, such as the geysers and hot springs of Yellowstone Park, where the hydrothermal activity erupts at the surface. However, the hydrothermal component is only about 3.4% of the total underground energy content, the remainder of the energy may be partially extracted by artificially injecting water into the hot, dry rock reservoir. According to Heiken et al. (1981), scientists at the Los Alamos National Laboratory, New Mexico have developed a method to extract energy from hot rocks by injecting water into the underground structure, fracturing the

rock, and then bringing hot water or steam to the surface. The method, illustrated in Figure 9, is simple in concept and difficult to implement. The Los Alamos scientists have been experimenting at Fenton Hill, New Mexico, on the western flank of the Valles Caldera, a large volcanic field where the hydrothermal component is only about 1%. They drilled an injection well to a depth of 2.93 km, where the temperature was 197°C. Then they drilled an extraction well whose withdrawal position was about 300 m above the input point of the injection well. Success of the operation depended upon the proper fracturing of the granitic reservoir rocks. Measurements indicated that large, multiple fractures did oc-

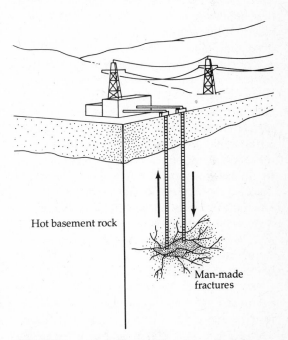

FIGURE 9. Artist's illustration of a method for extracting energy from hot rocks deep underground. Water that is injected into the hot rock structure fractures and penetrates the rocks. It is then recovered as hot water, which may be used to generate electricity, to heat homes, or to drive industrial processes.

cur and created a heat transfer area of about 50,000 m². Some water loss occurs between the injection and extraction wells, but it is expected to diminish as the pores of the surrounding rock become filled. Water withdrawn from the extraction well had a temperature of 153°C and only declined 8°C following 288 days of sustained operation. Encouraged by this experiment, the Los Alamos scientists are drilling wells into hotter sections of this volcanic field and more experiments are in progress.

Environmental and ecological considerations

The use of geothermal resources is not without potential environmental and ecological impacts. One primary concern is land subsidence following the withdrawal of hot water or steam from an underground field. Several potential pollutants are associated with geothermal sources, including hydrogen sulfide, carbon dioxide, ammonia, methane, and boric acid, along with trace amounts of mercury, arsenic, and other elements. In order to extract energy from hot, dry rocks, from molten magma, or from the normal temperature differences underground, it is necessary to force water down boreholes as a working fluid and to return it to the surface for use in a turbine or for direct heating. The chief environmental problem is that very large quantities of water are needed. The amount of water required is less if the geothermal reservoir is very hot so that it converts the water to steam. If the underground reservoir is highly permeable, there is no way to know how much water will need to be injected before a useful amount of steam or hot water is returned to the surface.

Axtmann (1975) has studied the environmental effects associated with the Wairakei, New Zealand geothermal power plant. He chose this one to study because it is a hot water rather than a dry steam geothermal source. He found that the Wairakei plant dis-charges approximately 6.5 times more heat, 5.5 times more water vapor, and one-half as much sulfur per unit of power produced as would a modern coal plant in New Zealand. It also contaminates the Waikato River with hydrogen sulfide, carbon dioxide, arsenic, and mercury at concentrations that have adverse, but not calamitous, effects. Because the Wairakei plant was initiated prior to expression of much of the current environmental concern it was allowed to have a greater environmental impact than would be acceptable or necessary today. Reinjection of the waste water into the geothermal field would reduce the plant's environmental impact greatly. Ground subsidence is not a serious problem at Wairakei but may prove to be so at the nearby Broadlands field. In studying the Wairakei site, Axtmann did not study impacts during the development of the borefield and construction of the plant, the effects of well blowouts, changes in the natural habitat, land use considerations, and aesthetic factors. All these issues need to be considered when using any geothermal resource.

The geothermal electric power plant that was to be built near the Salton Sea in the Imperial Valley of southern California might have produced a degradation of the air quality of the valley. It also might have caused an increase in the salinity of the Salton Sea, the increase resulting from the use of agricultural waters for power plant cooling. Some land subsidence might have resulted. Even earthquakes might have been generated by the extraction and injection of geothermal fluids underground. Hydrogen sulfide would have been released into the atmosphere when hot water from the geothermal reservoir deep underground was brought to the surface and flashed to produce the steam used to run a turbine generator. The amount of hydrogen sulfide released, without controls imposed, could have been as much as 160 g/MWh. If hydrogen sulfide emissions were not abated to below 30 g/MWh, the California air quality standard (42 $\mu g/m^3$ averaged over one hour) would have been exceeded over much of the Imperial Valley by the year 2010. Emissions would need to be reduced by 82% in order to meet the air quality standard. Unfortunately the odor of hydrogen sulfide can be detected by most people at very low concentrations—4 to 12 $\mu g/m^3$.

Hydrogen sulfide gas has a minor impact on most ecosystems except at concentrations exceeding ambient air quality standards, at which concentration it can have a fertilizing effect on crops. Shinn (1981) reported that the geothermal air pollutants, carbon dioxide and hydrogen sulfide, stimulated photosynthesis in sugar beets at low concentrations because these compounds provide essential elements for carbohydrate and amino acid production. The first significant decrease in photosynthesis took place when the hydrogen sulfide concentration exceeded about 3 ppm (4200 $\mu g/m^3$) in the mixture containing elevated amounts of carbon dioxide.

Geothermal water at The Geysers in northern California is cooled as it flows through cooling towers. Apparently this highly mineralized water is killing vegetation in deposition zones downwind from the power stations there. However, the area affected is only about 5% of the total area required for roads and wellheads. Shinn (1981) reported higher deposition of cooling tower minerals at a distance of 1200 m than would be expected from liquid aerosol transport and concluded that dry deposition is also playing a role. He also reported significant increases in the levels of minerals in runoff water collected at the confluence of streams draining the area.

HYDROELECTRIC POWER

The most concentrated form of solar energy is the water power of great rivers and of high-altitude lakes and streams. Solar heat evaporates water from ocean surfaces and from soil and plants on the land. The water vapor then condenses as precipitation and runs into the watersheds of the world. The hydrologic cycle is essential to all ecosystems. Hydropower comes from utilization of the potential energy contained in water behind dams. The release of the water through a spillway converts the potential energy to kinetic energy of fast moving water. The kinetic energy is converted to electrical energy by turning a turbine or to mechanical energy by rotating a water wheel or millwheel.

Hydropower was the first mechanical energy used by humans. With the advent of the electric generator, hydropower became a significant source of power. The potential for further hydroelectric development throughout the world is considerable. Participants in the 1980 World Energy Conference estimated that the current use of hydropower is about 17% of the overall world potential for this energy form. The total global water-power capacity is about 2,857,000 MW. Of this potential, North America has 11%, South America 20%, Africa 27%, Southeast Asia 16%, the Soviet Union and China 16%, and western Europe 6%. It is significant that the continents of Africa and South America, both deficient in coal, have the highest water-power potential of all the continents. Annual hydroelectric production is currently about 2% of the total world production of commercial energy and fuelwood (Sivard, 1981). Hydropower accounts for almost one-quarter of the electricity produced in the world; three times as much electricity as is generated by nuclear power. In developing countries only, the fraction of electricity produced hydroelectrically is about 45%. Projections of the composition of world energy supply in the future show hydroelectric to account for 5 to 6% of the total by the year 2000. Estimates for nuclear power range from 7 to 13% of the total world power needs by then.

Although hydroelectric power is a renewable resource, it is also one with considerable ecological impact and is not always acceptable to the people of the area involved. Hydroelectric power involves high capital costs associated with the construction of dam sites. These high industrial costs are offset by the longevity of the plant and by low operating costs. However, it means that many developing countries do not have the financial resources for the initial capital investment. There is also a safety risk to humans living downstream from hydroelectric dams, and this must be considered when undertaking more development of this power source.

Hydroelectric plants

Originally, water power was used purely mechanically; the water cascaded over a water wheel, which turned the machinery of a grist mill, a saw mill, a textile mill, or other industries. Such plants rarely exceeded a few hundred kilowatts in power capacity. Water power came into widespread use after the development of electricity and its transmission over large distances. Large and small hydroelectric plants have been built throughout the United States and Europe. In fact, in Europe, most of the best potential hydropower has been developed; because of the high density of population, they make effective use of what they have available. In the United States, about 30% of the potential water power resources have been developed (Figure 10). Further development will be difficult because of other uses of the valleys that might be dammed. Projects at some sites are prohibited by legislation; for example, the Colorado River Basin Project Act prohibits the Federal Power Commission from issuing licenses for projects on the Colorado River between Glen Canyon and Hoover Dam, a stretch of water containing 3500 MWe of potential capacity. The Wild and Scenic Rivers Act also prohibits the Federal Power Commission from licensing any projects in the 37 national wild and scenic rivers system, which contain another 9000 MWe of potential power.

On the Columbia River there are five hydroelectric power plants of more than 700 MWe each; and a 2000-MWe plant is located on the Niagara River. Hydroelectric power provides about 13% of the U.S. electrical generation today (coal provides 44%, oil and gas 30%, and nuclear 13%). Canada has developed substantial amounts of hydroelectric power and has the potential for considerably more. Many thousands of megawatts are available from the Churchill Falls projects in Labrador and other large projects in Quebec and New Brunswick, as well as from the Nelson River in Manitoba.

FIGURE 10. Hoover Dam hydroelectric generating plant.

Environmental and ecological impacts

Hydroelectric power development creates many environmental and ecological changes. The construction of a dam across a valley changes an ecosystem of rich diversity, of high scenic value, and of great natural productivity into a body of water attractive to fishermen, boating enthusiasts, and others. Here is a conflict of values. Many mountain valleys contain streams and ponds with spawning grounds for a number of fish species, habitat for many fur-bearing mammals, and winter feeding grounds for elk or deer. Conversion of this area into a lake eliminates habitats for many plants and animals and provides habitats to some species foreign to the mountain valley.

Often the construction of a big dam requires the resettlement of large numbers of people. With the construction of the Aswan High Dam in Egypt, as many as 120,000 demographic problems were created. In addition, serious health problems may arise for people living near the reservoir, particularly in tropical countries where schistosomiasis (also known as bilharzia) and malaria exist. The possibility of dam failure is always a potential hazard to people and to ecosystems downstream. When very large reservoirs are created, there may be a significant climate change nearby.

In Canada, the proposed James Bay project would flood 34,500 km^2, an area twice the size of England, and destroy the ecosystems upon which several thousand Indians depend. This project threatens to wipe out resting grounds for millions of migrating geese and one of the world's great salmon-spawning grounds.

An excellent summary of the environmental effects of hydroelectric power development is given by Jassby (1980).

Physical effects. The construction of a dam across a valley not only creates a reservoir and changes the physical character of the valley, it also creates an enormous increase in the weight supported by the underlying soil and rocks and greatly increases stresses. The increased fluid pressure produced by the water behind a dam may be sufficient to induce earthquakes, as has happened at approximately 30 sites throughout the world. Clearly the construction of many large dams has not led to earthquakes. However, at Koyna, India in 1976 an earthquake was induced by a reservoir and resulted in the loss of many lives and much property.

A deep reservoir will develop a temperature structure similar to those of lakes, with warm water on top and cold water at the bottom, particularly during the summer season. However, reservoirs differ from lakes in two respects. The heavier water of silt-laden streams may penetrate the lower cold water and upset the temperature structure of some parts of the reservoir. The colder, deeper water may be discharged downstream from a reservoir, whereas for a lake warm water usually flows off from the surface. Downstream ecosystems may be heavily impacted by this cold water discharge.

The increased surface area of a reservoir (compared with that of the stream that previously occupied the valley) leads to greatly increased evaporation. This loss of water may be serious in regions where water is scarce. The water loss from reservoirs of the U.S. western states is approximately 20 billion m^3/yr—enough water to supply the needs of 100 million people. Water loss from the reservoirs feeding the Colorado River alone may approach 1 billion m^3/yr.

Much siltation occurs in reservoirs. Siltation decreases the capacity of the reservoir and may seriously limit its lifetime. It is estimated that some reservoirs where siltation rates are high may have lifetimes of less than 50 years. If water levels vary a great deal, there will be zones around the periphery of the reservoir that are free of vegetation; and as a result, siltation rates will increase, banks will erode and become steeper, and more sediment will end up downstream. However, in most reservoirs, the average amount of sedimentation in the discharge waters is less than that in the original stream, and downstream waters may be clearer as a result. Less sediment going down a river can also mean a change in the river delta wherever it exits into a lake or an ocean. For example, the delta of the Nile River in the Mediterranean Sea has receded greatly since the construction of the Aswan High Dam.

Chemical effects. Increased evaporation from reservoirs results in an increased concentration of total dissolved solids. When water above a reservoir is used for irrigation, the salinity may increase many fold. Colorado River water, for example, is reused many times for irrigation, and its salinity level, about 50 ppm in the headwaters, is over 1000 ppm by the time the river reaches Mexico.

Mineral solution and precipitation may occur in a reservoir where the surrounding materials, such as gypsum or limestone, are soluble. At Lake Mead, for example, huge amounts of gypsum dissolved during the decade following dam construction. Mineral precipitation and the formation of marl on the bottom of a reservoir will result when algae and aquatic macrophytes alter the equilibrium concentration of dissolved inorganic carbon in the water.

Stratification of various compounds and nutrients will result from biological processes in the reservoir. Autotrophic metabolism in the upper layers will produce the following changes with increasing depth: lower pH, lower O_2, higher CO_2, and higher nutrient levels of phosphorus and nitrogen compounds. Often with the filling of a reservoir, there will be a transient flush of nutrients into the reservoir along with the initial siltation. Abundant waterfowl using a reservoir

will lead to nutrient enrichment through the decomposition of fecal matter.

Biological effects. As a result of the many physical and chemical changes occurring in a reservoir, there will be concomitant biological changes. Nutrient enrichment, warmer temperatures, and good light penetration will lead to large standing crops of phytoplankton, although these often decrease as the reservoir ages. Immediately downstream from a dam, algal growth is usually greater than in the same stream before the dam was built. Releases of water from the epilimnion (the warm upper layers) carry phytoplankton into the stream. Releases of water from the hypolimnion (the cold lower layers) provide nutrients and result in phytoplankton or algal growth in the stream. Further downstream, below where this plant growth may occur, there can be a drop of plankton levels because the nutrients have been taken out by the plants.

Aquatic macrophytes (water weeds) will respond to the favorable conditions in the reservoir in the same manner as the phytoplankton. In warm climates, the water hyacinth (*Eichornia crassipes*) is a great nuisance in reservoirs. Submerged macrophytes most often growing in reservoirs in the United States include species of *Ceratophyllum*, *Egeria*, *Elodea*, *Myriophyllum*, *Najas*, and *Potamogeton*.

The population of game fishes in a reservoir may increase at first as a result of the initial high phytoplankton productivity. Then, as conditions in the reservoir change with time, other competitive fishes come in and the number of game fishes declines. However, the total standing crop of fishes gradually increases as the reservoir ages.

The temperature structure of a reservoir may cause a shift to warm-water fish species. The production of low-oxygen conditions in the hypolimnion through organic decomposition forces fish to seek oxygen in the warm epilimnion, where temperatures may be above the thermal tolerance levels of cold-water fishes.

Hydroelectric dams often interfere with the migration of anadromous fish into spawning grounds. Along the Pacific Northwest coast of the United States, many dams have weirs to allow the anadromous fish to get above them.

Downstream from a reservoir, conditions may be highly variable, depending on discharge cycles, water quality, temperature, and turbidity. Nitrogen and oxygen supersaturation of the discharge water can occur as air is entrained with the discharge water. This increases the levels of dissolved oxygen in the body fluids of fish. Death may ensue from embolisms or air bubbles in the blood vessels. This is a frequent phenomenon along the Columbia River at certain times of year. In delta areas there is often a decrease in fish production resulting from a decrease of the nutrient flow because of upstream dams. There has been a significant drop of fish harvests in the eastern Mediterranean following construction of the Aswan High Dam.

Aswan High Dam

Probably the world's most classic example of unmitigated ecological degradation associated with the construction of a hydroelectric project is the High Dam at Aswan on the Nile River in Egypt, built in 1964. Designed to impound water in Lake Nasser for irrigation, to stem the flow of floods, to produce electricity, and to protect the Egyptian people against drought and famine, it is one of the largest and most expensive dams in the world. The problems associated with the High Dam at Aswan might have been avoided if proper environmental and ecological studies had been completed prior to its undertaking. Many things could have been done very differently

and very much better. Some of the problems associated with the Aswan Dam are

1. An unexpectedly high rate of water loss by evaporation and leakage through underground strata. A computational error underestimated evaporation by 50%, an estimate that did not properly allow for the very hot, strong, dry desert winds that blow over Lake Nasser. The impoundment of several smaller bodies of water behind several smaller dams might have greatly reduced this problem. The amount of water available behind the Aswan Dam is much less than expected.

2. The annual flooding associated with the free flow of the Nile carried huge amounts of nutrients into the lower Nile delta and into the eastern Mediterranean. Now the soils along the lower Nile are being impoverished. Holding back the waters of the Nile River has resulted in a one-third reduction of the plankton densities in the eastern Mediterranean. Because plankton is the food base for sardine, mackerel, lobster, and shrimp, a loss of plankton production resulted in a decrease of these marine populations. In fact, the annual sardine catch alone declined from 18,000 mt to 500 mt after the filling of Lake Nasser began. Some compensation for this loss may occur with the development of a freshwater fishery in Lake Nasser, but this has not yet occurred (George, 1972).

3. The salinity of soils is increasing rapidly in the region of the upper Nile, where once the annual flooding flushed out the salt accumulation. All arid lands have a serious problem with salt accumulation if they cannot be flushed out with a sufficient flow of fresh water. This now appears to be a serious problem in Egypt.

4. Lake Nasser and all its embayments are infested with snails carrying a most debilitating human disease—schistosomiasis. This

disease ranks along with river blindness and malaria as one of the worst of tropical or subtropical endemic diseases. There are more than 200 million people infected with schistosomiasis in the world today. Schistosomiasis has always been serious in Egypt, but the naturally flowing Nile with its swift currents flushed out the snail and restricted the disease to still waters along the edge of the river. The warm, stagnant waters of Lake Nasser are an ideal habitat for the snails. The people of Egypt live along the Nile. They bathe in the river, wash their clothes in it, empty their sewage into it, and use it in their daily lives in many ways. The parasite leaves the snail and becomes free-swimming in its search for a warm-blooded host into which it can burrow and lay its eggs. Because millions of Egyptians and millions of snails are comingling in the waters of Lake Nasser, it is inevitable that the human host and the snail intermediate host are each keeping one another infected. In the human body the *Schistosoma* parasite attacks the vital organs such as the liver, stomach, heart, and lungs. The effect on humans is complete debilitation and eventual death. In Egypt, 20 million people, or half of the total population, are infected with schistosomiasis.

5. About 80 km south of the southern shores of Lake Nasser, a virulent strain of malaria is transmitted by a particular species of mosquito found there. Health authorities are concerned that the waters of Lake Nasser may make it possible for this mosquito to spread into many parts of Egypt and to carry with it this fatal form of malaria.

The construction of the High Dam at Aswan may turn out to be one of the greatest disservices ever perpetrated on the Egyptian people. Today the consumption of electricity is such that only one of the six electrical generating units is needed. Lake Nasser is silting up, nutrients are not reaching the lower Nile and eastern Mediterranean, disease is rampant, and soil salinity is increasing in the upper Nile. Once the natural river ecosystem has been so drastically altered as it has here, it is going to require enormous expenditures of money for fertilizers and for the control of disease to compensate for these unprecedented negative impacts.

Kariba Dam

Construction of the Kariba Dam in central Africa has problems similar to those associated with the Aswan High Dam. Its construction displaced at least 50,000 people, including complete tribes, and many animals were exterminated. The Swende Valley, flooded by Lake Kariba, had a higher total faunal productivity of animal herds prior to flooding than the productivity of the fisheries that were established afterward. Attempts to transfer wildlife out of the area to be inundated were not very successful.

ENERGY FUTURE

Alternative energy sources—solar, wind, geothermal, ocean thermal, tides, ocean waves, hydro—can only provide a fraction of the total power required by a burgeoning world population living in an industrialized society. These are all dispersed energy sources and they cannot provide the concentration of power needed except at very low efficiencies. On the other hand, oil, gas, coal and uranium are highly concentrated energy forms but each has a finite supply. As previous chapters have documented, the notion of a glut of oil (or of coal) is a myth. There is no way in which a waste of nonrenewable energy can be justified.

The people of the world are in an environmental vise between carbon dioxide and acid rain on the one hand and radioactivity on the other. Populations overrunning food production are dying and millions of people are suffering from impoverished ecosystems today. Soil erosion is occurring at an alarming rate and no less so within the developed than among the developing nations. As soil deterioration occurs, more and more nutrients must be supplied to maintain crop productivity. However, nutrients, particularly nitrogen, require petroleum for their production, and petroleum shortages will eventually be commonplace.

Recently, new estimates have been made by the U.S. Geological Survey of the global oil resources to be recovered by conventional means; a total of 1718 billion barrels of oil. (This is in the middle of estimates given in Table 3 of Chapter 4.) Of this amount, 550 billion barrels of oil are in the undiscovered or yet to be discovered category. All of the oil known to exist in the ground today plus this new undiscovered oil will last the world about 60 years *at present rates of consumption.* Since the mid-1960s the amount of oil discovered each year in the world has dropped from about 38 billion barrels to about 10 billion barrels in 1980. More serious is the fact that beginning in the early 1970s more oil has been consumed from reserves each year than has been added by new discoveries and this trend continues today. The rate of annual discovery is declining.

Estimated quantities of oil, gas, coal and uranium resources will continue to change as new discoveries are made—or fail to be made. Even estimates concerning the amount of power derivable from renewable energy sources will change considerably as technology improves and our understanding of each source increases.

The need for intensive energy conservation, for strong development of alternative energy sources, and for breeder reactors will become critical fairly soon. If breeder reactors are rejected, the world's energy future is not very certain beyond the next few decades. We can choose to live in a fool's paradise for a while longer, but the day of energy reckoning will come and our children and grandchildren will pay for our profligate waste.

APPENDIX

Units and Conversion Factors

Length

$1\,m = 10^2\,cm = 3.28\,ft = 39.37\,in$

$1\,km = 10^3\,m = 3{,}280\,ft = 0.621\,mi$

$1\,mi = 5{,}280\,ft = 1{,}609\,m = 1.609\,km$

Area

$1\,m^2 = 10^4\,cm^2 = 10.76\,ft^2$

$1\,ha = 10^4\,m^2 = 10^{-2}\,km^2 = 2.47\,acre$

$1\,km^2 = 10^6\,m^2 = 247\,acre = 0.386\,mi^2$

$1\,mi^2 = 640\,acres = 2.59\,km^2 = 259\,ha$

$1\,acre = 43{,}560\,ft^2 = 4{,}047\,m^2 = 0.405\,ha$

Volume

$1\,cm^3 = 1.0\,ml = 0.061\,in^3 = 0.0338\,U.S.\,fl\,oz$

$1\,l = 10^3\,cm^3 = 61.0\,in^3 = 33.8\,U.S.\,fl\,oz = 1.057$
U.S. qt $= 0.264\,U.S.\,gal$

$1\,m^3 = 10^6\,cm^3 = 10^3\,l = 35.31\,ft^3 = 264\,U.S.\,gal$

$1\,ft^3 = 1728\,in^3 = 29.9\,U.S.\,qt = 7.48\,U.S.\,gal =$
$28.32\,l$

$1\,U.S.\,qt = 32\,U.S.\,fl\,oz = 0.25\,U.S.\,gal = 0.946\,l$

$1\,U.S.\,gal = 231\,in^3 = 128\,U.S.\,fl\,oz = 3.06 \times 10^{-6}$
acre ft $= 0.0238\,barrels\,(bbl)\,of\,oil = 3.785\,l$

$1\,barrel = 42\,U.S.\,gal$

Velocity

$1\,m\,s^{-1} = 3.6\,km\,hr^{-1} = 1.94\,knots = 2.237\,mi$
$hr^{-1}\,(mph)$

$1\,km\,hr^{-1} = 0.621\,mi\,hr^{-1}\,(mph)$

$1\,knot = 1.15\,mi\,hr^{-1} = 1.69\,ft\,s^{-1} = 0.515\,ms^{-1} =$
$1.85\,km\,hr^{-1}$

$1\,mi\,hr^{-1}\,(mph) = 0.868\,knot = 1.47\,ft\,s^{-1} = 0.447$
$m\,s^{-1} = 1.609\,km\,hr^{-1}$

Mass

$1\,g = 10^{-3}\,kg = 0.0022\,lb$

$1\,kg = 10^3\,g = 35.27\,oz = 2.20\,lb$

$1\,metric\,ton\,(mt) = 10^3\,kg = 2205\,lb = 1.102\,short$
tons

$1\,lb = 16\,oz = 0.4536\,kg$

$1\,short\,ton = 2000\,lb = 907.2\,kg = 0.9072\,mt$

Force

$1\,newton\,(N) = 1\,kg\,m\,s^{-2} = 10^5\,dynes = 0.2248$
pounds

$1\,pound = 32.2\,poundal = 4.448\,N$

Pressure

$1\,Pascal = 1\,N\,m^{-2} = 10^{-5}\,bar = 10^{-2}\,mbar\,(mb) =$
$9.87 \times 10^{-6}\,atm = 1.45 \times 10^{-4}\,psi$

$1\,psi = 0.069\,atm = 68.9\,mb = 51.7\,mm\,Hg$

$1\,atm = 1013\,mb = 14.7\,psi = 760\,mm\,Hg$

Energy

$1\,joule\,(J) = 10^7\,ergs = 2.78 \times 10^{-7}\,kW\text{-}hr = 0.2389$
cal $= 9.48 \times 10^{-4}\,Btu$

$1\,kW\text{-}hr = 3.6 \times 10^6\,J = 8.59 \times 10^5\,cal = 3410\,Btu$

$1\,cal = 10^{-3}\,kg\,cal\,(Kcal) = 4.186\,J = 3.968 \times$
$10^{-3}\,Btu$

$1\,Btu = 252\,cal = 1055\,J = 29.3 \times 10^{-5}\,kW\text{-}hr$

$1\,ft\,lb = 1.356\,J$

$1\,metric\,ton\,of\,coal\,equivalent\,(mtce) = 7 \times 10^9\,cal$
$= 29.3 \times 10^9\,J = 8145\,kW\text{-}hr = 27.8 \times 10^6\,Btu$

$1\,metric\,ton\,of\,oil\,equivalent\,(mtoe) = 1.43\,mtce =$
$41.9 \times 10^9\,J = 11{,}647\,kW\text{-}hr = 39.68 \times 10^6\,Btu$

$1\,barrel\,of\,oil\,(42\,gal) = 6.12 \times 10^9\,J = 1700\,kW\text{-}hr$
$= 194\,W\text{-}yr = 5.80 \times 10^6\,Btu$

$1\,cubic\,foot\,(ft^3)\,of\,natural\,gas = 1.05 \times 10^9\,J =$
$0.29\,kW\text{-}hr = 0.033\,W\text{-}yr = 1000\,Btu$

$1\,short\,ton\,of\,U_3O_8\,(in\,burner\,reactors) = 422 \times$
$10^{12}\,J = 400 \times 10^9\,Btu = 14.4 \times 10^3\,mtce$

$1\,short\,ton\,of\,U_3O_8\,(in\,breeder\,reactors) = 31.65 \times$
$10^{15}\,J = 30 \times 10^{12}\,Btu = 1.08 \times 10^6\,mtce$

$1\,quad = 1\,quadrillion\,Btu$

Energy Density

1 kJ kg^{-1} = 0.430 Btu lb^{-1}

1 kJ l^{-1} = 10^3 kJ m^{-3} = 26.8 Btu ft^{-3} = 3.59 Btu gal^{-1}

1 kJ m^{-3} = 10^{-3} kJ l^{-1} = 0.0268 Btu ft^{-3} = 3.59 × 10^{-3} Btu gal^{-1}

1 Btu lb^{-1} = 2.324 kJ kg^{-1}

1 Btu ft^{-3} = 0.134 Btu gal^{-1} = 37.28 kJ m^{-3}

1 Btu gal^{-1} = 7.46 Btu ft^{-3} = 0.279 kJ l^{-1} = 279 kJ m^{-3}

Power

1 Watt (W) = 1 J s^{-1} = 0.2389 cal s^{-1} = 0.0569 Btu min^{-1}

1 Btu min^{-1} = 17.58 W = 252 cal min^{-1}

1 horsepower = 550 ft lb s^{-1} = 745.7 W

FACTORS AND TERMS

Factor	Term	Prefix	Symbol
10^{15}	quadrillion	pecta	P
10^{12}	trillion	tera	T
10^9	billion	giga	G
10^6	million	mega	M
10^3	thousand	kilo	k
10^{-1}	tenth	deci	d
10^{-2}	hundredth	centi	c
10^{-3}	thousandth	milli	m
10^{-6}	millionth	micro	μ
10^{-9}	billionth	nano	n
10^{-12}	trillionth	pico	p

BIBLIOGRAPHY

Abelson, P.H. 1982. Energy and chemicals from trees. Science *215*, 1349.

Aber, J.D., G.R. Hendrey, A.J. Francis, D.B. Botkin and J.M. Melillo. 1982. Potential effects of acid precipitation on soil nitrogen and productivity of forest ecosystems. In F.M. D'Itri (ed.). *Acid Precipitation: Effects on Ecological Systems.* Ann Arbor Science Publ., Ann Arbor. 411–433.

Agee, E.M. 1980. Present climatic cooling and a proposed causative mechanism. Bull. Amer. Meteorol. Soc. *61*, 1356–1367.

Agnew, H.M. 1981. Gas-cooled nuclear power reactors. Sci. Amer. *244*, 55–63.

Almer, B., W. Dickson, C. Ekstrom and E. Hornstrom. 1978. Sulfur pollution and the aquatic ecosystem. In J.O. Nriagu (ed.). *Sulfur in the Environment, Part II: Ecological Impacts,* John Wiley and Sons, New York. 271–312.

American Institute of Physics. 1981. Studies revise dose estimates of bomb survivors. Physics Today *34*, 17–32.

Anderson, V.G. 1915. The influence of weather condition upon the amount of nitric acid and of nitrous acid in the rainfall at and near Melbourne, Australia. Quart. J. Ray. Meteorol. Soc. *41*. 99–122.

Angino, E.E. 1977. High-level and long-lived radioactive waste disposal. Science *198*, 885–890.

Atkins, W.R.G. 1922. Measurements of the acidity and alkalinity of natural waters: in their biological relationship. Salmon and Trout Magazine *30*, 184–198.

Atlas, R.M., A. Horowitz and M. Busdosh. 1978. Prudhoe crude oil in arctic marine ice, water and sediment ecosystems: degradation and interactions with microbial and benthic communities. J. Fish. Res. Bd. Can. *35*, 585–590.

Avery, W.H. and G.L. Dugger. 1980. Contributions of ocean thermal energy conversion to world energy needs. The International Journal of Ambient Energy *1*, 177–190.

Axtmann, R.C. 1975. Environmental impact of a geothermal power plant. Science *187*, 795–802.

Barber, R.T. and F.P. Chavez. 1982. Biological consequences of El Niño. Science *222*, 1203–1210.

Barnthouse, J. Boreman, S.W. Christensen, C.P. Goodyear, W. Van Winkle and D.G. Vaughan. 1984. Population biology in the courtroom: the Hudson River controversy. BioScience *34*, 14–19.

Barrett, E. and G. Brodin. 1955. The acidity of Scandinavian precipitation. Tellus *7*, 251–257.

Bartlett, A.A. 1978. Forgotten fundamentals of the energy crisis. Am. J. Phys. *46*, 876–888.

Batie, S.S. 1983. *Soil Erosion.* The Conservation Foundation, Washington, D.C.

Bauer, H.J. 1973. Ten years' studies of biocenological succession in the excavated mines of the Cologne lignite district. In R.J. Hutnick and G. Davis (eds.). *Ecology and Reclamation of Devastated Land.* Gordon and Breach, New York.

Bebbington, W.P. 1976. The reprocessing of nuclear fuel. Sci. Amer. *235*, 30–41.

Bennet, G. 1982. Big ships have their problems too. Cruising World *8*, 98–102.

Bennet, I. 1964. Monthly maps of mean daily insolation for the United States. Solar Energy *9*, 145–152.

Bick, H. and E.F. Drews. 1973. Sebstreinigung und Ciliaten-besiedlung. Hydrobiologia *42*, In Saurem Milieu (Modellversuche). 393–402.

Blumer, M., H.L. Sanders, J.F. Grassle and G.R. Hampson. 1971. Oil spill. Environment *13*, 3–21.

Boeck, William L. 1976. Meteorological consequences of atmospheric krypton-85. Science *193*, 195–198.

Bolin, B. 1977. Changes of land biota and their importance for the carbon cycle. Science *196*, 613–615.

Bormann, F.H., G.E. Likens, D.W. Fisher and R.S. Pierce. 1968. Nutrient loss accelerated by clearcutting of a forest ecosystem. Science *159*, 882–884.

Boughey, A.S. 1973. *Ecology of Populations,* 2nd Ed. Macmillan, New York.

Boyle, J.R., J.J. Phillips and A.R. Ek. 1973. Whole tree harvesting: nutrient budget evaluation. J. Forestry *71*, 760–762.

Brett, J.R., J.E. Shelbourn and C.T. Shoop. 1969. Growth rate and body composition of fingerling Sockeye Salmon, *Oncorhynchus nerka*: relationship to temperature and ration size. J. Fish. Res. Bd. of Canada *26*, 2363–2394.

Brezonik, P.L., E.S. Edgerton and C.D. Hendry. 1980. Acid precipitation and sulfate deposition in Florida. Science *208*, 1027–1029.

Brock, T.D. 1973. Lower pH limit for the existence of blue-green algae: evolutionary and ecological implications. Science *179*, 480–483.

Brown, J. and R.L. Berg. 1980. Environmental and ecological baseline investigations along the Yukon River-Prudhoe Bay haul road. Cold Regions Research and Engineering Laboratory Report 80-19, U.S. Army Corps of Engineers, Hanover, NH.

Bungay, H.R. 1982. Biomass refining. Science *18*, 643–646.

Cairns, J. 1956. Effects of increased temperatures on aquatic organisms. Ind. Wastes 1, 180–183.

Cairns, J. 1972. Coping with heated water discharges from steam-electric power plants. BioScience 22, 411–420.

Cairns, J. 1976. Heated waste-water effects on aquatic ecosystems. In G.W. Esch and R.W. McFarlane (eds.). Thermal Ecology II. Tech. Inf. Center, Springfield, VA. 32–36.

Cairns, J. 1979. Ecological considerations in relcaiming surface mined lands. Minerals and Environment 1. 83–89.

Cairns, J., A.L. Buikema, A.G. Heath and B.C. Parker. 1978. Effects of temperature on aquatic organism sensitivity to selected chemicals. Bulletin 106. Virginia Water Resources Center, Virginia Polytechnic Institute and State University, Blacksburg, VA.

Cairns, J., K.L. Dickson and A.W. Maki. 1979. Estimating the hazard of chemical substances to aquatic life. Hydrobiologia 64, 157–166.

Calvin, M. 1978. Green factories. Chem. Eng. News 56, 30–36.

Campbell, T. 1979. The Do-it-yourself Weather Book. Oxmoor House, Birmingham.

Chambers, R.S., R.A. Herendeen, J.J. Joyce and P.S. Penner. 1979. Gasohol: does it or doesn't it produce positive net energy? Science 206, 790–795.

Christy, E.J. and R.R. Sharitz. 1980. Characteristics of three populations of a swamp annual under different temperature regimes. Ecology 61, 454–460.

Clapham, W.B., Jr. 1973. Natural Ecosystems. Macmillan, New York.

Clark, J. 1969. Thermal pollution and aquatic life. Sci. Amer. 220, 19–27.

Clark, J. 1979. The world's biggest environmental hangover. Animal Kingdom 82, 7–14.

Clark, J. and W. Brownell. 1973. Electric power plants in the coastal zone. Amer. Littoral Society. Special Publ. No. 7. Highlands, NJ.

Colwell, R.R., A.L. Mills, J.D. Walker, P. Garcia-Tello and V. Compos-P. 1978. Microbial ecology studies of the Metula Spill in the Straits of Magellan. J. Fish. Res. Bd. Can. 35, 573–580.

CONAES. 1980. Energy in transition 1985–2010. Final Report of the Committee on Nuclear and Alternative Energy Systems, Natl. Res. Council. W.H. Freeman, San Francisco.

Conservation Foundation. 1981a. Toward a National Policy for Managing Low Level Radioactive Waste: Key Issues and Recommendations. The Conservation Foundation, Washington, D.C.

Conservation Foundation. 1981b. Managing the Nation's High Level Radioactive Waste: Key Issues and Recommendations. The Conservation Foundation, Washington, D.C.

Conway, H.W. and G.R. Hendry. 1981. Ecological effects of acid precipitation on primary producers. In Conference Proceedings: The Effects of Acid Precipitation on Ecological Systems in the Great Lakes Region of the United States. April 1–3, 1981, Michigan State University.

Cooper, C.F. 1978. What might man-induced climate change mean? Foreign Affairs, April, 500–520.

Council on Environmental Quality. 1973. Energy and the Environment. U.S. Govt. Printing Office, Washington, D.C.

Council on Environmental Quality. 1981. Global Energy Futures and the Carbon Dioxide Problem. U.S. Govt. Printing Office, Washington, D.C.

Coutant, C.C. 1978. Determining the ecological effects of power plant cooling. Fifth FAO/SIDA Workshop on Aquatic Pollution in Relation to Living Resources. Food and Agricultural Organization of the United Nations, Rome. TF-RAS-34 (SWE) Suppl. 1. 229–251.

Cowling, E.B. and R.A. Linthurst. 1981. The acid precipitation phenomenon and its ecological consequences. BioScience 31, 649–654.

Crowther, C. and H.G. Ruston. 1911. The nature, distribution and effects upon vegetation of atmospheric impurities in and near an industrial town. J. Agr. Sci. 4, 25–55.

Dahl, K. 1927. The effects of acid water on trout fry. Salmon and Trout Magazine 46, 35–43.

Davis, M.B. 1980. Holocene climate of New England. Quartern. Res. 14, 240–250.

Davis, R.B. and F. Berge. 1980. Atmospheric deposition in Norway during the last 100 years as recorded in SNSF lake sediments. II. Diatom stratigraphy and inferred pH. In D. Brablos and A. Tollan (eds.). Ecological Impact of Acid Precipitation: Proceedings of an International Conference, Sandefjord, Norway, March 11–14, 1980. Oslo-Aas, Norway.

Deely, D.J. and F.Y. Borden. 1973. High surface temperatures on strip-mine spoils. In R.J. Hutnick and G. Davis (eds.). Ecology and Reclamation of Devastated Land. Gordon and Breach, New York.

DeLaune, R.D., W.H. Patrick, Jr. and R.J. Buresh. 1979. Effect of crude oil on a Louisiana Spartina alterniflora salt marsh. Environ. Poll. 20, 21–31.

deMarsily, G., E. Dedoux, A. Barbreau and J. Margat. 1977. Nuclear waste disposal: Can the geologist guarantee isolation? Science 197, 519–526.

Duderstadt, J.J. and Kikuchi, C. 1979. Nuclear Power, Technology on Trial. Univ. of Michigan Press, Ann Arbor. 228.

Dvorak, A.J. 1978. Impacts of coal-fired power plants on fish, wildlife and their habitats. Fish and Wildlife Service, U.S. Dept. Interior FWS/OBS-78/29.

Ellers, F.S. 1982. Advanced offshore oil platforms. Sci. Amer. 246, 39–49.

Erickson, E. 1952. Composition of atmospheric precipitation: II sulfur, chloride and iodine compounds. Tellus 4, 280–303.

Erickson, E. 1959. The yearly circulation of chlorine and sulfur in nature: meteorological, geochemical and pedological implications. Part I. Tellus 11, 375–403.

Erickson, E. 1960. The yearly circulation of chlorine and sulfur in nature: meteorological, geochemical and pedological implications. Part II. Tellus 12, 63–109.

Etherington, J.R. 1982. Environment and Plant Ecology, 2nd Ed. John Wiley and Sons, New York.

Etkins, R. and E.S. Epstein. 1982. The rise of global mean sea level as an indication of climate change. Science *215*, 287–289.

Exxon. 1978. *The Offshore Search for Oil and Gas*, 3rd Ed. Exxon Background Series. July.

Exxon. 1979. *Exxon Company: USA's Energy Outlook 1980–2000*. Exxon Company, USA. Public Affairs Dept., Houston.

Exxon. 1980. *Middle East Oil*. Exxon Background Series. September.

Farrington, J.W. 1980. An overview of the biogeochemistry of fossil fuel hydrocarbons in the marine environment. In E. Petrakis and F.T. Weiss (eds.). *Petroleum in the Marine Environment*. Advances in Chemistry Series, No. *185*, Amer. Chem. Soc.

Flamm, B.R. and G.E. Bangay. 1981. *United States-Canada Memorandum of Intent on Transboundary Air Pollution, Impact Assessment, Interim Report*. Environmental Protection Service, Environment Canada, Ottawa.

Foltz, J.W. and C.R. Norden. 1977. Seasonal changes in food consumption and energy content of smelt (*Osmerus mordax*) in Lake Michigan. Trans. Amer. Fish. Soc. *106*, 230–234.

Foster, K.E. 1980. Environmental consequences of an industry based on harvesting the wild desert shrub Jojoba. BioScience *30*, 256–258.

Foster, M.S. and R.W. Holmes. 1977. The Santa Barbara oil spill: an ecological disaster? In J. Cairns, Jr., K.L. Dickson and E.E. Herricks (eds.). Recovery and Restoration of Damaged Ecosystems; Proceedings of the International Symposium on the Recovery of Damaged Ecosystems held at Virginia Polytechnic Institute and State University, Blacksburg, VA, March 1975. University Press of Virginia, Charlottesville.

Gates, D.M. 1980. *Biophysical Ecology*. Springer-Verlag, New York.

George, C.J. 1972. The role of the Aswan High Dam in changing the fisheries of the southeastern Mediterranean. In M.T. Farrar and J.P. Milton (eds.). *The Careless Technology*, Natural History Press, Garden City, NY. 159–178.

Geyer, W.A. 1973. Tree species performance on Kansas coal spoils. In R.J. Hutnick and G. Davis (eds.). *Ecology and Reclamation of Devastated Land*. Gordon and Breach, New York.

Gibbons, J.H. 1980. *Energy from Biological Processes*. Office of Technology Assessment, Supt. Documents, U.S. Govt. Printing Office, Washington, DC.

Gibbons, J.W. and R.R. Sharitz. 1981. Thermal ecology: environmental teachings of a nuclear reactor site. BioScience *31*, 293–298.

Gibbons, J.W., R.R. Sharitz and I.L. Brisbin. 1980. Thermal ecology research at the Savannah River Plant: a review. Nuclear Safety *21*, 367–379.

Glenn-Lewin, D.C. 1979. Natural revegetation of acid coal spoils in southeast Iowa. In M.K. Wali (ed.). *Ecology and Coal Resource Development. Vol. 2*. Pergamon Press, New York.

Golley, F.B. 1960. Energy dynamics of a food chain of an old-field community. Ecol. Monogr. *30*, 187–206.

Gorham, E. 1958. The influence and importance of daily weather conditions on the supply of chloride, sulphate and other ions to fresh waters from atmospheric precipitation. Roy. Soc. Lond. Phil. Trans., B. *241*, 147–178.

Gorham, E. 1982. What to do about acid rain? Tech. Rev. *85*, 3–12.

Gornitz, V., S. Lebedeff and J. Hansen. 1982. Global sea level trend in the past century. Science *215*, 1611–1614.

Gosner, K.L. and I.H. Black. 1957. The effects of acidity on the development and hatching of New Jersey frogs. Ecology *38*, 256–262.

Gundlach, E.R., P.D. Boehm, M. Marchand, R.M. Atlas, D.M. Ward and D.A. Wolfe. 1983. The fate of *Amoco Cadiz* oil. Science *221*, 122–129.

Gustavson, M.R. 1979. Limits to wind power utilization. Science *204*, 13–17.

Guthrie, R.K. and D.S. Cherry. 1976. Pollutant removal from coal-ash basin effluent. Water Res. Bull. *12*, 889–902.

Hammond, A.L. 1972. Geothermal energy: an emerging major resource. Science *177*, 978–980.

Hansen, J., D. Johnson, A. Lacis, S. Lebedeff, P. Lee, D. Rind and G. Russell. 1981. Climate impact of increasing atmospheric carbon dioxide. Science *213*, 957–966.

Hart, J.S. 1952. Geographic variations of some physiological and morphological characters in certain freshwater fish. Publ. Ontario Fish Res. Lab. LXXII, 79, Univ. of Toronto Biol. Ser. *60*, Toronto.

Hartline, B.K. 1980. Tapping sun-warmed ocean water for power. Science *209*, 794–796.

Heiken, G., H. Murphy, G. Nunz, R. Potter and C. Grigsby. 1981. Hot dry rock geothermal energy. Amer. Sci. *69*, 400–407.

Herricks, E.E. and V.O. Shanholtz. 1976. Predicting the environmental impact of mine drainage on stream biology. Trans. Amer. Soc. Agr. Eng. *19*, 271–283.

Herricks, E.E., V.O. Shanholtz and D.N. Contractor. 1975. Models to predict environmental impact of mine drainage on streams. Trans. Amer. Soc. Agr. Eng. *18*, 657–667.

Hill, A.C., S. Hill, C. Lamb and T.W. Barret. 1974. Sensitivity of native desert vegetation to SO_2 and to SO_2 and NO_2 combined. J. Air Pollut. Control Assoc. *24*, 153–157.

Hill, A.C., M.R. Pack, M. Treshow, R.T. Downs and L.G. Transtrum. 1961. Plant injury induced by ozone. Phytopathology *51*, 356–363.

Holdren, J.P., G. Morris and I. Mintzer. 1980. Environmental aspects of renewable energy sources. Ann. Rev. Energy *5*, 241–291.

Hopkinson, C.S. and J.W. Day. 1980. Net energy analysis of alcohol production from sugarcane. Science *207*, 302–303.

Hutchinson, T.C. and M. Havas (eds.). 1980. *Effects of Acid Precipitation on Terrestrial Ecosystems*. Plenum Press, New York. 654.

Inglis, D.R. 1978. *Wind Power and Other Energy Options*. Univ. of Michigan Press, Ann Arbor. 298.

Irving, P.M. and J.E. Miller. 1981. Productivity of field-grown soybeans exposed to acid rain and sulfur dioxide alone and in combination. J. Environ. Qual. *10*, 473–478.

Issaacs, J.D. and W.R. Schmitt. 1980. Ocean energy: forms and prospects. Science *207*, 265–273.

Jassby, A.D. 1980. The environmental effects of hydroelectric power development. In *Energy and the Fate of the Ecosystems*. National Academy Press, Washington, D.C.

Jensen, K.W. and W. Snekvik. 1972. Low pH levels wipe out salmon and trout populations in southernmost Norway. Ambio *1*, 223–225.

Johnson, D.W. 1968. Pesticides and fishes—a review of selected literature. Trans. Am. Fish. Soc. *97*, 398–424.

Jones, H.C. 1973. Investigation of alleged air pollution effects on yield of soybeans in the vicinity of the Shawnee Steam Plant. Tenn. Valley Auth. Publ. E-EB-73-3.

Keeling, C.D. 1978. Atmospheric carbon dioxide in the 19th century. Science *202*, 1109.

Kellogg, W.W. and R. Schware. 1981. *Climate Change and Society*. Westview Press, Boulder, CO.

Kelso, J., R.M. Milburn and G.S. Milburn. 1979. Entrainment and impingement of fish by power plants in the Great Lakes which use the once-through cooling process. J. Great Lakes Res. *5*, 182–194.

Kilburn, P.D. 1976. Environmental implications of oil-shale development. Environmental Conservation *3*, 101–115.

Klein, D.H. and P. Russell. 1973. Heavy metals: fallout around a power plant. Environ. Sci. Technol. *7*, 357–358.

Knabe, W. 1973. Investigations of soils and tree growth on five deep-mine refuse piles in the hard-coal region of the Ruhr. In R.J. Hutnick and G. Davis (eds.). *Ecology and Reclamation of Devastated Land*. Gordon and Breach, New York.

Kormondy, E.J. 1976. *Concepts of Ecology*. Prentice-Hall, Englewood Cliffs, NJ.

Kornberg, H.A. 1977. Concern overhead. EPRI Journal. June/July 6–13.

Krebs, C.T. and K.R. Burns. 1977. Long term effects of an oil spill on populations of the salt-marsh crab *Uca pugnax*. Science *197*, 484–487.

Kreith, F. and J.F. Kreider. 1978. *Principles of Solar Engineering*. McGraw-Hill, New York.

Krugman, H. and F. Von Hippel. 1977. Radioactive wastes: a comparison of U.S. military and civilian inventories. Science *197*, 883–885.

Kubo, A.S. and D.J. Rose. 1973. Disposal of nuclear wastes. Science *182*, 1205–1211.

Kukla, G. and J. Gavin. 1981. Summer ice and carbon dioxide. Science *214*, 497–503.

Ledbetter, M.C., P.W. Zimmerman and A.E. Hitchcock. 1959. The histopathological effects of ozone on plant foliage. Contrib. Boyce-Thompson Inst. *20*, 225–282.

Lee, R., W.G. Hutson and S.C. Hill. 1975. Energy exchange and plant survival on disturbed lands. In D.M. Gates and R.B. Schmerl (eds.). *Perspectives of Biophysical Ecology*. Springer-Verlag, New York.

Leivestad, H. and I.P. Muniz. 1976. Fish kill at low pH in a Norwegian River. Nature *359*, 391–392.

Lemon, E.R. (ed.) 1983. *CO_2 and Plants: The Response of Plants to Rising Levels of Atmospheric Carbon Dioxide*. Westview Press, Boulder, CO.

Likens, G. and T. Butler. 1981. Recent acidification of precipitation in North America. Atmos. Environ. *15*, 1103–1109.

Likens, G.E., F.H. Bormann, J.S. Eaton, R.S. Pierce and N.M. Johnson. 1967. Hydrogen ion input to the Hubbard Brook Experimental Forest, New Hampshire during the last decade. Water, Air, Soil Pollut. *6*, 435–445.

Likens, G.E., F.H. Bormann and J.S. Eaton. 1980. Variations in precipitation and streamwater chemistry at the Hubbard Brook Experimental Forest during 1964 and 1977. In T.C. Hutchinson and M. Havas (eds.) *Effects of Acid Precipitation on Terrestrial Ecosystems*. Plenum Press, New York. 443–464.

Likens, G.E., E.H. Bormann, R.S. Pierce, J.S. Eaton and N.M. Johnson. 1977. *Biogeochemistry of a Forested Ecosystem*. Springer-Verlag, New York.

Likens, G.E., R.F. Wright, J.N. Galloway and T.T. Butler. 1979. Acid rain. Sci. Amer. *241*, 43–51.

Linden. H.R. 1981. *U.S. Energy Outlook-1981*. Gas Research Institute, Chicago.

Linden. H.R. and J.D. Parent. 1980. *Perspectives on U.S. and World Energy Resources*. Gas Research Institute and Institute of Gas Technology, Chicago.

List, R.J. 1949. *Smithsonian Meteorological Tables*. Smithsonian Inst. Press., Washington, D.C.

Lockwood, J.G. 1979. *Causes of Climate*. John Wiley and Sons, New York.

Lonnroth, M., T.B. Johansson and P. Steen. 1980. Sweden beyond oil: nuclear commitments and solar options. Science, *208*, 557–563.

Lutgens. F.K. and E.J. Tarbuck. 1979. *The Atmosphere*. Prentice-Hall, Englewood Cliffs, NJ.

Manabe, S. and R.J. Stouffer. 1980. A CO_2 climate sensitivity study with a mathematical model of the global climate. Nature *282*, 491–493.

Manabe, S. and R.T. Wetherald. 1975. The effects of doubling the CO_2 concentration on the climate of a general circulation model. J. Atmos. Sci. *32*, 3–15.

Manabe, S. and R.T. Wetherald. 1980. On the distribution of climate change resulting from an increase in CO_2 content of the atmosphere. J. Atmos. Sci. *37*, 99–118.

Marland, G. 1978. Extracting energy from warm sea water. Endeavour *2*, 165–170.

Marshall, E. 1981. New A-bomb studies alter radiation estimates. Science *212*, 900–903.

Maugh, T.H. 1982. Solar with a grain of salt. Science *216*, 1213–1214.

McBride, J.P., R.E. Moore, J.P. Witherspoon and R.E. Blanco. 1977. Radiological impact of airborne effluents of coal-fired and nuclear power plants. ONRL-5315. Oak Ridge National Laboratory, Oak Ridge, TN, 43.

McCaull, J. 1973. Windmills. Environment *15*, 6–17.

McCollam, J. 1973. *The Yachtman's Weather Manual*. Dodd, Mead & Co., New York.

McNichols, J.L., W.S. Gineli and J.S. Cory. 1979. Thermoclines: a solar thermal energy resource for enhanced hydroelectric power production. Science *203*, 167–168.

Medvick, C. 1973. Selecting plant species for revegetating surface coal mined lands in Indiana—a forty-year record. In R.J. Hutnick and G. Davis (eds.). *Ecology and Reclamation of Devastated Land*. Gordon and Breach, New York.

Meinel, A.B. and M.P. Meinel. 1976. *Applied Solar Energy: An Introduction*. Addison-Wesley, Reading, MA. 651.

Mendonca, B.G., K. Hanson and J.J. DeLuisi. 1978. Volcanically related secular trends in atmospheric transmission at Mauna Loa Observatory, Hawaii. Science *202*, 513–515.

Metz, W.D. 1977a. Solar thermal electricity: power tower dominates research. Science *197*, 353–356.

Metz, W.D. 1977b. Ocean thermal energy: the biggest gamble in solar power. Science *198*, 178–180.

Miles, V.C., R.W. Ruble and R.L. Bond. 1973. Performance of plants in relation to spoil classification in Pennsylvania. In R.J. Hutnick and G. Davis (eds.). *Ecology and Reclamation of Devastated Land*. Gordon and Breach, New York.

Miller, M.W. and G.E. Kaufman. 1978. High voltage overhead. Environment *20*, 6–36.

Muller, R.N., J.E. Miller and D.G. Sprugel. 1979. Photosynthetic response of field-grown soybeans to fumigations with sulfur dioxide. Jour. Appl. Ecol. *16*, 567–576.

Murray, R.L. 1982. *Understanding Radioactive Waste*. Batelle Press, Columbus, OH.

Mynatt, F.R. 1982. Nuclear reactor safety research since Three Mile Island. Science *216*, 131–135.

National Academy of Sciences. 1969. *Resources and Man*. W.H. Freeman, San Francisco. 259.

National Research Council. 1980. *Energy in Transition 1985–2010*. Natl. Acad. Sci. Washington, D.C. W.H. Freeman, San Francisco, 677.

Nelson-Smith, A. 1977. Recovery of some British rocky seashores from oil spills and cleanup operations. In J. Cairns, Jr., K.L. Dickson and E.E. Herricks (eds.). *Recovery and Restoration of Damaged Ecosystems; Proceedings of the International Symposium on the Recovery of Damaged Ecosystems held at Virginia Polytechnic Institute and State University, Blacksburg, VA, March 1975*. University Press of Virginia, Charlottesville. 190–207.

Neyman, J. 1977. Public health hazards from electricity-producing plants. Science *195*, 754–758.

Nielsen, P.E., H. Nishimura, J.W. Otvos and M. Calvin. 1977. Plant crops as a source of fuel and hydrocarbon-like materials. Science *198*, 942–944.

Oden, S. 1976. The acidity problem—an outline of concepts. Water, Air and Soil Pollution *6*, 137–166.

Odum, H.T. 1973. Energy, ecology and economics. Ambio *2*, 220–227.

Patrick, R. 1963. The structures of diatom communities under varying ecological conditions. Ann. N.Y. Acad. Sci. *108*, 353–358.

Patrick. R., V.P. Binetti and S.G. Halterman, 1981. Acid lakes from natural and anthropogenic causes. Science *211*, 446–448.

Perry, H. 1983. Coal in the United States: a status report. Science *222*, 377–384.

Piementel, D., M.A. Moran, S. Fast, G. Weber, R. Bukantis, L. Balliett, P. Boveng, C. Cleveland, S. Hindman and M. Young. 1981. Biomass energy from crop and forest residues. Science *212*, 1110–1115.

Pough, F.H. 1976. Acid precipitation and embryonic mortality of spotted salamanders, *Ambystoma maculatum*. Science *192*, 68–70.

Pough, F.H. and R.E. Wilson. 1977. Acid precipitation and reproductive success of *Ambystoma* salamanders. Water, Air, Soil Pollut. *7*, 307–316.

Precht, H., J. Christopherson, H. Hensel and W. Larcher (eds.). 1973. *Temperature and Life*. Springer-Verlag, New York.

Radford, E.P. 1981. Radiation dosimetry. Science *213*, 602.

Ramanthan, V., R.D. Lian and R.D. Cess. 1979. Increased atmospheric CO_2: zonal seasonal estimates of the effect on the radiation energy balance and surface temperature. J. Geophys. Res. *84*, 494–495.

Rampino, M.R. and S. Self. 1979. Can rapid climate change cause volcanic eruptions? Science *206*, 826–828.

Rand, P.J. 1982. *Land and Water Issues Related to Energy Development*. Ann Arbor Science Publ., Ann Arbor. 469.

Reed, T.B. and R.M. Lerner. 1973. Methanol: a versatile fuel for immediate use. Science *182*, 1299–1304.

Richardson, C.J. and J.A. Lund. 1975. Effects of clear cutting on nutrient losses in aspen forests on three soil types in Michigan. In F.G. Howell, J.B. Gentry and M.H. Smith (eds.). *Mineral Cycling in Southeastern Ecosystems, ERDA Symposium Series*. CONF-740513.

Ricklefs, R.E. 1979. *Ecology*, 2nd Ed. Chiron Press, Newton, MA.

Roberts, W.O. and H. Landsford. 1979. *The Climate Mandate*. W.H. Freeman, San Francisco.

Rose, A.H. (ed.). 1967. *Thermobiology*. Academic Press, London and New York.

Samson, A.L., J.H. Vandermeulen, P.G. Wells and C. Moyse. 1980. A selected bibliography on the fate and effects of oil pollution relevant to the Canadian marine environment, 2nd Ed. Res. and Dev. Div., Environ. Protection Service, Environment Canada. Rept. No. EPS 3-EC-80-5.

Schofield, C.L. 1976. Acid precipitation: effect on fish. Ambio *5*, 228–230.

Schull, W.J., M.O. Take and J.V. Neel. 1981. Genetic effects of the atomic bombs: a reappraisal. Science *213*, 1220–1227.

Schurr, S.H., J. Darmstadter, H. Perry, W. Ramsay and M. Russell. 1979. *Energy in America's Future*. The Johns Hopkins University Press, Baltimore.

Sharitz, R.R. and J.W. Gibbons. 1981. Effects of thermal effluents on a lake: enrichment and stress. In G.W. Barrett and R. Rosenberg (eds.). *Stress Effects on Natural Ecosystems.* John Wiley and Sons, New York.

Sherwood, L. and D. Meadows. 1978. The fuel requirements of a 50-MW wood fired electrical generating facility in northern Vermont. Resource Policy Center. Thayer School of Engineering, Dartmouth College. Paper DSD #106.

Sivard, R.L. 1981. *World Energy Survey.* World Priorities, Leesburg, VA.

Skinner, B.J. 1969. *Earth Resources.* Prentice-Hall, Englewood Cliffs, NJ.

Sorenson, B. 1981. Turning to the wind. Amer. Sci. *69*, 500–508.

Southern, W.E. 1975. Orientation of gull chicks exposed to Project Sanguine's electromagnetic field. Science *189*, 143–145.

Southern, W.E. 1977. Migrating birds respond to Project Seafarer's electromagnetic field. Science *195*, 777–779.

Sprugel, D.G., J.E. Miller, R.N. Miller, H.J. Smith and P.B. Xerikos. 1980. Sulfur dioxide effects on yield and seed quality in field-grown soybeans. Phytopathology *70*, 1129–1133.

Starr, C. and R.P. Hammond. 1972. Nuclear waste storage. Science *177*, 744–745.

Stensland, G.J. and R.G. Semonin. 1982. Another interpretation of the pH trend in the United States. Bull. Amer. Meteorol. Soc. *63*, 2177–2183.

Streeter, R.G., R.T. Moore, J.J. Skinner, S.G. Margin, T.L. Terrel, W.D. Llimstra, J. Tate, Jr. and M.J. Nolde. 1979. Energy mining impacts and wildlife management: which way to turn. Paper presented at 44th North American Wildlife and Natural Resources Conf., Toronto. March 26–28.

Striffler, W.D. 1973. Surface mining disturbance and water quality in eastern Kentucky. In R.J. Hutnick and G. Davis (eds.). *Ecology and Reclamation of Devastated Land.* Gordon and Breach, New York.

Strijbosch, H. 1979. Habitat selection of amphibians during their aquatic phase. Oikos *33*, 363–372.

Stuiver, M. 1978a. Atmospheric carbon dioxide and carbon reservoir changes. Science *199*, 253–258.

Stuiver, M. 1978b. Atmospheric carbon dioxide in the 19th century. Science *202*, 1109.

Sundquist, E.T. and G.A. Miller. 1980. Oil shales and carbon dioxide. Science *208*, 740–741.

Svensson, B.H. and R. Soderlund (eds.). 1975. Nitrogen, phosphorus and sulphur-global cycles. Ecological Bulletins No. 22. SCOPE Report 7. Swedish Natural Science Research Council, NFR, Stockholm.

Teal, J.M., K. Burns and J.W. Farrington. 1978. Analyses of aromatic hydrocarbons in intertidal sediments resulting from two spills of No. 2 fuel oil in Buzzards Bay, MA. J. Fish Res. Bd. Can. *35*, 510–520.

Tilly, L.J. 1975. Changes in water chemistry and primary productivity of a reactor cooling reservoir (Par Pond). In F.G. Howell, J.B. Gentry and M.H. Smith (eds.). *Mineral Cycling in Southeastern Ecosystems.* ERDA

Sump. Ser. (CONF-740513), National Technical Information Center, Oak Ridge, TN. 394–407.

Toribara, T.Y., M.W. Miller and P.E. Marrow (eds.). 1980. *Polluted Rain.* Plenum Press, New York. 502.

Traaen, T.S. 1980. Ecological impact of acid precipitation. In D. Drablos and A. Tollan (eds.). *Ecological Impact of Acid Precipitation, Proc. Int. Conf., Sandefjord, Norway, March 11–14, 1980.* 340–341.

Travers, W.B. and P.R. Luney. 1976. Drilling, tankers and oil spills on the Atlantic outer continental shelf. Science *194*, 791–796.

Upton, A.C. 1982. The biological effects of low-level ionizing radiation. Sci. Amer. *246*, 41–49.

Vogel, W.G. and W.A. Berg. 1973. Fertilizer and herbaceous cover influence establishment of direct-seeded black locust on coal-mine spoils. In R.J. Hutnick and G. Davis (eds.). *Ecology and Reclamation of Devastated Land.* Gordon and Breach, New York.

Vogelmann, H.W. 1982. Catastrophe on Camel's Hump. Natural History *91*, 8–14.

von Hippel, F. 1982. Potassium iodide for thyroid protection. Science *218*, 1174–1175.

Wakeham, S.G. and J.W. Farrington. 1980. Hydrocarbons in contemporary aquatic sediments. In R.A. Baker (ed.). *Contaminants and Sediments, Vol. I.* 3–33.

Walker, D.A., P.J. Webber, K.R. Everett and J. Brown. 1978. Effects of crude and diesel oil spills on plant communities at Prudhoe Bay, Alaska and the derivation of oil spill sensitivity maps. Arctic *31*, 242–259.

Walton, S. 1981. Academy looks again at petroleum in the marine environment. BioScience *31*, 93–96.

Warner, R.W. 1973. Acid coal mine drainage effects on aquatic life. In R.J. Hutnick and G. Davis (eds.). *Ecology and Reclamation of Devastated Land.* Gordon and Breach, New York.

Webber, P.J. and J.D. Ives. 1978. Damage and recovery of tundra vegetation. Environmental Conservation *5*, 171–182.

Webber, P.J., P.C. Miller, F.S. Chapin and B.H. McCowan. 1980. The vegetation: pattern and succession. In J. Brown, P.C. Miller, L.L. Tieszen and F.L. Bunnell (eds.). *An Arctic Ecosystem: The Coastal Tundra at Barrow, Alaska.* Dowden, Hutchinson and Ross, Stroudsberg, PA. 186–218.

Whipple, C. 1980. The energy impacts of solar heating. Science *208*, 262–266.

White, D.E. 1965. Geothermal energy, U.S. Geol. Survey Circ. 519.

Whittaker, R.H. 1975. *Communities and Ecosystems,* 2nd Ed. Macmillan, New York.

Wigg, E.E. 1974. Methanol as a gasoline extender: a critique. Science *186*, 785–795.

Wigley, T.M., P.D. Jones and P.M. Kelly. 1980. Scenario for a warm, high-CO_2 world. Nature *283*, 17–21.

Williams, P.J. 1979. *Pipelines and Permafrost: Physical Geography and Development in the Circumpolar North.* Longman Group, London.

Wilson, E.O. and W.H. Bossert. 1971. *A Primer of Population Biology.* Sinauer Associates, Sunderland, MA.

Winograd, I.J. 1981. Radioactive waste disposal in thick

unsaturated zones. Science *212*, 1457–1464.

Witherspoon, P.A., N.G.W. Cook and J.E. Gale. 1981. Geologic storage in radioactive waste: field studies in Sweden. Science *211*, 894–900.

Woodson, R.D. 1971. Cooling towers. Sci. Amer: *224*, 70–78.

Woodwell, G.M., R.H. Whittaker, W.A. Reiners, G.E. Likens, C.C. Delwiche and D.B. Botkin. 1978. The biota and the world carbon budget. Science *199*, 141–146.

Wrenn, W.B. 1980. Effects of elevated temperature on growth and survival of smallmouth bass. Trans. Amer. Fish. Soc. *109*, 617–625.

Yan, N.D. 1979. Phytoplankton community of an acidified, heavymetal contaminated lake near Sudbury, Ontario: 1973–1977. Water, Air, Soil Pollut. *11*, 43–55.

INDEX

367